Constitutive Equations of Nonlinear Electromagnetic–Elastic Crystals

E. Kiral
A. Cemal Eringen

Constitutive Equations of Nonlinear Electromagnetic–Elastic Crystals

With 17 Illustrations

Springer-Verlag
New York Berlin Heidelberg
London Paris Tokyo Hong Kong

E. Kiral
Professor of Engineering Mechanics
Faculty of Engineering
Cukurova University
Balcali
Adana
Turkey

A. Cemal Eringen
Princeton University
Princeton, NJ 08544
U.S.A.

Library of Congress Cataloging in Publication Data
Kiral, Ehran.
Constitutive equations of nonlinear electromagnetic-elastic crystals / E. Kiral, A.C. Eringen.
p. cm.
Includes bibliographical references.
ISBN 0-387-97120-3
1. Crystallography, Mathematical. 2. Magnetic crystals. 3. Group theory. I. Eringen, A. Cemal. II. Title.
QD911.K48 1990
548'.7—dc20 89-26331
CIP

Printed on acid-free paper.

Typeset by Asco Trade Typesetting Ltd., Hong Kong.
Printed and bound by Edwards Brothers, Ann Arbor, Michigan
Printed in the United States of America.

9 8 7 6 5 4 3 2 1

ISBN 0-387-97120-3 Springer-Verlag New York Berlin Heidelberg
ISBN 3-540-97120-3 Springer-Verlag Berlin Heidelberg New York

Preface

Continuum physics is concerned with the predictions of deformations, stress, temperature, and electromagnetic fields in deformable and fluent bodies. To that extent, mathematical formulation requires the establishment of basic balance laws and constitutive equations. Balance laws are the union of those of continuum thermomechanics and Maxwell's equations, as collected in Chapter 1. To close the theory it is necessary to formulate equations for the material response to external stimuli. These equations bring into play the material properties of the media under consideration. In their simplest forms these are the constitutive laws, such as Hooke's law of classical elasticity, Stokes' law of viscosity of viscous fluids, Fourier's law of heat conduction, Ohm's law of electric conduction, etc. For large deformations and fields in material media, the constitutive laws become very complicated, involving all physical effects and material symmetry.

The present work is concerned with the material symmetry regulations arising from the crystallographic symmetry of magnetic crystals. While there exist some works on the thirty-two conventional crystal classes, excluding the linear case, there exists no study on the nonlinear constitutive equations for the ninty magnetic crystal classes. Yet the interaction of strong electromagnetic fields with deformable solids cannot be explained without the material symmetry regulations relevant to magnetic crystals.

In this monograph, we present a thorough discussion of magnetic symmetry by means of group theory. We consider only one scalar function which depends on one symmetric second-order tensor (e.g., the strain tensor), one vector (e.g., the electric field), and one axial vector (e.g., the magnetic field). However, our method is valid for more tensors and vectors, leading to much lengthier results. Ultimately, such studies will have to be done by means of computers. To make the material accessible to those who are not familiar with group theory, we have also provided essentials of group theory in the Appendices.

The lists and tables given in this monograph are essential to the discussion of the material properties of magnetic crystal classes and for research into the nonlinear electromagnetic theory of elastic solids. Nonlinear optics, nonlinear magnetism, elastic solids subject to large fields, and plasma physics are but a few areas where we need nonlinear constitutive equations which are in final form. Therefore, we believe that the methods and material provided here should be useful to material scientists, engineers, and physicists working in these fields. We have also provided several examples of applications to demonstrate the use of the theory.

The major part of the present monograph was completed in 1977, but its publication was postponed in the hope that we could verify our results with the aid of a computer. This turned out to be a major task. Yet no other work in this area has appeared in the intervening years, so we have decided to publish our work, in order that it might provide a source for researchers.

In spite of our efforts, no doubt, some errors and/or misprints may have escaped our attention. We would be grateful if these are relayed to us by readers who note them.

This work was initiated while Dr. Kiral was visiting Princeton University in 1976–1977. We express our thanks to the Department of Civil Engineering and to the administration of Princeton University who allowed us this opportunity. Professor G.F. Smith, who has contributed extensively to the field, has read and criticized some parts of the manuscript. We gratefully express our appreciation to him.

Contents

List of Tables

Introduction

Although magnetic crystals have been the subject of research and conjecture for some time, it is only relatively recently that this field has become an active one for experimental and theoretical work. The basic and broad approach to the study of magnetic materials is the investigation of their constitutive behavior. Constitutive equations are the relations between dependent (response) and independent (state) tensors. For a given physical situation, the problem is to determine the restrictions to be imposed on the form of these constitutive relations as dictated by the symmetry of the material.

When the constitutive equations are not linear, the problem is usually discussed by means of a Taylor series expansion (polynomial approximation) of the response functions, in terms of state variables and the determination of the restrictions to be imposed by the material symmetry on the constant coefficient tensors (material moduli). For example, in piezomagnetoelectricity, the magnetization, **M**, may be assumed to depend on the electric field, **E**, the magnetic field, **H**, and the strain, **e**, i.e., $\mathbf{M} = \mathbf{M}(\mathbf{E}, \mathbf{H}, \mathbf{e})$. The possible effects are then described by the relation

$$\begin{aligned} M_i = {} & a_{ij}H_j + b_{ij}E_j + c_{ijk}H_jH_k + d_{ijk}H_jE_k + e_{ijk}E_jE_k + f_{ijk}e_{jk} \\ & + h_{ijkm}H_jH_kH_m + k_{imjk}H_jH_kE_m + m_{ikpj}H_jE_kE_p \\ & + n_{ijkp}E_jE_kE_p + p_{ijkm}H_je_{km} + q_{ijkm}E_je_{km}, \qquad (1) \end{aligned}$$

where only material tensors up to the fourth order are included. Although this procedure of approximating by polynomials enables us to consider material tensors of much higher degree (higher-order effects), it has the disadvantage that the problem is not solved in closed form. Application of the theory of invariants, however, eliminates this limitation to some extent.

Unfortunately, the representation theorems available for the nonmagnetic crystals are not valid for magnetic crystals. This is because, in addition to the geometrical symmetries present in the lattice structure of the crystals, the atoms of the lattice in magnetic materials are endowed with atomic magnetic moments (spins). The usual spatial symmetry operations, rotations, and rotation-inversions, while preserving the geometrical properties of the lattice, may reverse the orientation of the spins. Therefore, for magnetic crystals, additional symmetry properties must be taken into account. With the incorporation of the spin symmetry, thirty-two crystallographic point groups are enlarged to ninety magnetic point groups.

Certain magnetic properties and various cross effects have been shown to be possible, after recognizing the magnetic symmetry classes (see Laundau and Lifshitz [1957] and Dzyaloshinskii [1960]), and their existence has received experimental confirmation (see Borovik-Romanov [1959] and Astrov [1961]). For an extensive review the reader is referred to O'Dell [1970, Chap. 1]. Certain crystals become magnetically polarized when placed in an electric field. For example, b_{ij} in (1) characterizes this effect, called magnetoelectric susceptibility, the existence of which depends on the magnetic symmetry of crystals. The magnetoelectric effect was first proposed by Curie [1908], but the experimental verification came much later from Astrov [1960, 1961], Folen et al. [1961], and Rado and Folen [1962]. On the basis of the erroneous assumption that all crystals are time-symmetric, Zocher and Török [1953] concluded that pyromagnetism and piezomagnetism (f_{ijk} in (1)) are forbidden effects. Since magnetic materials are not time-symmetric, this conclusion is wrong, and these effects were observed by Borovik-Romanov (1959) in the antiferromagnetic crystals CoF_2 and MnF_2.

The foundation of magnetic crystallography were laid down by Shubnikov [1951], Zamorzaev [1957, 1958], and Belov et al. [1957]. The same topic was developed by Tavger and Zaitsev [1956], where it was specifically applied to magnetic crystals. For an extensive review of these works, see, e.g., Birss [1962, 1963], Koptsik [1966, 1968], and Bradley and Cracknell [1972].

The interaction of electromagnetic fields with deformable bodies has been studied extensively during the last two decades. The balance laws, thermodynamics, and the constitutive theory for electromagnetic solids have been formulated (see Grot and Eringen [1966], Maugin and Eringen [1977], and Eringen [1980]). However, when electromagnetic effects are present, the symmetry of electromagnetic crystalline solids should, rationally, be based on space–time (magnetic) symmetry groups. As far as nonlinear constitutive equations are concerned, very little research has been done in this field.

For the symmetry of a magnetic material, the antisymmetry operator, τ, is the operator that reverses a magnetic moment. Alternatively, we may regard τ as reversing the sense of the direction of time. Besides being polar (true) or axial (pseudo), the tensors are further classified into two types by considering the effect of time-reversal, τ, on the magnetic crystals. Hence, appropriate transformation rules must be used, according to the nature of a given tensor.

Let $\{M\} = \{\mathbf{M}^1, \mathbf{M}^2, \ldots, \mathbf{M}^g\}$ be the group of symmetry transformations (ordinary $\mathbf{M}^\alpha = \mathbf{S}^\alpha$, or complementary $\mathbf{M}^\alpha = \tau\mathbf{S}^\alpha$) which carry the material from its initial configuration into a final indistinguishable configuration. Consider a general constitutive relation given by

$$E_{i_1 \ldots i_n} = E_{i_1 \ldots i_n}(A_{i_1 \ldots i_r}, B_{i_1 \ldots i_p}, \ldots, C_{i_1 \ldots i_q}), \tag{2}$$

where **E**, **A**, **B**, ..., **C** are tensors of the order indicated by their indices. For example, **E** may be the stress tensor (a second-order polar tensor); **A** may be the strain tensor (also of second order); and **B** and **C** may be electric and magnetic field vectors (time-symmetric and time-antisymmetric vectors, respectively). Due to the principle of material invariance, the relation (2) must have the same form under each transformation by the symmetry elements, $\mathbf{M}^\alpha$, associated with the material. Therefore, we write

$$\bar{E}_{i_1 \ldots i_n} = E_{i_1 \ldots i_n}(\bar{A}_{i_1 \ldots i_r}, \bar{B}_{i_1 \ldots i_p}, \ldots, \bar{C}_{i_1 \ldots i_q}), \tag{3}$$

where tensors carrying an overbar are the components of those without a bar in the new coordinate systems obtained by the group of transformation $\mathbf{M}^\alpha$, for each member of $\{M\}$. It is then said that the tensor-valued function $\mathbf{E}(\mathbf{A}, \mathbf{B}, \ldots, \mathbf{C})$ is invariant under $\{M\}$. In the case of a scalar function $W(\mathbf{A}, \mathbf{B}, \ldots, \mathbf{C})$, the invariance requirement is

$$W(\mathbf{A}, \mathbf{B}, \ldots, \mathbf{C}) = W(\bar{\mathbf{A}}, \bar{\mathbf{B}}, \ldots, \bar{\mathbf{C}}) \qquad \text{for all} \quad \mathbf{M}^\alpha \text{ in } \{M\}, \tag{4}$$

which is a special case of (3). In this case, a solution to (4) is furnished by giving a set of scalar-valued invariants $I_p(\mathbf{A}, \mathbf{B}, \ldots, \mathbf{C})$. The invariants $I_p(\mathbf{A}, \mathbf{B}, \ldots, \mathbf{C})$, $p = 1, 2, \ldots$, are said to form an *integrity basis* for functions of **A**, **B**, ..., **C** which are invariant under $\{M\}$. We may omit from such an integrity basis any element which can be expressed as a polynomial in the remaining ones. The set of invariants so obtained is called a *minimal* integrity basis. In the case of tensor-valued functions, the problem can be reduced to that of a scalar, as in (4), with the inclusion of an appropriate extra tensor into the argument list of W (cf. eq. (1.38)). Thus, the main problem of the constitutive theory for the ninety magnetic crystals is to determine the minimal integrity basis of the vectors and tensors that affect an absolute scalar function (e.g., the free energy function).

The relevant research works available in the literature on nonlinear anisotropic constitutive equations are solely confined to nonmagnetic crystalline solids. The integrity bases for a single symmetric tensor for all conventional crystal classes have been obtained by Smith and Rivlin [1958]. Their results were reproduced by Green and Adkins [1960]. The integrity bases for a single symmetric second-order tensor and a single absolute vector for all nonmagnetic classes were given by Smith et al. [1963]. The appropriate integrity bases for an arbitrary number of vectors were provided by Smith and Rivlin [1964] and Smith [1967, 1968], again for each of the nonmagnetic crystal classes. Smith and Kiral [1969] have derived the integrity bases for an arbitrary number of symmetric second-order tensors for each of the conventional crystal classes.

The form of the energy function $W = W(\mathbf{C}, \mathbf{D})$, where **C** is a second-order symmetric tensor and **D** is an asymmetric second-order traceless tensor, was given by Huang [1968, 1969]. Kiral and Smith [1974] derived the most general integrity bases for the scalar polynomial function $W = W(\mathbf{A}, \mathbf{B}, \ldots, \mathbf{C})$, where the number and type of the arguments **A**, **B**, ..., **C** are arbitrary. The crystallographic groups considered by them are those associated with the triclinic, monoclinic, orthorhombic, tetragonal, trigonal, and hexagonal crystal systems, with the exception of the class $D_{6h} = 6/mmm$. Integrity bases for an arbitrary number of arguments were given by Kiral [1972] for the crystal classes T, T_d, and 0 of the cubic system.

The purpose of this investigation is to determine and tabulate the forms of the nonlinear constitutive equations which are restricted by the magnetic symmetry of electromagnetic solids. The independent electromagnetic constitutive variables to be chosen are dictated by the physical phenomenon that we are interested in. We consider electromagnetic solids which are not heat- and electric-conducting. We also exclude viscoelastic effects, and dielectric and magnetic dissipation. Moreover, we do not include magnetic spin phenomena, polarization gradient effects, and electric and magnetic quadrupoles. For such materials, the internal energy may be expressed as (see Chapter 1):

$$\Sigma = \Sigma(E_{KL}, P_K, M_K, \theta) \tag{5}$$

with similar expressions for the entropy, the electric field, the magnetic induction, the heat flux, and the stress tensor. These materials are conservative, and dependent (response) quantities are derivable from the internal energy, which is a true scalar function of the independent (state) variables E_{KL}, P_K, M_K, and θ. These arguments are the material description of the strain tensor (a symmetric, true, and time-symmetric tensor), polarization (a polar and time-symmetric vector), magnetization (an axial time-asymmetric vector), and temperature (a true scalar). Constitutive equations for electric and heat conductions can also be obtained by the present method when one of $\mathbf{P}$ and $\mathbf{M}$ is not present in (5). This is done as indicated by eq. (1.38).

Conventional crystal classes are reviewed in Chapter 2. Chapter 3 deals briefly with magnetic crystal classes. Although most of these results are available in the literature, they are scattered and various sets of notations are used by different authors. The thirty-two conventional point groups are listed in Tables 2.1–2.3 and 2.5. Subgroups of the conventional point groups are given in Table 2.6. The fifty-eight magnetic point groups are listed in Table 3.1. The magnetic point groups and their irreducible representations are presented in Table 3.3. In Chapter 4 we decompose the spaces of the mechanical and electromagnetic tensors; and the linear combination of their components that form the carrier spaces of the irreducible representation of the magnetic point groups are listed in Table 4.3(1–58). It is assumed that the reader has a basic working knowledge of the theory of groups and of group representation; however, we have summarized the relevant parts of this theory in Appendix A.

Chapter 5 is devoted to explaining how these linear combinations (basic quantities) may be used to determine the forms of nonlinear constitutive equations due to material symmetry, in a general context. Chapter 6, in particular, deals with linear constitutive equations. The application of the symmetry arguments to linear constitutive equations for crystalline solids is, of course, not new, but emphasis is given to magnetic crystals in the present work. Moreover, we introduce a new method for determining the forms of tensors characterizing the physical properties of crystals. This method is based on the basic quantities of the relevant tensors, and completely eliminates the algebraic computations involved in the existing methods. It is equally applicable to both conventional and magnetic crystals, and no complications arise when trigonal and hexagonal classes are considered.

Chapter 7 is devoted to nonlinear polynomial constitutive equations of the form $W = W(E_{KL}, P_K, M_K, \theta)$. Irreducible integrity bases for E_{KL}, P_K, and M_K, which are invariant under the magnetic symmetry of the electromagnetic solids, are listed for each of the fifty-eight purely magnetic crystal classes in Sections 7.1–7.58. The importance of these results is that the internal energy (or free energy) will be expressed as a linear combination of the elements of the integrity bases appropriate to the magnetic symmetry of the crystal under consideration. The constitutive equation for the dependent quantities, such as stress, electric and magnetic fields, etc., will then be readily determined from the partial derivatives of the internal energy, with respect to the appropriate independent variables (cf. eq. (1.40)).

In particular, in the case of linear interactions, the internal energy is restricted to the second-order terms. For example, magnetoelectric effects are studied by retaining second-order elements of the integrity basis in **P** and **M** (Rado and Folen [1962] and Rado [1962]). Piezomagnetoelectric effects are obtained as a special case of the results presented in Chapter 7, when third-order terms in E_{KL}, P_K, and M_K are considered such that each term is linear in **E**, **P**, and **M** (see Borovik-Romanov [1960] and Rado [1962]).

In each magnetic symmetry class, the results are presented in seven groups that are associated with *elastic*, *rigid dielectric*, *rigid magnetic*, *elastic dielectric*, *elastic magnetic*, *rigid polarizable and magnetizable solids*, and *elastic polarizable and magnetizable solids*.

In order to demonstrate the uses of the theory, and the tables constructed in Chapter 8, we present several applications. Symmetry regulations have been obtained for various nonlinear electromagnetic elastic crystals. These include piezomagnetoelectricity and magnetoelectric effects in Cr_2O_3 ($\underline{\bar{3}m}$); piezomagnetism in MnF_2, CoF_2, and FeF_2 ($\underline{4}/m\underline{mm}$); and rigid nonconducting electromagnetic solids and magnetoelasticity of nonconducting solids.

CHAPTER 1

Electromagnetic Theory

In this chapter we present a summary of the basic equations of the electromagnetic theory of deformable and fluent bodies. These consist of balance laws and constitutive equations. The purpose of this summary is to close the theory, without a thorough discussion of the derivations which would require a long, separate account. Interested readers may consult Eringen [1980, Chap. 10] and Eringen and Maugin [1989] for deformable bodies; De Groot and Suttorp [1972] for the statistical foundations of electromagnetic theory; and Jackson [1975] and Landau and Lifshitz [1960] for the classical theories.

1.1. Deformation and Motion

Consider a material manifold (called body $\mathscr{B}$) embedded in a three-dimensional Euclidean space R^3, which is covered by a rectangular coordinate network. At its reference state the body possesses volume V enclosed by its surface ∂V. A material point P in the body, at the reference state at time $t = 0$, is located by its rectangular coordinates X_K, $K = 1, 2, 3$, or the position vector $\mathbf{X}$. Upon the application of external loads and fields, the body deforms to the spatial configuration, having volume $\mathscr{V}$ enclosed within the surface $\partial\mathscr{V}$. The material point $\mathbf{X}$ is now located at a spatial place $\mathbf{x}$ with the rectangular coordinates x_k, $k = 1, 2, 3$ (Fig. 1.1). We employ majuscule letters and indices for quantities referring to the reference frame X_K and minuscule letters and indices for those referring to x_k.

The motion of the material point $\mathbf{X}$ is a one-parameter family of transformations expressed by

$$x_k = x_k(X_K, t) \qquad \text{or} \qquad \mathbf{x} = \mathbf{x}(\mathbf{X}, t). \tag{1.1}$$

We assume that these three functions possess continuous partial derivatives with respect to X_K throughout V, and that the Jacobian J is positive, i.e.,

$$J = \det\left(\frac{\partial x_k}{\partial X_K}\right) > 0. \tag{1.2}$$

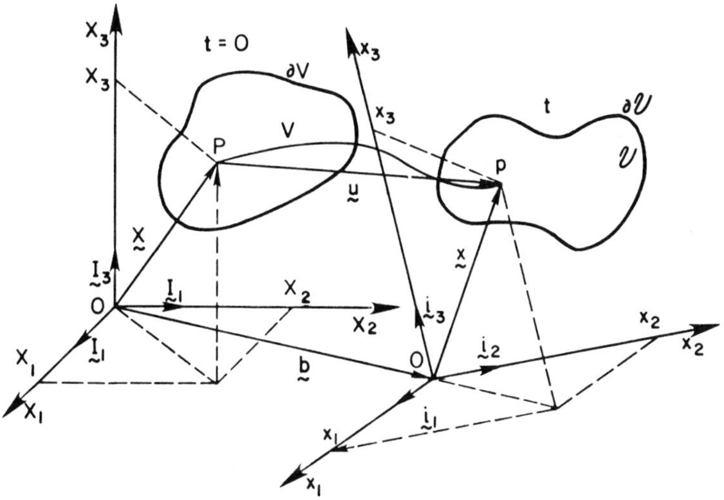

Figure 1.1. Deformation.

From the implicit function theorem of calculus, it then follows that (1.1) possesses a unique inverse in a neighborhood of $\mathbf{x} \in \mathscr{V}$:

$$X_K = X_K(x_k, t) \qquad \text{or} \qquad \mathbf{X} = \mathbf{X}(\mathbf{x}, t). \tag{1.3}$$

The velocity $\mathbf{v}$ and acceleration $\mathbf{a}$ of the material point $\mathbf{X}$ are defined by

$$\mathbf{v} = \frac{\partial \mathbf{x}(\mathbf{X}, t)}{\partial t}, \qquad \mathbf{a} = \frac{\partial^2 \mathbf{x}(\mathbf{X}, t)}{\partial t^2}. \tag{1.4}$$

By substituting from (1.3) into $\mathbf{v}(\mathbf{X}, t)$, we also have

$$\mathbf{v}(\mathbf{X}, t) = \mathbf{v}(\mathbf{X}(\mathbf{x}, t), t) = \hat{\mathbf{v}}(\mathbf{x}, t). \tag{1.5}$$

In this form $\hat{\mathbf{v}}$ is the velocity field at $\mathbf{x} \in \mathscr{V}$, at time t. The identity of the material point at $\mathbf{x}$ is not known. A material point, in passing through the spatial place $\mathbf{x}$ at time t, acquires the velocity $\mathbf{v}$. This is the Eulerian point of view.

The acceleration field at $(\mathbf{x}, t)$ can be calculated by using $\hat{\mathbf{v}}$:

$$\mathbf{a}(\mathbf{x}, t) = \frac{\partial \hat{\mathbf{v}}}{\partial t} + \frac{\partial \hat{\mathbf{v}}}{\partial x_k} \hat{v}_k. \tag{1.6}$$

Henceforth, we shall drop the hat (ˆ) on $\mathbf{v}$ and express (1.6) in the form

$$\mathbf{a} = \frac{D\mathbf{v}}{Dt} = \dot{\mathbf{v}} = \frac{\partial \mathbf{v}}{\partial t} + \mathbf{v}_{,k} v_k, \tag{1.7}$$

where an overdot or D/Dt represents the material derivative, as defined by the partial derivative with respect to t with $\mathbf{X} =$ fixed. We also employ indices after a comma to denote partial derivatives with respect to x_k or X_K, e.g.,

$$\dot{\mathbf{v}} = \left.\frac{\partial \mathbf{v}}{\partial t}\right|_{\mathbf{X}}, \qquad \mathbf{v}_{,k} = \frac{\partial \mathbf{v}}{\partial x_k}, \qquad x_{k,K} = \frac{\partial x_k}{\partial X_K}.$$

As usual, summation convention is understood on repeated indices, e.g.,

$$v_{l,k}v_k = v_{l,1}v_1 + v_{l,2}v_2 + v_{l,3}v_3.$$

Green and Cauchy deformation tensors are defined by

$$\begin{aligned} C_{KL}(\mathbf{X}, t) &= x_{k,K}x_{k,L} \qquad &\text{(Green)},\\ c_{kl}(\mathbf{x}, t) &= X_{K,k}X_{K,l} \qquad &\text{(Cauchy)}. \end{aligned} \tag{1.8}$$

The inverses B_{KL} and b_{kl} of these tensors are called the Piola and Finger deformation tensors, respectively,

$$\begin{aligned} B_{KL} &= \overset{-1}{C}_{KL} = X_{K,k}X_{L,k} \qquad &\text{(Piola)},\\ b_{kl} &= \overset{-1}{c}_{kl} = x_{k,K}x_{l,K} \qquad &\text{(Finger)}. \end{aligned} \tag{1.9}$$

The Lagrangian and Eulerian strain tensors are given by

$$\begin{aligned} E_{KL} &= \tfrac{1}{2}(C_{KL} - \delta_{KL}) \qquad &\text{(Lagrange)},\\ e_{kl} &= \tfrac{1}{2}(\delta_{kl} - c_{kl}) \qquad &\text{(Euler)}, \end{aligned} \tag{1.10}$$

where δ_{kl} is the Kronecker symbol which is 1 when $k = l$ and zero when $k \neq l$. If we employ the displacement vector $\mathbf{u}$ (Fig. 1.1),

$$\mathbf{U} = \mathbf{x} - \mathbf{X} - \mathbf{b} \tag{1.11}$$

in (1.8) and (1.10), we obtain

$$\begin{aligned} E_{KL} &= \tfrac{1}{2}(U_{K,L} + U_{L,K} + U_{M,K}U_{M,L}),\\ e_{kl} &= \tfrac{1}{2}(u_{k,l} + u_{l,k} - u_{m,k}u_{m,l}), \end{aligned} \tag{1.12}$$

where U_K and u_k are, respectively, components of $\mathbf{u}$ in X_K and x_k.

The deformation rate tensor $\mathbf{d}$ and the vorticity tensor $\mathbf{w}$ are defined by

$$\begin{aligned} d_{kl} &= \tfrac{1}{2}(v_{k,l} + v_{l,k}),\\ w_{kl} &= \tfrac{1}{2}(v_{k,l} - v_{l,k}). \end{aligned} \tag{1.13}$$

A useful lemma, in calculating the material derivative of various tensors, is

$$\frac{D}{Dt}(x_{k,K}) = v_{k,l}x_{l,K}; \qquad \frac{D}{Dt}(dx_k) = v_{k,l}\,dx_l, \tag{1.14}$$

which follows from the fact that the operators D/Dt and $\partial/\partial X_K$ are commutative.

Using this result we find that

$$\frac{D}{Dt}(ds^2) = 2d_{kl}\,dx_k\,dx_l = \dot{C}_{KL}\,dX_K\,dX_L. \tag{1.15}$$

Consequently, *the motion is locally rigid if and only if* $\mathbf{d} = \mathbf{0}$ (Killing's theorem).

Cauchy's stress tensor is denoted by t_{kl} and the Piola–Kirchhoff stress tensors by T_{Kl} and T_{KL}. The heat vector in the spatial and material frames are denoted, respectively, by q_k and Q_K. We have the relations

$$
\begin{aligned}
T_{KL} &= \frac{\rho_0}{\rho} X_{K,k} X_{L,l} t_{kl} = T_{Kl} X_{L,l}, \\
Q_K &= \frac{\rho_0}{\rho} X_{K,k} q_k.
\end{aligned}
\tag{1.16}
$$

Electromagnetic fields in a fixed reference frame will be denoted by

$\mathbf{E}$ = electric vector, $\mathbf{H}$ = magnetic vector,

$\mathbf{P}$ = polarization vector, $\mathbf{M}$ = magnetization vector,

$\mathbf{J}$ = current vector, q_e = charge density,

$\mathbf{D} = \mathbf{E} + \mathbf{P}$ = dielectric displacement vector,

$\mathbf{B} = \mathbf{H} + \mathbf{M}$ = magnetic induction vector.

We use Lorentz–Heaviside units.

In a frame of reference x_k, moving with velocity $\mathbf{v}$, the electromagnetic fields are given by

$$
\begin{aligned}
\mathscr{E} &= \mathbf{E} + \frac{1}{c}\mathbf{v} \times \mathbf{B}, & \mathscr{H} &= \mathbf{H} - \frac{1}{c}\mathbf{v} \times \mathbf{D}, \\
\mathbf{P} &= \mathbf{P}, & \mathscr{M} &= \mathbf{M} + \frac{1}{c}\mathbf{v} \times \mathbf{P}, \\
\mathscr{J} &= \mathbf{J} - q_e \mathbf{v}.
\end{aligned}
\tag{1.17}
$$

The material and spatial components of these vectors are related to each other by

$$
\begin{aligned}
\mathscr{E}_K &= \mathscr{E}_k x_{k,K}, & B_K &= B_k x_{k,K}, \\
\Pi_K &= \frac{\rho_0}{\rho} X_{K,k} P_k, & M_K &= \frac{\rho_0}{\rho} X_{K,k} \mathscr{M}_k, \\
\mathscr{J}_K &= \frac{\rho_0}{\rho} X_{K,k} \mathscr{J}_k, & \frac{\rho_0}{\rho} &= J = \det(x_{k,K}).
\end{aligned}
\tag{1.18}
$$

1.2. Balance Laws (in $\mathscr{V} - \sigma$)

In order to consider the jump conditions, we assume that the body $\mathscr{B}$ is swept by a discontinuity surface σ with velocity $\mathbf{v}$, at the spatial configurations. The image of σ, in the material frame, is denoted by Σ. We write $V - \Sigma$ and

$\mathscr{V} - \sigma$ for the volumes occupied by the body, excluding the points on the discontinuity surface; and $\partial V - \Sigma$ and $\partial \mathscr{V} - \sigma$ for the surface of the body, excluding the points of intersections of Σ and σ with surfaces ∂V and $\partial \mathscr{V}$, e.g.,

$$\partial \mathscr{V} - \sigma = \partial \mathscr{V} - \partial \mathscr{V} \cap \sigma.$$

The balance laws are expressed in $\mathscr{V} - \sigma$ and the jump conditions on σ.

A. *Maxwell's Equations* (in $\mathscr{V} - \sigma$)

$$\begin{gathered} \nabla \cdot \mathbf{D} = q_e, \qquad \nabla \times \mathbf{E} + \frac{1}{c}\frac{\partial \mathbf{B}}{\partial t} = \mathbf{0}, \\ \nabla \cdot \mathbf{B} = 0, \qquad \nabla \times \mathbf{H} - \frac{1}{c}\frac{\partial \mathbf{D}}{\partial t} = \frac{1}{c}\mathbf{J}, \\ \frac{\partial q_e}{\partial t} + \nabla \cdot \mathbf{J} = 0. \end{gathered} \tag{1.19}$$

B. *Mechanical Balance Laws* (in $\mathscr{V} - \sigma$)

$$\begin{gathered} \rho_0 = \rho J \qquad \text{or} \qquad \dot{\rho} + \rho \nabla \cdot \mathbf{v} = 0, \\ t_{kl,k} + \rho(f_l - \dot{v}_l) + {}_{\mathrm{M}}f_l = 0, \\ t_{[kl]} = \mathscr{E}_{[k} P_{l]} + B_{[k} \mathscr{M}_{l]}, \\ \rho \dot{\varepsilon} - t_{kl} v_{l,k} - \nabla \cdot \mathbf{q} - \rho h - \rho \mathscr{E} \cdot (\mathbf{P}/\rho)^{\cdot} + \mathscr{M} \cdot \dot{\mathbf{B}} - \mathscr{J} \cdot \mathscr{E} = 0, \\ \rho \dot{\eta} - \nabla \cdot \left(\frac{\mathbf{q}}{\theta}\right) - \frac{\rho h}{\theta} \geq 0. \end{gathered} \tag{1.20}$$

The first line of (1.20) is the expression for mass conservation; the second line is for linear momentum; the third line is the expression for the balance of angular momentum; the fourth line is for the energy balance; and the last inequality is the expression for the second law of thermodynamics. Brackets enclosing the indices indicate the antisymmetry, e.g.,

$$t_{[kl]} \equiv \tfrac{1}{2}(t_{kl} - t_{lk}).$$

The physical meanings of various symbols are:

ρ_0 = mass density in $V - \Sigma$,
ρ = mass density in $\mathscr{V} - \sigma$,
f_l = mechanical body force density,
${}_{\mathrm{M}}f_l$ = electromagnetic body force density (see the expression $(1.23)_1$ below),
ε = internal energy density,
h = heat source density,
η = entropy density,
θ = absolute temperature ($\theta > 0$, $\inf \theta = 0$).

1.3. Jump Conditions (on σ)

Corresponding to (1.19) and (1.20) (except for the angular momentum balance), we have the jump conditions at the discontinuity surface σ. When σ coincides with the surface $\partial\mathscr{V}$ of the body these equations give boundary conditions. The jump discontinuity across σ is denoted by boldface brackets, e.g.,

$$[A] = A^+ - A^-,$$

where A^+ and A^- are, respectively, the values of A at σ, taken from the positive and negative directions of the unit normal $\mathbf{n}$ of σ.

A. *Electromagnetic Jump Conditions* (*on* σ):

$$\begin{aligned} &\mathbf{n}\cdot[\mathbf{D}] = w_e, \qquad \mathbf{n}\times\left[\mathbf{E} + \frac{1}{c}\mathbf{v}\times\mathbf{B}\right] = \mathbf{0}, \\ &\mathbf{n}\cdot[\mathbf{B}] = 0, \qquad \mathbf{n}\times\left[\mathbf{H} - \frac{1}{c}\mathbf{v}\times\mathbf{D}\right] = \mathbf{0}, \\ &\mathbf{n}\cdot[\mathbf{J} - q_e\mathbf{v}] = 0. \end{aligned} \tag{1.21}$$

B. *Mechanical Jump Conditions* (*on* σ):

$$\begin{aligned} &[\rho(\mathbf{v} - \mathbf{\nu})]\cdot\mathbf{n} = 0, \\ &[\rho v_l(v_k - \nu_k) - t_{kl} - {}_{\mathrm{M}}t_{kl} - \nu_k G_l]n_k = 0, \\ &[\{\rho\varepsilon + \tfrac{1}{2}\rho v^2 + \tfrac{1}{2}(E^2 + B^2)\}(v_k - \nu_k) \\ &\quad - (t_{kl} + {}_{\mathrm{M}}t_{kl} + \nu_k G_l)v_l - q_k + \mathscr{S}_k]n_k = 0, \\ &\left[\rho\eta(\mathbf{v} - \mathbf{\nu}) - \frac{\mathbf{q}}{\theta}\right]\cdot\mathbf{n} \geq 0. \end{aligned} \tag{1.22}$$

The electromagnetic body force ${}_{\mathrm{M}}\mathbf{f}$, the electromagnetic stress tensor ${}_{\mathrm{M}}\mathbf{t}$, the electromagnetic momentum $\mathbf{G}$, and the Poynting vector $\mathscr{S}$ are given by

$$\begin{aligned} {}_{\mathrm{M}}\mathbf{f} &= q_e\mathbf{E} + \frac{1}{c}\mathbf{J}\times\mathbf{B} + (\nabla\mathbf{E})\cdot\mathbf{P} + (\nabla\mathbf{B})\cdot\mathbf{M} + \frac{1}{c}[(\mathbf{P}\times\mathbf{B})v_k]_{,k} \\ &\quad + \frac{1}{c}\frac{\partial}{\partial t}(\mathbf{P}\times\mathbf{B}), \\ {}_{\mathrm{M}}t_{kl} &= P_k\mathscr{E}_l - B_k\mathscr{M}_l + E_kE_l + B_kB_l - \tfrac{1}{2}(E^2 + B^2 - 2\mathscr{M}\cdot\mathbf{B})\delta_{kl}, \\ G_k &= \frac{1}{c}(\mathbf{E}\times\mathbf{B})_k, \\ \mathscr{S}_k &= c(\mathbf{E}\times\mathbf{H})_k + [{}_{\mathrm{M}}t_{kl} + v_kG_l - \tfrac{1}{2}(E^2 + B^2 + 2\mathbf{E}\cdot\mathbf{B})\delta_{kl}]v_l. \end{aligned} \tag{1.23}$$

1.4. Constitutive Equations of Electromagnetic Elastic Solids

It is convenient to replace the nonsymmetrical stress tensor t_{kl} by a symmetric one. This is done by introducing

$$ {}_{\mathrm{E}}t_{kl} = {}_{\mathrm{E}}t_{lk} = t_{kl} + P_k \mathscr{E}_l + \mathscr{M}_k B_l. \tag{1.24}$$

This eliminates the necessity of using the angular momentum balance, since it reduces to ${}_{\mathrm{E}}t_{kl} = {}_{\mathrm{E}}t_{lk}$. Corresponding to the Piola–Kirchhoff stress tensor reads

$$ {}_{\mathrm{E}}T_{KL} = {}_{\mathrm{E}}T_{LK} = \frac{\rho_0}{\rho} X_{K,k} X_{L,l} \, {}_{\mathrm{E}}t_{kl}. \tag{1.25}$$

The constitutive equations of heat and electric conducting electromagnetic elastic solids are expressed in the form

$$ \rho_0 \Psi \equiv \rho_0(\varepsilon - \theta\eta) - \Pi_K \mathscr{E}_K = \Sigma(C_{KL}, \theta, \theta_{,K}, \mathscr{E}_K, B_K, \mathbf{X}) \tag{1.26}$$

The same functional dependence is assumed for η, ${}_{\mathrm{E}}T_{KL}$, Q_K, Π_K, M_K, and $\mathscr{J}_K$.

In order to obtain the restrictions arising from the second law of thermodynamics, we first eliminate $\rho h/\theta$ from the entropy inequality $(1.20)_6$ by substituting ρh from the energy equation $(1.20)_5$. This leads to the generalized Clausius–Duhem (C–D) inequality

$$ -\rho(\dot{\Psi} + \dot{\theta}\eta) + t_{kl} v_{l,k} + \frac{1}{\theta} q_k \theta_{,k} - P_k \dot{\mathscr{E}}_k - \mathscr{M}_k \dot{B}_k + \mathscr{J}_k \mathscr{E}_k \geq 0, \tag{1.27}$$

where

$$ \Psi = \frac{\Sigma}{\rho_0} = \varepsilon - \theta\eta - \frac{1}{\rho} \mathscr{E}_k P_k = \varepsilon - \theta\eta - \frac{1}{\rho_0} \Pi_K \mathscr{E}_K. \tag{1.28}$$

By the use of (1.8), (1.15), $(1.16)_3$, (1.24), and (1.25) this inequality may be transformed to

$$ -(\dot{\Sigma} + \rho_0 \eta \dot{\theta}) + \tfrac{1}{2} {}_{\mathrm{E}}T_{KL} \dot{C}_{KL} + \frac{1}{\theta} Q_K \theta_{,K} - \Pi_K \dot{\mathscr{E}}_K - M_K \dot{B}_K + \mathscr{J}_K \mathscr{E}_K \geq 0. \tag{1.29}$$

Upon substituting $\dot{\Sigma}$, calculated from (1.26), into (1.29) we have

$$ -\left(\frac{\partial \Sigma}{\partial \theta} + \rho_0 \eta\right)\dot{\theta} + \left(\tfrac{1}{2} T_{KL} - \frac{\partial \Sigma}{\partial C_{KL}}\right)\dot{C}_{KL} - \left(\Pi_K + \frac{\partial \Sigma}{\partial \mathscr{E}_K}\right)\dot{\mathscr{E}}_K $$

$$ - \left(M_K + \frac{\partial \Sigma}{\partial B_K}\right)\dot{B}_K + \frac{\partial \Sigma}{\partial \theta_{,K}} \dot{\theta}_{,K} + \frac{1}{\theta} Q_K \theta_{,K} + \mathscr{J}_K \mathscr{E}_K \geq 0. \tag{1.30}$$

This inequality is linear in $\dot{\theta}$, $\dot{\theta}_{,K}$, $\dot{C}_{KL}$, $\dot{\mathscr{E}}_K$, and $\dot{B}_K$. The necessary and sufficient conditions, for the inequality not to be violated, and for all independent variations of these quantities, are

$$ \frac{\partial \Sigma}{\partial \theta_{,K}} = 0, $$

$$
{}_{\mathrm{E}}T_{KL} = 2\frac{\partial \Sigma}{\partial C_{KL}}, \qquad \eta = -\frac{1}{\rho_0}\frac{\partial \Sigma}{\partial \theta},
$$
$$
\Pi_K = -\frac{\partial \Sigma}{\partial \mathscr{E}_K}, \qquad M_K = -\frac{\partial \Sigma}{\partial B_K}, \tag{1.31}
$$

and

$$
\frac{1}{\theta} Q_K \theta_{,K} + \mathscr{J}_K \mathscr{E}_K \geq 0. \tag{1.32}
$$

From the continuity of $\mathbf{Q}$ and $\mathscr{J}$ in $\nabla\theta$ and $\mathscr{E}$ it also follows that

$$
\mathbf{Q} = \mathscr{J} = \mathbf{0}, \qquad \text{when} \quad \nabla\theta = \mathscr{E} = \mathbf{0}. \tag{1.33}
$$

The energy equation is now reduced to

$$
\rho\theta\dot{\eta} - \nabla\cdot\mathbf{q} - \mathscr{J}\cdot\mathscr{E} - \rho h = 0. \tag{1.34}
$$

Equations (1.31) show that Σ is independent of $\theta_{,K}$; and ${}_{\mathrm{E}}\mathbf{T}$, η, $\mathbf{\Pi}$, and $\mathbf{M}$ are determined when the function Σ is known. It remains to write the constitutive equations for $\mathbf{Q}$ and $\mathscr{J}$, as functions of the same list of independent variables that appear in (1.26), i.e.,

$$
\begin{aligned}
\mathscr{J} &= \mathscr{J}(C_{KL}, \theta, \theta_{,K}, \mathscr{E}_K, B_K, \mathbf{X}),\\
\mathbf{Q} &= \mathbf{Q}(C_{KL}, \theta, \theta_{,K}, \mathscr{E}_K, B_K, \mathbf{X}).
\end{aligned} \tag{1.35}
$$

These equations are then restricted by the C–D inequality (1.32).

Spatial forms of the constitutive equations follow from (1.18) and (1.25)

$$
\begin{aligned}
\rho_0\Psi &= \Sigma(C_{KL}, \theta, \mathscr{E}_K, B_K, \mathbf{X}),\\
\eta &= -\frac{1}{\rho_0}\frac{\partial \Sigma}{\partial \theta},\\
{}_{\mathrm{E}}t_{kl} &= 2\frac{\rho}{\rho_0}\frac{\partial \Sigma}{\partial C_{KL}} x_{k,K} x_{l,L},\\
P_k &= -\frac{\rho}{\rho_0}\frac{\partial \Sigma}{\partial \mathscr{E}_K} x_{k,K},\\
\mathscr{M}_k &= -\frac{\rho}{\rho_0}\frac{\partial \Sigma}{\partial B_K} x_{k,K},\\
\mathscr{J}_k &= \frac{\rho}{\rho_0}\mathscr{J}_K x_{k,K},\\
q_k &= \frac{\rho}{\rho_0} Q_K x_{k,K}.
\end{aligned} \tag{1.36}
$$

To discuss the material symmetry regulations it proves convenient to study the invariance of scalar functions such as Σ. This can be done for $\mathscr{J}$ and $\mathbf{Q}$ by including one additional vector variable A_K to the list of vector and tensor variables $(C_{KL}, \mathscr{E}_K, B_K)$.

$$Q = Q_K A_K, \qquad \mathscr{J} = \mathscr{J}_K A_K. \tag{1.37}$$

Here Q and $\mathscr{J}$ are scalar functions that depend on C_{KL}, $\mathscr{E}_K$, B_K, and A_K. Once the invariance requirements of Q and $\mathscr{J}$ are determined, we can calculate Q_K and $\mathscr{J}_K$ by

$$Q_K = \left.\frac{\partial Q}{\partial A_K}\right|_{\mathbf{A}=\mathbf{0}}, \qquad \mathscr{J}_K = \left.\frac{\partial \mathscr{J}}{\partial A_K}\right|_{\mathbf{A}=\mathbf{0}}. \tag{1.38}$$

By means of a Legendre transformation, alternative forms of constitutive equations may be written in terms of the independent variables E_{KL}, Π_K, and M_K. To this end, we introduce a new energy function W by

$$W(E_{KL}, \theta, \Pi_K, M_K, \mathbf{X}) = \Sigma + \mathbf{\Pi}\cdot\mathscr{E} + \mathbf{M}\cdot\mathbf{B}. \tag{1.39}$$

From the C–D inequality (1.29) there follows

$$\begin{aligned} \eta &= -\frac{1}{\rho_0}\frac{\partial W}{\partial \theta}, \\ {}_{\mathrm{E}}T_{KL} &= \frac{\partial W}{\partial E_{KL}}, \\ \mathscr{E}_K &= \frac{\partial W}{\partial \Pi_K}, \\ B_K &= \frac{\partial W}{\partial M_K}, \end{aligned} \tag{1.40}$$

and the inequality (1.32). Equations (1.40) replace equations (1.31).

The following chapters present discussions of the magnetic symmetry regulations for single scalar functions Σ or W which may be replaced by Q or $\mathscr{J}$.

1.5. Constitutive Equations of Electromagnetic Fluids

For electromagnetic fluids, the starting constitutive assumption is

$$\Psi = \Psi(\rho^{-1}, d_{kl}, \theta, \theta_{,k}, \mathscr{E}_k, B_k). \tag{1.41}$$

The C–D inequality (1.27) leads to

$$\begin{aligned} &\frac{\partial \Psi}{\partial \mathbf{d}} = \mathbf{0}, && \frac{\partial \Psi}{\partial(\nabla\theta)} = \mathbf{0}, && t_{[kl]} = 0, \\ &\eta = -\frac{\partial \Psi}{\partial \theta}, && P_k = -\rho\frac{\partial \Psi}{\partial \mathscr{E}_k}, && \mathscr{M}_k = -\rho\frac{\partial \Psi}{\partial B_k}, \end{aligned} \tag{1.42}$$

and

$$\rho\gamma \equiv {}_{\mathrm{D}}t_{kl}d_{lh} + \frac{1}{\theta}q_k\theta_{,k} + \mathscr{J}_k\mathscr{E}_k \geq 0, \tag{1.43}$$

where ${}_{\mathrm{D}}\mathbf{t}$ is the symmetric dissipative stress tensor defined by

$$ {}_{\mathrm{D}}t_{kl} = t_{kl} + \pi\delta_{kl}; \qquad \pi \equiv -\frac{\partial\Psi}{\partial\rho^{-1}}. \tag{1.44}$$

Here π is the thermodynamic pressure. Clearly Ψ and π are independent of $\mathbf{d}$ and $\nabla\theta$.

$$\Psi = \Psi(\rho^{-1}, \theta, \mathscr{E}_k, B_k). \tag{1.45}$$

From the continuity of ${}_{\mathrm{D}}\mathbf{t}$, $\mathbf{q}$, and $\mathscr{J}$ in $\mathbf{d}$, $\nabla\theta$, and $\mathscr{E}$, it follows that

$$ {}_{\mathrm{D}}\mathbf{t} = \mathbf{0}, \quad \mathbf{q} = \mathbf{0}, \quad \mathscr{J} = \mathbf{0} \qquad \text{when} \quad \mathbf{d} = \mathbf{0}, \quad \nabla\theta = \mathbf{0}, \quad \mathscr{E} = \mathbf{0}. \tag{1.46}$$

Again, separate constitutive equations will have to be written for ${}_{\mathrm{D}}\mathbf{t}$, $\mathbf{q}$, and $\mathscr{J}$ in the form of (1.41) and subjected to the C–D inequality (1.43).

Since Ψ must be invariant under *arbitrary time-dependent rotations*, it will depend on $\mathscr{E}$ and $\mathbf{B}$ only through their invariants, i.e.,

$$\Psi = \Psi(\mathrm{I}_1, \mathrm{I}_2, \mathrm{I}_3, \theta, \rho^{-1}), \tag{1.47}$$

where

$$\mathrm{I}_1 = \mathscr{E}\cdot\mathscr{E}, \qquad \mathrm{I}_2 = \mathbf{B}\cdot\mathbf{B}, \qquad \mathrm{I}_3 = (\mathscr{E}\cdot\mathbf{B})^2. \tag{1.48}$$

This leads to

$$\begin{aligned}
\eta &= -\frac{\partial\Psi}{\partial\theta}, \\
\mathbf{P} &= -2\rho\left[\frac{\partial\Psi}{\partial\mathrm{I}_1}\mathscr{E} + \frac{\partial\Psi}{\partial\mathrm{I}_3}(\mathscr{E}\cdot\mathbf{B})\mathbf{B}\right], \\
\mathscr{M} &= -2\rho\left[\frac{\partial\Psi}{\partial\mathrm{I}_2}\mathbf{B} + \frac{\partial\Psi}{\partial\mathrm{I}_3}(\mathscr{E}\cdot\mathbf{B})\mathscr{E}\right].
\end{aligned} \tag{1.49}$$

From the nonlinear constitutive equations obtained, we can derive various other approximate equations. Usually, polynomial equations are written in terms of the powers of $\mathbf{C}$, $\mathscr{E}$, $\mathbf{B}$, and $\nabla\theta$. The linear theory involves only the first power of these quantities. For these we refer the reader to Eringen [1980, Chap. 10].

CHAPTER 2

Conventional Crystallographic Point Groups

A three-dimensional point group is a group of symmetry operators which acts at a fixed point O, and leaves invariant all distances and angles in a three-dimensional Euclidean space. The symmetry operators having these properties are rotations about the axes through O, or products of such rotations and the inversion. Such products, of course, include reflections in planes through O.

If the group contains all possible rotations and no other elements, it is called the three-dimensional *proper rotation group*. It is isomorphic with the group $O^+(3)$ for all 3×3 orthogonal matrices with determinant $(+1)$.

If a group contains all possible rotations and their products, with the inversion it is called the three-dimensional *rotation group*. It is isomorphic with the group $O(3)$ for all 3×3 orthogonal matrices. Operators whose matrices having determinant (-1) are called *improper rotations*. They are products of the proper rotation and the inversion. Note that the inversion commutes with all rotations.

Every subgroup of the three-dimensional rotation group is a point group. Subgroups of $O^+(3)$ are called *proper point groups*. Proper point groups of finite order are classified as

$$\begin{gathered}\text{Cyclic } (Cn = n); \qquad \text{Dihedral } (D_n = n22,\ n \text{ even},\ D_n = n2,\ n \text{ odd}); \\ \text{Tetrahedral } (T = 23); \qquad \text{and} \qquad \text{Octahedral } (O = 432).\end{gathered} \tag{2.1}$$

We are concerned with the crystallographic point groups. We have an extra requirement that an operator is compatible with the translational symmetry of crystalline solids. Hence, the appropriate symmetry operations are the identity E, the inversion C, reflections in certain planes σ, and rotations about certain axes of orders $n = 1, 2, 3, 4$, or 6. The last symmetry operators are denoted by C_{nr}, which means an anticlockwise rotation through $2\pi/n$ radians about the same axis. The eleven proper point groups obtained from (2.1) are listed in Table 2.1, together with their symmetry elements.

In Table 2.1, $j = 1, 2, 3, 4$; $m = x, y, z$; $p = a, b, c, d, e, f$; and $r = 1, 2, 3$; and the labels of the symmetry operations can be identified from Figures 2.1–2.3.

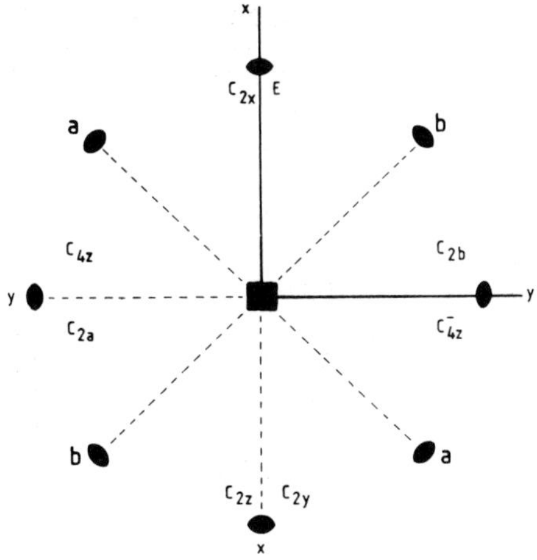

Figure 2.1. Symmetry elements: triclinic, monoclinic, rhombic, and tetragonal systems.

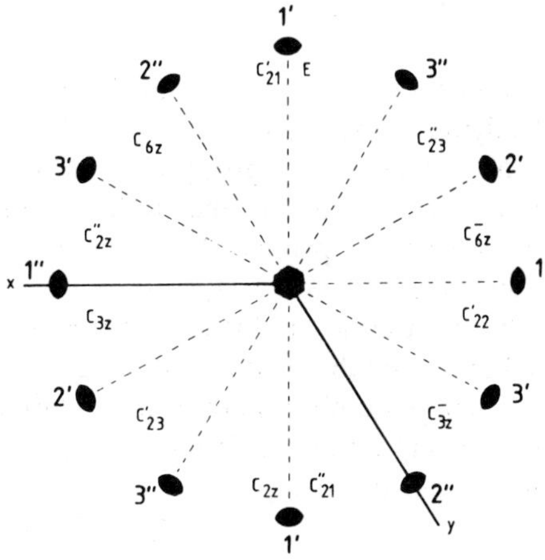

Figure 2.2. Symmetry elements: trigonal and hexagonal systems.

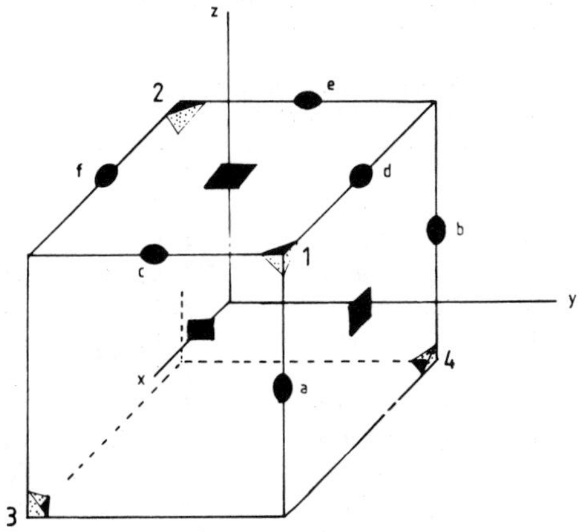

Figure 2.3. Symmetry elements: cubic system.

Table 2.1. Crystallographic pure rotation groups.

Cyclic groups	Symmetry elements
$C_1 = 1$	E
$C_2 = 2$	E, C_{2z}
$C_3 = 3$	E, C_{3z}, C_{3z}^-
$C_4 = 4$	$E, C_{4z}, C_{4z}^-, C_{2z}$
$C_6 = 6$	$E, C_{6z}, C_{6z}^-, C_{3z}, C_{3z}^-, C_{2z}$
Dihedral groups	
$D_2 = 222$	$E, C_{2x}, C_{2y}, C_{2z}$
$D_3 = 32$	$E, C_{3z}, C_{3z}^-, C'_{21}, C'_{22}, C'_{23}$
$D_4 = 422$	$E, C_{4z}, C_{4z}^-, C_{2x}, C_{2y}, C_{2z}, C_{2a}, C_{2b}$
$D_6 = 622$	$E, C_{6z}, C_{6z}^-, C_{3z}, C_{3z}^-, C_{2z}, C'_{2r}, C''_{2r}$
Tetrahedral group	
$T = 23$	$E, C_{2m}, C_{3j}, C_{3j}^-$
Octahedral group	
$O = 432$	$E, C_{2m}, C_{3j}, C_{3j}^-, C_{2p}, C_{4m}, C_{4m}^-$

In Figures 2.1 and 2.2, the labels of the symmetry operations are placed on the figure in the position in which the letter E is taken by that operation.

Now consider the group $C_i = \{E, C\}$ consisting of the identity and inversion. Then, given a proper point group $\{P\}$, it is always possible to form the outer direct product $\{P\} \times \{E, C\}$. This process gives a new set of point groups that are subgroups of $O(3)$. From the proper point groups listed in Table 2.1, the eleven distinct extra point groups are generated in this manner, and they are tabulated in Table 2.2.

There is another method for obtaining new point groups from a proper point group $\{P\}$, provided that $\{P\}$ has an invariant subgroup $\{H\}$ of index 2. In this case the group

$$\{P\} = \{H\} + C\{P - H\} \tag{2.2}$$

Table 2.2. Direct product point groups $\{P'\} = \{P\} + C\{P\}$.

Proper point groups	Direct product groups $\{P'\} = \{P\} + C\{P\}$	Symmetry operators
C_1	$S_2 = C_i = \bar{1}$	E, C
C_2	$C_{2h} = 2/m$	E, C, C_{2z}, σ_z
C_3	$S_6 = C_{3i} = \bar{C}_3 = \bar{3}$	$E, C, C_{3z}, C_{3z}^-, S_{6z}^-, S_{6z}$
C_4	$C_{4h} = 4/m$	$E, C_{4z}, C_{4z}^-, C_{2z}, C, S_{4z}, S_{4z}^-, \sigma_z$
C_6	$C_{6h} = 6/m$	$E, C_{6z}, C_{6z}^-, C_{3z}, C_{3z}^-, C_{2z}, C, S_{3z}, S_{3z}^-, S_{6z}, S_{6z}^-, \sigma_h$
D_2	$D_{2h} = mmm$	$E, C, C_{2x}, C_{2y}, C_{2z}, \sigma_x, \sigma_y, \sigma_z$
D_3	$D_{3v} = \bar{3}m$	$E, C, C_{3z}, C_{3z}^-, S_{6z}, S_{6z}^-, C'_{2r}, \sigma_{dr}$
D_4	$D_{4h} = 4/mmm$	$E, C, C_{4z}, C_{4z}^-, S_{4z}, S_{4z}^-, C_{2m}, C_{2a}, C_{2b}, \sigma_m, \sigma_{da}, \sigma_{db}$
D_6	$D_{6h} = 6/mmm$	$E, C, C_{6z}, C_{6z}^-, C_{3z}, C_{3z}^-, C_{2z}, C'_{2r}, C''_{2r}, S_{3z}, S_{3z}^-, C_{6z}^{\mp}, \sigma_h, \sigma_{dr}, \sigma_{vr}$
T	$T_h = m3$	$E, C, C_{2m}, \sigma_m, C_{3j}, C_{3j}^-, S_{6j}, S_{6j}^-$
O	$O_h = m3m$	$E, C, C_{2m}, C_{2p}, C_{3j}, C_{3j}^-, C_{4m}, C_{4m}^-, \sigma_m, \sigma_{dp}, S_{6j}^-, S_{6j}, S_{4m}^-, S_{4m}$

Note that the operator S denotes rotation reflection and $CC_{3z}^- = S_{6z}$, $CC_{3z} = S_{6z}^-$, $CC_{4z}^- = S_{4z}$, etc.

Table 2.3. Point groups $\{P'\} = \{H\} + C\{P - H\}$.

Proper point groups	Invariant subgroup	New point groups $\{P'\}$	Symmetry operators
C_2	1	$C_s = m$	E, σ_z
C_4	2	$S_4 = \bar{4}$	$E, C_{2z}, S_{4z}, S_{4z}^-$
C_6	3	$C_{3h} = \bar{6}$	$E, S_{3z}, S_{3z}^-, C_{3z}, C_{3z}^-, \sigma_h$
D_2	2	$C_{2v} = 2mm$	$E, C_{2z}, \sigma_x, \sigma_y$
D_3	3	$C_{3v} = 3m$	$E, C_{3z}, C_{3z}^-, \sigma_{dr}$
D_4	4	$C_{4v} = 4mm$	$E, C_{4z}, C_{4z}^-, C_{2z}, \sigma_x, \sigma_y, \sigma_{da}, \sigma_{db}$
	222	$D_{2d} = \bar{4}2m$	$E, S_{4z}, S_{4z}^-, C_{2x}, C_{2y}, C_{2z}, \sigma_{da}, \sigma_{db}$
D_6	6	$C_{6v} = 6mm$	$E, C_{6z}, C_{6z}^-, C_{3z}, C_{3z}^-, C_{2z}, \sigma_{dr}, \sigma_{vr}$
	32	$D_{3h} = \bar{6}m2$	$E, S_{3z}, S_{3z}^-, C_{3z}, C_{3z}^-, \sigma_h, C_{2r}', \sigma_{vr}$
O	23	$T_d = \bar{4}3m$	$E, C_{2m}, C_{3j}, C_{3j}^-, \sigma_{dp}, S_{4m}, S_{4m}^-$

is also a point group. The ten distinct point groups derived in this manner are listed in Table 2.3.

The thirty-two point groups listed in Tables 2.1–2.3 are further classified into seven crystal systems according to their order on the principal axis. There are five crystal systems with a single principal axis of order 1, 2, 3, 4, or 6, namely, triclinic, monoclinic, trigonal, tetragonal, and hexagonal systems, respectively. The rhombic system has three mutually perpendicular rotation axes of order two, and the cubic system has four rotation axes directed towards the vertices of a regular tetrahedron, each of order three.

The symmetry operation describing the symmetry properties of each of the thirty-two crystallographic classes are represented by the 3×3 matrices in a given three-dimensional basis. The fifteen matrices listed below, together with their products, produces the symmetry elements given in Tables 2.1–2.3.

Symmetry operators and the corresponding representation matrices are listed in Table 2.4.

Symmetry operation matrices describing the symmetry property of each of the thirty-two crystal classes are given in Table 2.5. The classes are arranged

Table 2.4. Symmetry operators and their representation matrices.

Symmetry elements	Representing matrices	Symmetry elements	Representing matrices
E	$\mathbf{I}$	$C_{31}^-, C_{32}^-, C_{33}^-, C_{34}^-$	$\mathbf{M}_1, \mathbf{D}_1\mathbf{M}_1, \mathbf{D}_2\mathbf{M}_1, \mathbf{D}_3\mathbf{M}_1$
C	$\mathbf{C}$		
C_{2x}, C_{2y}, C_{2z}	$\mathbf{D}_1, \mathbf{D}_2, \mathbf{D}_3$	$C_{31}, C_{32}, C_{33}, C_{34}$	$\mathbf{M}_2, \mathbf{D}_2\mathbf{M}_2, \mathbf{D}_3\mathbf{M}_2, \mathbf{D}_1\mathbf{M}_2$
$\sigma_x, \sigma_y, \sigma_z$	$\mathbf{R}_1, \mathbf{R}_2, \mathbf{R}_3$	$S_{61}, S_{62}, S_{63}, S_{64}$	$\mathbf{CM}_1, \mathbf{R}_1\mathbf{M}_1, \mathbf{R}_2\mathbf{M}_1, \mathbf{R}_3\mathbf{M}_1$
σ_h	$\mathbf{R}_3$		
$\sigma_{df}, \sigma_{de}, \sigma_{db}$	$\mathbf{T}_1, \mathbf{T}_2, \mathbf{T}_3$	$S_{61}^-, S_{62}^-, S_{63}^-, S_{64}^-$	$\mathbf{CM}_2, \mathbf{R}_2\mathbf{M}_2, \mathbf{R}_3\mathbf{M}_2, \mathbf{R}_1\mathbf{M}_2$
$\sigma_{da}, \sigma_{dc}, \sigma_{dd}$	$\mathbf{D}_3\mathbf{T}_3, \mathbf{D}_2\mathbf{T}_2, \mathbf{D}_1\mathbf{T}_1$	C_{4x}, C_{4y}, C_{4z}	$\mathbf{R}_2\mathbf{T}_1, \mathbf{R}_3\mathbf{T}_2, \mathbf{R}_1\mathbf{T}_3$
C_{2f}, C_{2e}, C_{2b}	$\mathbf{CT}_1, \mathbf{CT}_2, \mathbf{CT}_3$	$C_{4x}^-, C_{4y}^-, C_{4z}^-$	$\mathbf{R}_3\mathbf{T}_1, \mathbf{R}_1\mathbf{T}_2, \mathbf{R}_2\mathbf{T}_3$
C_{2a}, C_{2c}, C_{2d}	$\mathbf{R}_3\mathbf{T}_3, \mathbf{R}_2\mathbf{T}_2, \mathbf{R}_1\mathbf{T}_1$		
$S_{4x}^-, S_{4y}^-, S_{4z}^-$	$\mathbf{D}_2\mathbf{T}_1, \mathbf{D}_3\mathbf{T}_2, \mathbf{D}_1\mathbf{T}_3$	S_{4x}, S_{4y}, S_{4z}	$\mathbf{D}_3\mathbf{T}_1, \mathbf{D}_1\mathbf{T}_2, \mathbf{D}_2\mathbf{T}_3$
C_{3z}^-, C_{3z}	$\mathbf{S}_1, \mathbf{S}_2$	S_{6z}, S_{6z}^-	$\mathbf{CS}_1, \mathbf{CS}_2$

Table 2.5. The thirty-two conventional crystal classes.

System	Class number	Class name		Symmetry transformations	Order
Triclinic	1	Pedial	$C_1 = 1$	I	1
	2	Pinacoidal	$C_i = \bar{1}$	I, C	2
Monoclinic	3	Sphenoidal	$C_2 = 2$	I, D_3	2
	4	Domatic	$C_s = m$	I, R_3	2
	5	Prismatic	$C_{2h} = 2/m$	I, C, R_3, D_3	4
Orthorhombic	6	Rhombic–disphenonidal	$D_2 = 222$	I, D_1, D_2, D_3	4
	7	Rhombic–pyromidal	$C_{2v} = 2mm$	I, R_1, R_2, D_3	4
	8	Rhombic–dipyramidal	$D_{2h} = mmm$	$I, C, D_1, D_2, D_3, R_1, R_2, R_3$	8
Tetragonal	9	Tetragonal–pyramidal	$C_4 = 4$	$I, D_3, R_1 T_3, R_2 T_3$	4
	10	Tetragonal–disphenoidal	$\bar{C}_2 = \bar{4}$	$I, D_3, D_1 T_3, D_2 T_3$	4
	11	Tetragonal–dipyramidal	$C_{4h} = 4/m$	$I, D_3, D_1 T_3, D_2 T_3, R_1 T_3, R_2 T_3, C, R_3$	8
	12	Tetragonal–trapezahedral	$D_4 = 422$	$I, D_1, D_2, D_3, CT_3, R_1 T_3, R_2 T_3, R_3 T_3$	8
	13	Ditetragonal–pyramidal	$C_{4v} = 4mm$	$I, R_1, R_2, D_3, T_3, R_1 T_3, R_2 T_3, D_3 T_3$	8
	14	Tetragonal–scalenohedral	$D_{2v} = \bar{4}2m$	$I, D_1, D_2, D_3, T_3, D_1 T_3, D_2 T_3, D_3 T_3$	8
	15	Ditetragonal–dipyramidal	$D_{4h} = 4/mmm$	$I, D_1, D_2, D_3, CT_3, R_1 T_3, R_2 T_3, R_3 T_3, C, R_1, R_2, R_3, T_3, D_1 T_3, D_2 T_3, D_3 T_3$	16

Trigonal	16	Trigonal–pyramidal	$C_3 = 3$	I, S_1, S_2	3
	17	Rhombohedral	$\bar{C}_3 = \bar{3}$	$I, S_1, S_2, C, CS_1, CS_2$	6
	18	Trigonal–trapezohedral	$D_3 = 32$	$I, S_1, S_2, D_1, D_1S_1, D_1S_2$	6
	19	Ditrigonal–pyramidal	$C_{3v} = 3m$	$I, S_1, S_2, R_1, R_1S_1, R_1S_2$	6
	20	Hexagonal–scalenohedral	$D_{3v} = \bar{3}m$	$I, S_1, S_2, C, CS_1, CS_2, R_1, R_1S_1, R_1S_2, D_1, D_1S_1, D_1S_2$	12
Hexagonal	21	Hexagonal–pyramidal	$C_6 = 6$	$I, S_1, S_2, D_3, D_3S_1, D_3S_2$	6
	22	Trigonal–dipyramidal	$C_{3h} = \bar{6}$	$I, S_1, S_2, R_3, R_3S_1, R_3S_2$	6
	23	Hexagonal–dipyramidal	$C_{6h} = 6/m$	$I, S_1, S_2, R_3, R_3S_1, R_3S_2, C, CS_1, CS_2, D_3, D_3S_1, D_3S_2$	12
	24	Hexagonal–trapezohedral	$D_6 = 622$	$I, S_1, S_2, D_3, D_3S_1, D_3S_2, D_1, D_1S_1, D_1S_2, D_2S_1, D_2S_2, D_2$	12
	25	Dihexagonal–pyramidal	$C_{6v} = 6mm$	$I, S_1, S_2, D_3, D_3S_1, D_3S_2, R_1, R_1S_1, R_1S_2, R_2, R_2S_1, R_2S_2$	12
	26	Ditrigonal–dipyramidal	$D_{3h} = \bar{6}m2$	$I, S_1, S_2, R_3, R_3S_1, R_3S_2, D_1, D_1S_1, D_1S_2, R_2, R_2S_1, R_2S_2$	12
	27	Dihexagonal–dipyramidal	$D_{6h} = 6/mm$	$I, S_1, S_2, C, CS_1, CS_2, D_1, D_1S_1, D_1S_2, D_2, D_2S_1, D_2S_2, R_1, R_1S_1,$ $R_1S_2, R_2, R_2S_1, R_2S_2, R_3, R_3S_1, R_3S_2, D_3, D_3S_1, D_3S_2$	24
Cubic	28	Tetartoidal	$T = 23$	$I, D_1, D_2, D_3, C_{3j}, C_{3j}^{-}$	12
	29	Diploidal	$T_h = m3$	$I, D_1, D_2, D_3, C, R_1, R_2, R_3, C_{3j}, C_{3j}^{-}, S_{6j}, S_{6j}^{-}$	24
	30	Gyroidal	$O = 432$	$I, D_1, D_2, D_3, C_{2p}, C_{3j}, C_{3j}^{-}, C_{4m}, C_{4m}^{-}$	24
	31	Hextetrahedral	$Td = \bar{4}3m$	$I, D_1, D_2, D_3, \sigma_2, C_{3j}, C_{3j}^{-}, S_{4m}, S_{4m}^{-}$	24
	32	Hexoctohedral	$O_h = m3m$	$I, D_1, D_2, D_3, C_{2p}, C_{3j}, C_{3j}^{-}, C_{4m}, C_{4m}^{-}, C, R_1, R_2, R_3, \sigma_{dp}, S_{6j}, S_{6j}^{-},$ S_{4m}, S_{4m}^{-}	48

Table 2.6. Subgroups of the thirty-two conventional point groups.

Point groups	Subgroups (halving subgroups are enclosed by parenthesis)
Triclinic	
1	
$\bar{1}$	(1)
Monoclinic	
2	(1)
m	(1)
2/*m*	1, ($\bar{1}$), (2), (*m*)
Rhombic	
222	1, (2)
2*mm*	1, (2), (*m*)
mmm	1, $\bar{1}$, 2, *m*, (2/*m*), (222), (2*mm*)
Tetragonal	
4	1, (2)
$\bar{4}$	1, (2)
4/*m*	1, $\bar{1}$, 2, *m*, (2/*m*), (4), ($\bar{4}$)
422	1, 2, (222), (4)
4*mm*	1, 2, *m*, (2*mm*), (4)
$\bar{4}$2*m*	1, 2, *m*, (222), (2*mm*), ($\bar{4}$)
4/*mmm*	1, $\bar{1}$, 2, *m*, 2/*m*, 222, 2*mm*, (*mmm*), 4, $\bar{4}$, (4/*m*), (422), (4*mm*), ($\bar{4}$2*m*)
Trigonal	
3	
$\bar{3}$	(3)
32	(3)
3*m*	(3)
$\bar{3}$*m*	3, ($\bar{3}$), (32), (3*m*)
Hexagonal	
6	(3)
$\bar{6}$	(3)
6/*m*	3, ($\bar{3}$), (6), ($\bar{6}$)
622	3, (32), (6)
6*mm*	3, (3*m*), (6)
$\bar{6}$*m*2	3, (32), (3*m*), ($\bar{6}$)
6/*mmm*	3, $\bar{3}$, 32, 3*m*, ($\bar{3}$*m*), 6, $\bar{6}$, (6/*m*), (622), (6*mm*), ($\bar{6}$*m*2)
Cubic	
23	3
*m*3	3, $\bar{3}$, (23)
432	3, 32, (23)
$\bar{4}$3*m*	3, 3*m*, (23)
*m*3*m*	3, $\bar{3}$, 32, 3*m*, $\bar{3}$*m*, 23, (*m*3), (432), ($\bar{4}$3*m*)

according to the seven crystal systems, and their Schönflies and International symbols are also indicated. The arbitrary class numbers in column 2 are those of Koster et al. [1963]. Finally, in Table 2.6, we list subgroups of the thirty-two conventional point groups for future use.

Matrices for the symmetry operations are:

$$\mathbf{I} = \begin{pmatrix} 1 & 0 & 0 \\ 0 & 1 & 0 \\ 0 & 0 & 1 \end{pmatrix}, \qquad \mathbf{C} = \begin{pmatrix} -1 & 0 & 0 \\ 0 & -1 & 0 \\ 0 & 0 & -1 \end{pmatrix},$$

$$\mathbf{R}_1 = \begin{pmatrix} -1 & 0 & 0 \\ 0 & 1 & 0 \\ 0 & 0 & 1 \end{pmatrix}, \quad \mathbf{R}_2 = \begin{pmatrix} 1 & 0 & 0 \\ 0 & -1 & 0 \\ 0 & 0 & 1 \end{pmatrix}, \quad \mathbf{R}_3 = \begin{pmatrix} 1 & 0 & 0 \\ 0 & 1 & 0 \\ 0 & 0 & -1 \end{pmatrix},$$

$$\mathbf{D}_1 = \begin{pmatrix} 1 & 0 & 0 \\ 0 & -1 & 0 \\ 0 & 0 & -1 \end{pmatrix}, \quad \mathbf{D}_2 = \begin{pmatrix} -1 & 0 & 0 \\ 0 & 1 & 0 \\ 0 & 0 & -1 \end{pmatrix}, \quad \mathbf{D}_3 = \begin{pmatrix} -1 & 0 & 0 \\ 0 & -1 & 0 \\ 0 & 0 & 1 \end{pmatrix},$$

$$\mathbf{T}_1 = \begin{pmatrix} 1 & 0 & 0 \\ 0 & 0 & 1 \\ 0 & 1 & 0 \end{pmatrix}, \quad \mathbf{T}_2 = \begin{pmatrix} 0 & 0 & 1 \\ 0 & 1 & 0 \\ 1 & 0 & 0 \end{pmatrix}, \quad \mathbf{T}_3 = \begin{pmatrix} 0 & 1 & 0 \\ 1 & 0 & 0 \\ 0 & 0 & 1 \end{pmatrix},$$

$$\mathbf{M}_1 = \begin{pmatrix} 0 & 1 & 0 \\ 0 & 0 & 1 \\ 1 & 0 & 0 \end{pmatrix}, \quad \mathbf{M}_2 = \begin{pmatrix} 0 & 0 & 1 \\ 1 & 0 & 0 \\ 0 & 1 & 0 \end{pmatrix},$$

$$\mathbf{S}_1 = \begin{pmatrix} -1/2 & \sqrt{3}/2 & 0 \\ -\sqrt{3}/2 & -1/2 & 0 \\ 0 & 0 & 1 \end{pmatrix}, \qquad \mathbf{S}_2 = \begin{pmatrix} -1/2 & -\sqrt{3}/2 & 0 \\ \sqrt{3}/2 & -1/2 & 0 \\ 0 & 0 & 1 \end{pmatrix},$$

where $\mathbf{I}$ is the identity and $\mathbf{C}$ is the central inversion. $\mathbf{R}_1, \mathbf{R}_2, \mathbf{R}_3$ are reflections in the planes whose normals are along the $x_1 = x$-, $x_2 = y$-, and $x_3 = z$-directions, respectively. $\mathbf{D}_1, \mathbf{D}_2, \mathbf{D}_3$ are rotations through π radians about the x_1-, x_2-, and x_3-axes, respectively. $\mathbf{T}_1$ is a reflection through a plane which bisects the x_2- and x_3-axes and contains the x_1-axis. $\mathbf{T}_2$ and $\mathbf{T}_3$ are analogously defined. $\mathbf{M}_1$ and $\mathbf{M}_2$ are rotations through $2\pi/3$ clockwise and anticlockwise, about an axis making equal acute angles with the axes x_1, x_2, and x_3. $\mathbf{S}_1$ and $\mathbf{S}_2$ are rotations through $2\pi/3$ clockwise and anticlockwise, respectively, about the $x_3 = z$-axis.

CHAPTER 3

Crystallographic Magnetic Point Groups

In addition to the geometrical symmetries present in the lattice structure of the crystals, the atoms of the lattice in magnetic materials are endowed with atomic magnetic moments (spins). The usual spatial symmetry operations, rotations and rotation-inversions, while preserving the geometrical properties of the lattice, may reverse the orientation of the spins.

For the symmetry of real magnetic materials, the antisymmetry operator τ is the operator that reverses a magnetic moment. Alternatively, we may regard τ as reversing the direction of an electric current, since a change in magnetic moment can be caused by a reversal of the direction of the electric current which produces the magnetic moment. This is then equivalent to a reversal of the sense of time, since $i = dq/dt$. It is possible to have compound operations of antisymmetry corresponding to the performance of both an element of the ordinary point group $\{G\}$ together with the operation of time-reversal τ. Operations of these types produce the magnetic (black and white) point groups.

We illustrate the derivation of the magnetic point groups from the ordinary point groups $\{G\}$ by considering the example of the black and white point groups that are derived from $C_{3v} = 3m$, which is the point group of the symmetry operations of an equilateral triangle. An equilateral triangle has the symmetry operations E, C_{3z}, C_{3z}^-, σ_{d1}, σ_{d2}, σ_{d3}, (see Table 2.3 and Figure A.1 in Appendix A). If irregular black and white patches are drawn at random on the triangle, these symmetry operations, other than E, will be destroyed. However, if half of the triangle is colored black and the other half is colored white in some regular way, such as in Figure 3.1, then three of these symmetry operations E, C_{3z}, C_{3z}^- survive and may be symmetry operations of the colored triangle. The remaining operations σ_{d1}, σ_{d2}, and σ_{d3} are no longer symmetry operations, since reflection in the vertical plane passing, say, through axis 1 leaves a black patch where there was a white patch before, and vice versa. But if each of these three operations is combined with τ (color-changing operation), to produce $\tau\sigma_{d1}$, $\tau\sigma_{d2}$, and $\tau\sigma_{d3}$, these operations now become symmetry operations of the colored triangle. The symmetry elements of Figure 2.1(b) are therefore

$$E, C_{3z}, C_{3z}^-, \tau\sigma_{d1}, \tau\sigma_{d2}, \tau\sigma_{d3},$$

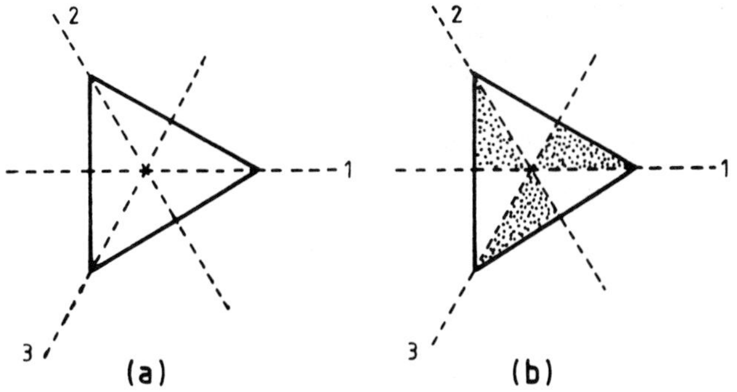

Figure 3.1. Black and white triangles to illustrate $3\underline{m}$.

which still form a group, and this group is denoted by $3\underline{m}$. m is underlined since mirror images are combined with the color-changing operator. Note that the uncolored operators $\{E, C_{3z}, C_{3z}^{-}\}$ form a group of index 2.

Now suppose that $\{G\}$ is one of the ordinary point groups discussed in the previous chapter. Altogether we have:

(i) The ordinary point groups $\{M\} = \{G\}$: 32.

The operation of the antisymmetry τ is not present in these groups. They are the conventional crystallographic point groups.

(ii) The grey point groups $\{M\} = \{G\} + \tau\{G\}$: 32.

The difference between the ordinary and the grey point groups is that in the latter the antisymmetry is an operator of the group. Note that $\{M\} = \{G\} \times \{E, \tau\}$ since $\tau^2 = E$, and τ commutes with all the elements of $\{G\}$.

(iii) The magnetic (black and white) point groups

$$\{M\} = \{H\} + \tau\{G - H\}: 58, \tag{3.1}$$

where $\{H\}$ is a subgroup of index 2 of the conventional point group $\{G\}$.

Here τ itself is not an element of the group $\{M\}$, but half of the elements of $\{G\}$ are now in combination with τ. Out of the thirty-two conventional point groups $\{G\}$, fifty-eight distinct magnetic point groups are derived. They are listed by Tavger and Zaitsev [1956]. Reproduction of these tables can be found in Birss [1963, 1964], Opechowski and Guccione [1965], and Bhagavantam [1966]. Stereographic projections of magnetic point groups are given by Koptsik [1966] and McMillan [1967]. All these tables are reproduced in the book by Bradley and Cracknell [1972]. Due to notational differences, we list the fifty-eight magnetic point groups in Table 3.1 for ease of reference. In Table 3.1, in column 3 we identify $\{H\}$, the halving subgroup of $\{G\}$, in the international and Schönflies notation. The actual symmetry elements in $\{H\}$ can be identified from Table 2.5. The elements in column 5 are to be understood as multiplied by τ.

Table 3.1. Magnetic point groups.

No.	Magnetic point group $\{M\}$	Classical subgroup $\{H\}$ International	Classical subgroup $\{H\}$ Schönflies	G-H
1	$\underline{\bar{1}}$	1	C_1	C
2	$\underline{2}$	1	C_1	D_3
3	$\underline{m}$	1	C_1	R_3
4	$2/\underline{m}$	2	C_2	C, R_3
5	$\underline{2}/m$	m	$C_{1h} = C_s$	C, D_3
6	$\underline{2}/\underline{m}$	$\bar{1}$	C_i	D_3, R_3
7	$\underline{2}\underline{2}2$	2	C_2	D_1, D_2
8	$2\underline{m}\underline{m}$	2	C_2	R_1, R_2
9	$\underline{2}m\underline{m}$	m	C_s	D_3, R_1
10	$\underline{m}\underline{m}\underline{m}$	222	D_2	C, R_1, R_2, R_3
11	$\underline{m}mm$	$2mm$	C_{2v}	C, D_1, D_2, R_3
12	$m\underline{m}\underline{m}$	$2/m$	C_{2h}	D_1, D_2, R_1, R_2
13	$\underline{4}$	2	C_2	$R_2 T_3, R_1 T_3$
14	$\underline{\bar{4}}$	2	C_2	$D_2 T_3, D_1 T_3$
15	$4\underline{2}\underline{2}$	4	C_4	$D_1, D_2, C T_3, R_3 T_3$
16	$\underline{4}2\underline{2}$	222	D_2	$R_2 T_3, R_1 T_3, C T_3, R_3 T_3$
17	$4/\underline{m}$	4	C_4	$C, R_3, D_2 T_3, D_1 T_3$
18	$\underline{4}/\underline{m}$	$\bar{4}$	S_4	$C, R_3, R_2 T_3, R_1 T_3$
19	$\underline{4}/m$	$2/m$	C_{2h}	$R_2 T_3, R_1 T_3, D_2 T_3, D_1 T_3$
20	$4\underline{m}\underline{m}$	4	C_4	$R_1, R_2, T_3, D_3 T_3$
21	$\underline{4}m\underline{m}$	$2mm$	C_{2v}	$R_2 T_3, R_1 T_3, T_3, D_3 T_3$
22	$\bar{4}\underline{2}\underline{m}$	$\bar{4}$	S_4	$D_1, D_2, T_3, D_3 T_3$
23	$\underline{\bar{4}}2\underline{m}$	222	D_2	$D_2 T_3, D_1 T_3, T_3, D_3 T_3$
24	$\underline{\bar{4}}\underline{2}m$	$2mm$	C_{2v}	$D_1, D_2, D_2 T_3, D_1 T_3$
25	$4/\underline{m}\underline{m}\underline{m}$	422	D_4	$C, R_1, R_2, R_3, D_2 T_3, D_1 T_3, T_3, D_3 T_3$
26	$4/\underline{m}mm$	$4mm$	C_{4v}	$C, R_3, D_2 T_3, D_1 T_3, D_1, D_2, C T_3, R_3 T_3$
27	$\underline{4}/mm\underline{m}$	mmm	D_{2h}	$R_2 T_3, R_1 T_3, C T_3, R_3 T_3, D_2 T_3, D_1 T_3,$ $T_3, D_3 T_3$
28	$\underline{4}/\underline{m}m\underline{m}$	$\bar{4}2m$	D_{2v}	$C, R_1, R_2, R_3, R_2 T_3, R_1 T_3, C T_3, R_3 T_3$
29	$\underline{4}/m\underline{m}\underline{m}$	$4/m$	C_{4h}	$D_1, D_2, R_1, R_2, C T_3, R_3 T_3, T_3, D_3 T_3$
30	$3\underline{2}$	3	C_3	$D_1, D_1 S_1, D_1 S_2$
31	$3\underline{m}$	3	C_3	$R_1, R_1 S_1, R_1 S_2$
32	$\underline{\bar{6}}$	3	C_3	$R_3, R_3 S_1, R_3 S_2$
33	$\bar{6}\underline{m}\underline{2}$	$\bar{6}$	C_{3h}	$R_2, R_2 S_1, R_2 S_2, D_1, D_1 S_1, D_1 S_2$
34	$\underline{\bar{6}}m\underline{2}$	$3m$	C_{3v}	$D_1, D_1 S_1, D_1 S_2, R_3, R_3 S_1, R_3 S_2$
35	$\underline{\bar{6}}\underline{m}2$	32	D_3	$R_3, R_3 S_2, R_3 S_1, R_2, R_1 S_2, R_2 S_2$
36	$\underline{6}$	3	C_3	$D_3, D_3 S_2, D_3 S_1$
37	$\underline{\bar{3}}$	3	C_3	$C, C S_1, C S_2$
38	$\bar{3}\underline{m}$	$\bar{3}$	C_{3i}	$D_1, D_1 S_1, D_1 S_2, R_1, R_1 S_1, R_1 S_2$
39	$\underline{\bar{3}}m$	$3m$	C_{3v}	$D_1, D_1 S_1, D_1 S_2, C, C S_1, C S_2$
40	$\underline{\bar{3}}\underline{m}$	32	D_3	$C, C S_1, C S_2, R_1, R_1 S_1, R, S_2$
41	$6\underline{2}\underline{2}$	6	C_6	$D_1, D_1 S_1, D_1 S_2, D_2, D_2 S_1, D_2 S_2$
42	$\underline{6}2\underline{2}$	32	D_3	$D_3, D_3 S_2, D_3 S_1, D_2, D_2 S_1, D_2 S_2$
43	$6/\underline{m}$	6	C_6	$C, C S_1, C S_2, R_3, R_3 S_2, R_3 S_1$
44	$\underline{6}/\underline{m}$	$\bar{3}$	C_{3i}	$D_3, D_3 S_2, D_3 S_1, R_3, R_3 S_2, R_3 S_1$
45	$\underline{6}/m$	$\bar{6}$	C_{3h}	$C, C S_1, C S_2, D_3, D_3 S_2, D_3 S_1$
46	$6\underline{m}\underline{m}$	6	C_6	$R_1, R_1 S_1, R_1 S_2, R_2, R_2 S_1, R_2 S_2$
47	$\underline{6}m\underline{m}$	$3m$	C_{3v}	$D_3, D_3 S_2, D_3 S_1, R_2, R_2 S_1, R_2 S_2$

(continued)

Table 3.1 (*continued*)

No.	Magnetic point group $\{M\}$	Classical subgroup $\{H\}$ International	Classical subgroup $\{H\}$ Schönflies	G-H
48	$\underline{6}/mm\underline{m}$	$\bar{6}2m$	D_{3h}	$C, CS_1, CS_2, D_3, D_3S_2, D_3S_1, D_2, D_2S_1, D_2S_2, R_1, R_1S_1, R_1S_2$
49	$\underline{6}/\underline{m}m\underline{m}$	$\bar{3}m$	D_{3d}	$D_3, D_3S_2, D_3S_1, D_2, D_2S_1, D_2S_2, R_3, R_3S_1, R_3S_2, R_2, R_2S_1, R_2S_2$
50	$6/\underline{m}\underline{m}\underline{m}$	622	D_6	$C, CS_1, CS_2, R_3, R_3S_1, R_3S_2, R_1, R_1S_1, R_1S_2, R_2, R_2S_1, R_2S_2$
51	$6/\underline{m}mm$	$6mm$	C_{6v}	$D_1, D_1S_1, D_1S_2, D_2, D_2S_1, D_2S_2, C, CS_1, CS_2, R_3, R_3S_1, R_3S_2$
52	$6/m\underline{m}\underline{m}$	$6/m$	C_{6h}	$D_1, D_1S_1, D_1S_2, D_2, D_2S_2, R_1, R_1S_1, R_1S_2, R_2, R_2S_1, R_2S_2, D_2S_1$
53	$\underline{m}3$	23	T	$C, S_{6j}, S_{6j}^{-}, R_1, R_2, R_3$
54	$\underline{\bar{4}}3\underline{m}$	23	T	$\sigma_{dp}, S_{4m}, S_{4m}^{-}$
55	$\underline{4}3\underline{2}$	23	T	$C_{2p}, C_{4m}, C_{4m}^{-}$
56	$\underline{m}3\underline{m}$	432	O	$C, S_{6j}, S_{6j}^{-}, R_1, R_2, R_3, \sigma_{dp}, S_{4m}, S_{4m}^{-}$
57	$\underline{m}3m$	$\bar{4}3m$	T_d	$C, S_{6j}, S_{6j}^{-}, R_1, R_2, R_3, C_{2p}, C_{4m}, C_{4m}^{-}$
58	$m3\underline{m}$	$m3$	T_h	$C_{2p}, C_{4m}, C_{4m}^{-}, \sigma_{dp}, S_{4m}, S_{4m}^{-}$

The ordinary point group $\{G\}$ and the magnetic points generated from it, by means of the prescription (3.1) from all the halving subgroups of $\{G\}$, are said to be the family of $\{G\}$. The magnetic groups which belong to the same family are all isomorphic (Indenbom [1960] and Mert [1975]). This implies that they all have the same irreducible representations, as has the generating ordinary point group $\{G\}$. Since there are eighteen nonisomorphic ordinary point groups, the number of the nonisomorphic family of magnetic groups will be the same. The isomorphic family of magnetic groups and the generating ordinary point groups are tabulated in Table 3.2. Note that the crystal classes $C_1 = 1$, $C_3 = 3$, and $T = 23$, numbered (1), (9), and (15), respectively, in Table 3.2, do not generate magnetic classes, since they do not have a subgroup of index 2 (see Table 2.6).

Irreducible representations of fifteen families of isomorphic magnetic point groups are listed in Table 3.3(2–18). The symmetry elements in $\tau\{G - H\}$ are said to be complementary, and they are underlined in these tables, e.g., $\tau R_1 = \underline{R_1}$. The two-dimensional irreducible representations are defined in terms of the matrices **E**, **A**, ..., **L** listed below:

$$\mathbf{E} = \begin{pmatrix} 1 & 0 \\ 0 & 1 \end{pmatrix}, \quad \mathbf{A} = \begin{pmatrix} -1/2 & \sqrt{3}/2 \\ -\sqrt{3}/2 & -1/2 \end{pmatrix}, \quad \mathbf{B} = \begin{pmatrix} -1/2 & -\sqrt{3}/2 \\ \sqrt{3}/2 & -1/2 \end{pmatrix},$$

$$\mathbf{F} = \begin{pmatrix} 1 & 0 \\ 0 & -1 \end{pmatrix}, \quad \mathbf{G} = \begin{pmatrix} -1/2 & \sqrt{3}/2 \\ \sqrt{3}/2 & 1/2 \end{pmatrix}, \quad \mathbf{H} = \begin{pmatrix} -1/2 & -\sqrt{3}/2 \\ -\sqrt{3}/2 & 1/2 \end{pmatrix},$$

$$\mathbf{K} = \begin{pmatrix} 0 & 1 \\ 1 & 0 \end{pmatrix}, \quad \mathbf{L} = \begin{pmatrix} 0 & 1 \\ -1 & 0 \end{pmatrix}.$$

Table 3.2. Family of magnetic point groups.

Number	Isomorphic classical classes	Isomorphic family of magnetic classes
1	$C_1 = 1$	
2	$C_i = \bar{1}$	$\underline{\bar{1}}$
	$C_s = m$	$\underline{m}$
	$C_2 = 2$	$\underline{2}$
3	$C_{2h} = 2/m$	$\underline{2}/m, 2/\underline{m}, \underline{2}/\underline{m}$
	$C_{2v} = 2mm$	$\underline{2}m\underline{m}, 2\underline{m}\underline{m}$
	$D_2 = 222$	$\underline{2}\underline{2}2$
4	$D_{2h} = mmm$	$\underline{m}mm, \underline{m}\underline{m}\underline{m}, m\underline{m}\underline{m}$
5	$C_4 = 4$	$\underline{4}$
	$S_4 = \bar{C}_2 = \bar{4}$	$\underline{\bar{4}}$
6	$C_{4h} = 4/m$	$\underline{4}/\underline{m}, 4/\underline{m}, \underline{4}/m$
7	$C_{4v} = 4mm$	$4\underline{m}\underline{m}, \underline{4}m\underline{m}$
	$D_{2d} = \bar{4}2m$	$\underline{\bar{4}}2\underline{m}, \underline{\bar{4}}\underline{2}m, \bar{4}\underline{2}\underline{m}$
	$D_4 = 422$	$4\underline{2}\underline{2}, \underline{4}2\underline{2}$
8	$D_{4h} = 4/mmm$	$\underline{4}/\underline{m}m\underline{m}, 4/\underline{m}mm, 4/m\underline{m}\underline{m}, 4\underline{m}\underline{m}\underline{m}, \underline{4}/mm\underline{m}$
9	$C_3 = 3$	
10	$C_{3v} = 3m$	$3\underline{m}$
	$D_3 = 32$	$3\underline{2}$
11	$C_{3i} = S_6 = \bar{3}$	$\underline{\bar{3}}$
	$C_{3h} = \bar{6}$	$\underline{\bar{6}}$
	$C_6 = 6$	$\underline{6}$
12	$C_{6h} = 6/m$	$6/\underline{m}, \underline{6}/m, \underline{6}/\underline{m}$
13	$D_{3v} = \bar{3}m$	$\bar{3}\underline{m}, \underline{\bar{3}}m, \underline{\bar{3}}\underline{m}$
	$D_{3h} = \bar{6}m2$	$\bar{6}\underline{m}\underline{2}, \underline{\bar{6}}m\underline{2}, \underline{\bar{6}}\underline{m}2$
	$C_{6v} = 6mm$	$6\underline{m}\underline{m}, \underline{6}m\underline{m}$
	$D_6 = 622$	$6\underline{2}\underline{2}, \underline{6}2\underline{2}$
14	$D_{6h} = 6/mmm$	$6/\underline{m}mm, 6/m\underline{m}\underline{m}, 6/\underline{m}\underline{m}\underline{m}, \underline{6}/mm\underline{m}, \underline{6}/\underline{m}m\underline{m}$
15	$T = 23$	
16	$T_h = m3$	$\underline{m}3$
17	$T_d = \bar{4}3m$	$\underline{\bar{4}}3\underline{m}$
	$O = 432$	$\underline{4}3\underline{2}$
18	$Oh = m3m$	$\underline{m}3\underline{m}, m3\underline{m}, \underline{m}3m$

Table 3.3. Irreducible representations of the magnetic point groups.

(2)

Pinacoidal	$\bar{1}$	I	C
(C_i)	$\underline{\bar{1}}$	I	$\underline{C}$
Domatic	m	I	R_3
(C_s)	$\underline{m}$	I	$\underline{R_3}$
Sphenoidal	2	I	D_3
(C_2)	$\underline{2}$	I	$\underline{D_3}$
	Γ_1	1	1
	Γ_2	1	−1

(3)

Prismatic	$2/m$	I	D_3	R_3	C
(C_{2h})	$\underline{2}/m$	I	$\underline{D_3}$	R_3	$\underline{C}$
	$2/\underline{m}$	I	D_3	$\underline{R_3}$	$\underline{C}$
	$\underline{2}/\underline{m}$	I	$\underline{D_3}$	$\underline{R_3}$	C
Rhombic–pyramidal	$2mm$	I	D_3	R_1	R_2
(C_{2v})	$\underline{2}m\underline{m}$	I	$\underline{D_3}$	$\underline{R_1}$	R_2
	$2\underline{m}\underline{m}$	I	D_3	$\underline{R_1}$	$\underline{R_2}$
Rhombic–disphenoidal	222	I	D_1	D_2	D_3
(D_2)	$\underline{2}\underline{2}2$	I	$\underline{D_1}$	$\underline{D_2}$	D_3
	Γ_1	1	1	1	1
	Γ_2	1	1	−1	−1
	Γ_3	1	−1	1	−1
	Γ_4	1	−1	−1	1

(4)

Rhombic–dypramidal	mmm	I	D_1	D_2	D_3	C	R_1	R_2	R_3
(D_{2h})	$\underline{m}mm$	I	$\underline{D_1}$	$\underline{D_2}$	D_3	$\underline{C}$	R_1	R_2	$\underline{R_3}$
	$\underline{m}\underline{m}\underline{m}$	I	D_1	D_2	D_3	$\underline{C}$	$\underline{R_1}$	$\underline{R_2}$	$\underline{R_3}$
	$m\underline{m}\underline{m}$	I	$\underline{D_1}$	$\underline{D_2}$	D_3	C	$\underline{R_1}$	$\underline{R_2}$	R_3
	Γ_1	1	1	1	1	1	1	1	1
	Γ_2	1	1	−1	−1	1	1	−1	−1
	Γ_3	1	−1	1	−1	1	−1	1	−1
	Γ_4	1	−1	−1	1	1	−1	−1	1
	Γ_1'	1	1	1	1	−1	−1	−1	−1
	Γ_2'	1	1	−1	−1	−1	−1	1	1
	Γ_3'	1	−1	1	−1	−1	1	−1	1
	Γ_4'	1	−1	−1	1	−1	1	1	−1

(5)

Tetragonal–disphenoidal	$\bar{4}$	I	D_3	$D_1 T_3$	$D_2 T_3$
($\bar{C}_2 = S_4$)	$\underline{\bar{4}}$	I	D_3	$\underline{D_1 T_3}$	$\underline{D_2 T_3}$
Tetragonal–pyramidal	4	$\underline{I}$	D_3	$R_1 T_3$	$R_2 T_3$
(C_4)	$\underline{4}$	I	D_3	$\underline{R_1 T_3}$	$\underline{R_2 T_3}$
	Γ_1	1	1	1	1
	Γ_2	1	1	−1	−1
	Γ_3	1	−1	i	$-i$
	Γ_4	1	−1	$-i$	i

Table 3.3 (*continued*)

(6)

Tetragonal–dipyramidal (C_{4h})	$4/m$	I	D_3	R_1T_3	R_2T_3	C	R_3	D_1T_3	D_2T_3
	$\underline{4}/\underline{m}$	I	D_3	$\underline{R_1T_3}$	$\underline{R_2T_3}$	$\underline{C}$	$\underline{R_3}$	D_1T_3	D_2T_3
	$4/\underline{m}$	I	D_3	R_1T_3	R_2T_3	$\underline{C}$	$\underline{R_3}$	$\underline{D_1T_3}$	$\underline{D_2T_3}$
	$\underline{4}/m$	I	D_3	$\underline{R_1T_3}$	$\underline{R_2T_3}$	C	R_3	$\underline{D_1T_3}$	$\underline{D_2T_3}$
	Γ_1	1	1	1	1	1	1	1	1
	Γ_2	1	1	−1	−1	1	1	−1	−1
	Γ_3	1	−1	i	$-i$	1	−1	i	$-i$
	Γ_4	1	−1	$-i$	i	1	−1	$-i$	i
	Γ_1'	1	1	1	1	−1	−1	−1	−1
	Γ_2'	1	1	−1	−1	−1	−1	1	1
	Γ_3'	1	−1	i	$-i$	−1	1	$-i$	i
	Γ_4'	1	−1	$-i$	i	−1	1	i	$-i$

(7)

Ditetragonal–pyramidal (C_{4v})	$4mm$	I	R_2	R_1	D_3	T_3	R_2T_3	R_1T_3	D_3T_3
	$4\underline{m}\underline{m}$	I	$\underline{R_2}$	$\underline{R_1}$	D_3	$\underline{T_3}$	R_2T_3	R_1T_3	$\underline{D_3T_3}$
	$\underline{4}m\underline{m}$	I	R_2	R_1	D_3	$\underline{T_3}$	$\underline{R_2T_3}$	$\underline{R_1T_3}$	$\underline{D_3T_3}$
Tetragonal–trapezohedral (D_4)	422	I	D_1	D_2	D_3	R_3T_3	R_2T_3	R_1T_3	CT_3
	$4\underline{2}\underline{2}$	I	$\underline{D_1}$	$\underline{D_2}$	D_3	$\underline{R_3T_3}$	R_2T_3	R_1T_3	$\underline{CT_3}$
	$\underline{4}2\underline{2}$	I	D_1	D_2	D_3	$\underline{R_3T_3}$	$\underline{R_2T_3}$	$\underline{R_1T_3}$	$\underline{CT_3}$
Tetragonal–scalenohedral (D_{2v})	$\bar{4}2m$	I	D_1	D_2	D_3	T_3	D_1T_3	D_2T_3	D_3T_3
	$\underline{\bar{4}}2\underline{m}$					—	—	—	—
	$\underline{4}\underline{2}m$		—	—			—	—	
	$4\underline{2}\underline{m}$		—	—		—			—
	Γ_1	1	1	1	1	1	1	1	1
	Γ_2	1	−1	−1	1	−1	1	1	−1
	Γ_3	1	−1	−1	1	1	−1	−1	1
	Γ_4	1	1	1	1	−1	−1	−1	−1
	Γ_5	E	F	$-F$	$-E$	K	L	$-L$	$-K$

(8)

Ditetragonal	$4/mmm$	I	D_1	D_2	D_3	CT_3	R_1T_3	R_2T_3	R_3T_3
	$\underline{4}/\underline{m}m\underline{m}$	I	D_1	D_2	D_3	$\underline{CT_3}$	$\underline{R_1T_3}$	$\underline{R_2T_3}$	$\underline{R_3T_3}$
Dipyramidal	$4/\underline{m}mm$		—	—		—			—
(D_{4h})	$\underline{4}/m\underline{m}\underline{m}$		—	—		—			—
	$4/\underline{m}\underline{m}\underline{m}$								
	$\underline{4}/mm\underline{m}$					—	—	—	—
	Γ_1	1	1	1	1	1	1	1	1
	Γ_2	1	−1	−1	1	−1	1	1	−1
	Γ_3	1	−1	−1	1	1	−1	−1	1
	Γ_4	1	1	1	1	−1	−1	−1	−1
	Γ_5	E	F	$-F$	$-E$	$-K$	$-L$	L	K
	Γ_1'	1	1	1	1	1	1	1	1
	Γ_2'	1	−1	−1	1	−1	1	1	−1
	Γ_3'	1	−1	−1	1	1	−1	−1	1
	Γ_4'	1	1	1	1	−1	−1	−1	−1
	Γ_5'	E	F	$-F$	$-E$	$-K$	$-L$	L	K

Table 3.3 (*continued*)

Ditetragonal	$4/mmm$	C	R_1	R_2	R_3	T_3	D_1T_3	D_2T_3	D_3T_3
	$\underline{4}/\underline{m}m\underline{m}$	$\underline{C}$	$\underline{R_1}$	$\underline{R_2}$	$\underline{R_3}$	T_3	D_1T_3	D_2T_3	D_3T_3
Dipyramidal	$4/\underline{m}mm$	–			–		–	–	
(D_{4h})	$\underline{4}/m\underline{m}\underline{m}$		—	—		—			—
	$4/\underline{m}\underline{m}\underline{m}$	—	—	—	—	—	—	—	—
	$\underline{4}/mm\underline{m}$					—	—	—	—
	Γ_1	1	1	1	1	1	1	1	1
	Γ_2	1	−1	−1	1	−1	1	1	−1
	Γ_3	1	−1	−1	1	1	−1	−1	1
	Γ_4	1	1	1	1	−1	−1	−1	−1
	Γ_5	E	F	$-F$	$-E$	$-K$	$-L$	L	K
	Γ_1'	−1	−1	−1	−1	−1	−1	−1	−1
	Γ_2'	−1	1	1	−1	1	−1	−1	1
	Γ_3'	−1	1	1	−1	−1	1	1	−1
	Γ_4'	−1	−1	−1	−1	1	1	1	1
	Γ_5'	$-E$	$-F$	F	E	K	L	$-L$	$-K$

(10)

Ditrigonal–pyramidal	$3m$	I	S_1	S_2	R_1	R_1S_1	R_1S_2
(C_{3v})	$3\underline{m}$	I	S_1	S_2	$\underline{R_1}$	$\underline{R_1S_1}$	$\underline{R_1S_2}$
Trigonal–trapezohedral	32	I	S_1	S_2	D_1	D_1S_1	D_1S_2
(D_3)	$3\underline{2}$	I	S_1	S_2	$\underline{D_1}$	$\underline{D_1S_1}$	$\underline{D_1S_2}$
	Γ_1	1	1	1	1	1	1
	Γ_2	1	1	1	−1	−1	−1
	Γ_3	E	A	B	$-F$	$-G$	$-H$

(11)

Rhombohedral	$\bar{3}$	I	S_1	S_2	C	CS_1	CS_2
($\bar{C}_3$)	$\underline{\bar{3}}$	I	S_1	S_2	$\underline{C}$	$\underline{CS_1}$	$\underline{CS_2}$
Trigonal–dipyramidal	$\bar{6}$	I	S_1	S_2	R_3	R_3S_1	R_3S_2
(C_{3h})	$\underline{\bar{6}}$	I	S_1	S_2	$\underline{R_3}$	$\underline{R_3S_1}$	$\underline{R_3S_2}$
Hexagonal–pyramidal	6	I	S_1	S_2	D_3	D_3S_1	D_3S_2
(C_6)	$\underline{6}$	I	S_1	S_2	$\underline{D_3}$	$\underline{D_3S_1}$	$\underline{D_3S_2}$
	Γ_1	1	1	1	1	1	1
	Γ_2	1	w	w^2	1	w	w^2
	Γ_3	1	w^2	w	1	w^2	w
	Γ_4	1	1	1	−1	−1	−1
	Γ_5	1	w	w^2	−1	$-w$	$-w^2$
	Γ_6	1	w^2	w	−1	$-w^2$	$-w$

where

$$w = -\tfrac{1}{2} + i\sqrt{3}/2 = e^{2\pi i/3},$$

$$w^2 = -\tfrac{1}{2} - i\sqrt{3}/2,$$

$$w^3 = 1.$$

Table 3.3 (*continued*)

(12)													
Hexagonal–dipyramidal (C_{6h})	$6/m$	I	S_1	S_2	D_3	D_3S_1	D_3S_2	C	CS_1	CS_2	R_3	R_3S_1	R_3S_2
	$6/\underline{m}$							—	—	—	—	—	—
	$\underline{6}/m$				—	—	—	—	—	—			
	$\underline{6}/\underline{m}$				—	—	—				—	—	—
	Γ_1	1	1	1	1	1	1	1	1	1	1	1	1
	Γ_2	1	w	w^2	1	w	w^2	1	w	w^2	1	w	w^2
	Γ_3	1	w^2	w	1	w^2	w	1	w^2	w	1	w^2	w
	Γ_4	1	1	1	-1	-1	-1	1	1	1	-1	-1	-1
	Γ_5	1	w	w^2	-1	$-w$	$-w^2$	1	w	w^2	-1	$-w$	$-w^2$
	Γ_6	1	w^2	w	-1	$-w^2$	$-w$	1	w^2	w	-1	$-w^2$	$-w$
	Γ_1'	1	1	1	1	1	1	-1	-1	-1	-1	-1	-1
	Γ_2'	1	w	w^2	1	w	w^2	-1	$-w$	$-w^2$	-1	$-w$	$-w^2$
	Γ_3'	1	w^2	w	1	w^2	w	-1	$-w^2$	$-w$	-1	$-w^2$	$-w$
	Γ_4'	1	1	1	-1	-1	-1	-1	-1	-1	1	1	1
	Γ_5'	1	w	w^2	-1	$-w$	$-w^2$	-1	$-w$	$-w^2$	1	w	w^2
	Γ_6'	1	w^2	w	-1	$-w^2$	$-w$	-1	$-w^2$	$-w$	1	w^2	w
(13)													
Ditrigonal–dipyramidal (D_{3h})	$\bar{\underline{6}}m2$	I	S_1	S_2	R_3	R_3S_1	R_3S_2	R_1	R_1S_1	R_1S_2	D_2	D_2S_1	D_2S_2
	$\bar{6}\underline{m}\underline{2}$							—	—	—	—	—	—
	$\bar{\underline{6}}m\underline{2}$				—	—	—				—	—	—
	$\bar{\underline{6}}\underline{m}2$				—	—	—	—	—	—			
Hexagonal–scalenohedral (D_{3v})	$\bar{3}m$	I	S_1	S_2	C	CS_1	CS_2	D_1	D_1S_1	D_1S_2	R_1	R_1S_1	R_1S_2
	$\bar{3}\underline{m}$							—	—	—	—	—	—
	$\bar{\underline{3}}m$				—	—	—	—	—	—			
	$\bar{\underline{3}}\underline{m}$				—	—	—				—	—	—
Hexagonal–trapezohedral (D_6)	622	I	S_1	S_2	D_3	D_3S_1	D_3S_2	D_1	D_1S_1	D_1S_2	D_2	D_2S_1	D_2S_2
	$6\underline{2}\underline{2}$							—	—	—	—	—	—
	$\underline{6}2\underline{2}$				—	—	—				—	—	—

Dihexagonal–pyramidal	$6mm$	I	S_1	S_2	D_3	D_3S_1	D_3S_2	R_2	R_2S_1	R_2S_2	R_1	R_1S_1	R_1S_2
(C_{6v})	$6\underline{mm}$							—	—	—	—	—	—
	$\underline{6}m\underline{m}$				—	—	—	—	—	—			
	Γ_1	1	1	1	1	1	1	1	1	1	1	1	1
	Γ_2	1	1	1	1	1	1	−1	−1	−1	−1	−1	−1
	Γ_3	1	1	1	−1	−1	−1	1	1	1	−1	−1	−1
	Γ_4	1	1	1	−1	−1	−1	−1	−1	−1	1	1	1
	Γ_5	E	A	B	$-E$	$-A$	$-B$	F	G	H	$-F$	$-G$	$-H$
	Γ_6	E	A	B	E	A	B	$-F$	$-G$	$-H$	$-F$	$-G$	$-H$

(14)

Dihexagonal–dipyramidal	$6/mmm$	I	S_1	S_2	D_1	D_1S_1	D_1S_2	D_2	D_2S_1	D_2S_2	D_3	D_3S_1	D_3S_2
(D_{6h})	$6/\underline{m}mm$				—	—	—	—	—	—			
	$6/m\underline{mm}$				—	—	—	—	—	—			
	$6/\underline{mmm}$												
	$\underline{6}/mm\underline{m}$							—	—	—	—	—	—
	$\underline{6}/\underline{mm}m$							—	—	—	—	—	—
	Γ_1	1	1	1	1	1	1	1	1	1	1	1	1
	Γ_2	1	1	1	−1	−1	−1	−1	−1	−1	1	1	1
	Γ_3	1	1	1	1	1	1	−1	−1	−1	−1	−1	−1
	Γ_4	1	1	1	−1	−1	−1	1	1	1	−1	−1	−1
	Γ_5	E	A	B	F	G	H	$-F$	$-G$	$-H$	$-E$	$-A$	$-B$
	Γ_6	E	A	B	$-F$	$-G$	$-H$	$-F$	$-G$	$-H$	E	A	B
	Γ_1'	1	1	1	1	1	1	1	1	1	1	1	1
	Γ_2'	1	1	1	−1	−1	−1	−1	−1	−1	1	1	1
	Γ_3'	1	1	1	1	1	1	−1	−1	−1	−1	−1	−1
	Γ_4'	1	1	1	−1	−1	−1	1	1	1	−1	−1	−1
	Γ_5'	E	A	B	F	G	H	$-F$	$-G$	$-H$	$-E$	$-A$	$-B$
	Γ_6'	E	A	B	$-F$	$-G$	$-H$	$-F$	$-G$	$-H$	E	A	B

Table 3.3 (*continued*)

(15)

Dihexagonal–dipyramidal	$6/mmm$	C	CS_1	CS_2	R_1	R_1S_1	R_1S_2	R_2	R_2S_1	R_2S_2	R_3	R_3S_1	R_3S_2
(D_{6h})	$6/\underline{m}mm$	—	—	—							—	—	—
	$6/m\underline{m}\underline{m}$				—	—	—	—	—	—			
	$6/\underline{m}\underline{m}\underline{m}$	—	—	—	—	—	—	—	—	—	—	—	—
	$\underline{6}/mm\underline{m}$	—	—	—	—	—	—						
	$\underline{6}/\underline{m}m\underline{m}$							—	—	—	—	—	—
	Γ_1	1	1	1	1	1	1	1	1	1	1	1	1
	Γ_2	1	1	1	−1	−1	−1	−1	−1	−1	1	1	1
	Γ_3	1	1	1	1	1	1	−1	−1	−1	−1	−1	−1
	Γ_4	1	1	1	−1	−1	−1	1	1	1	−1	−1	−1
	Γ_5	E	A	B	F	G	H	$-F$	$-G$	$-H$	$-E$	$-A$	$-B$
	Γ_6	E	A	B	$-F$	$-G$	$-H$	$-F$	$-G$	$-H$	E	A	B
	Γ_1'	−1	−1	−1	−1	−1	−1	−1	−1	−1	−1	−1	−1
	Γ_2'	−1	−1	−1	1	1	1	1	1	1	−1	−1	−1
	Γ_3'	−1	−1	−1	−1	−1	−1	1	1	1	1	1	1
	Γ_4'	−1	−1	−1	1	1	1	−1	−1	−1	1	1	1
	Γ_5'	$-E$	$-A$	$-B$	$-F$	$-G$	$-H$	F	G	H	E	A	B
	Γ_6'	$-E$	$-A$	$-B$	F	G	H	F	G	H	$-E$	$-A$	$-B$

Table 3.3 (*continued*)

(16)

						$3D$						$3R$	
Diploidal	$m3$	I	C_{3j}^{-}	C_{3j}	D_1	D_2	D_3	C	S_{6j}	S_{6j}^{-}	R_1	R_2	R_3
(T_h)	$\underline{m}3$	I	C_{3j}^{-}	C_{3j}	D_1	D_2	D_3	$\underline{C}$	$\underline{S_{6j}}$	$\underline{S_{6j}^{-}}$	$\underline{R_1}$	$\underline{R_2}$	$\underline{R_3}$
	Γ_1	1	1	1	1	1	1	1	1	1	1	1	1
	Γ_2	1	w	w^2	1	1	1	1	w	w^2	1	1	1
	Γ_3	1	w^2	w	1	1	1	1	w^2	w	1	1	1
	Γ_4	I	C_{3j}^{-}	C_{3j}	D_1	D_2	D_3	I	C_{3j}^{-}	C_{3j}	D_1	D_2	D_3
	Γ_1'	1	1	1	1	1	1	-1	-1	-1	-1	-1	-1
	Γ_2'	1	w	w^2	1	1	1	-1	$-w$	$-w^2$	-1	-1	-1
	Γ_3'	1	w^2	w	1	1	1	-1	$-w^2$	$-w$	-1	-1	-1
	Γ_4'	I	C_{3j}^{-}	C_{3j}	D_1	D_2	D_3	C	S_{6j}	S_{6j}^{-}	R_1	R_2	R_3

(17)

Hextetrahedral	$\bar{4}3m$	I	$3D$	σ_{dp}	C_{3j}^{-}	C_{3j}	S_{4m}	S_{4m}^{-}
(Td)	$\underline{\bar{4}}3\underline{m}$			—			—	—
Gyroidal	432	I	$3D$	C_{2p}	C_{3j}^{-}	C_{3j}	C_{4m}^{-}	C_{4m}
(O)	$\underline{4}3\underline{2}$	I	$3D$	$\underline{C_{2p}}$	C_{3j}^{-}	C_{3j}	$\underline{C_{4m}^{-}}$	$\underline{C_{4m}}$
	Γ_1	1	1	1	1	1	1	1
	Γ_2	1	1	-1	1	1	-1	-1
	Γ_3	I	I	$\Gamma_3(\sigma_{dp})$	$\Gamma_3(C_{3j}^{-})$	$\Gamma_3(C_{3j})$	$\Gamma_3(C_{4m}^{-})$	$\Gamma_3(C_{4m}^{-})$
	$T_d = \Gamma_4$	I	$3D$	σ_{dp}	C_{3j}^{-}	C_{3j}	S_{4m}	S_{4m}^{-}
	$O = \Gamma_5$	0	$3D$	C_{2p}	C_{3j}^{-}	C_{3j}	C_{4m}^{-}	C_{4m}

where

$\sigma_{dp} = T_1, T_2, T_3, D_1T_1, D_2T_2, D_3T_3,$

$\Gamma_3(\sigma_{dp}) = F, H, G, F, H, G,$

$C_{3j}^{-} = M_1, D_1M_1, D_2M_1, D_3M_1,$

$\Gamma_3(C_{3j}^{-}) = A, A, A, A,$

$C_{3j} = M_2, D_1M_2, D_2M_2, D_3M_2,$

$\Gamma_3(C_{3j}) = B, B, B, B,$

$C_{2p} = CT_1, CT_2, CT_3, R_1T_1, R_2T_2, R_3T_3,$

$S_{4m} = D_3T_1, D_1T_2, D_2T_3,$

$\Gamma_3(S_{4m}) = F, H, G,$

$S_{4m}^{-} = D_2T_1, D_3T_2, D_1T_3,$

$\Gamma_3(S_{4m}^{-}) = F, H, G,$

$C_{4m}^{-} = R_3T_1, R_1T_2, R_2T_3,$

$C_{4m} = R_2T_1, R_3T_2, R_1T_3.$

(18)

Hexoctohedral	$m3m$	I	$3D$	C_{2p}	C_{3j}^{-}	C_{3j}	C_{4m}^{-}	C_{4m}
(O_h)	$m3\underline{m}$			—			—	—
	$\underline{m}3\underline{m}$							
	$\underline{m}3m$			—			—	—
	Γ_1	1	1	1	1	1	1	1
	Γ_2	1	1	-1	1	1	-1	-1
	Γ_3	I_2	I_2	$\Gamma_3(\sigma_{dp})$	$\Gamma_3(C_{3j}^{-})$	$\Gamma_3(C_{3j})$	$\Gamma_3(C_{4m}^{-})$	$\Gamma_3(C_{4m})$
	Γ_4	I	$3D$	σ_{dp}	C_{3j}^{-}	C_{3j}	S_{4m}	S_{4m}^{-}
	Γ_5	I	$3D$	C_{2p}	C_{3j}^{-}	C_{3j}	C_{4m}^{-}	C_{4m}
	Γ_1'	1	1	1	1	1	1	1
	Γ_2'	1	1	-1	1	1	-1	-1
	Γ_3'	I_2	I_2	$\Gamma_3(\sigma_{dp})$	$\Gamma_3(C_{3j}^{-})$	$\Gamma_3(C_{3j})$	$\Gamma_3(C_{4m}^{-})$	$\Gamma_3(C_{4m})$
	Γ_4'	I	$3D$	σ_{dp}	C_{3j}^{-}	C_{3j}	S_{4m}	S_{4m}^{-}
	Γ_5'	I	$3D$	C_{2p}	C_{3j}^{-}	C_{3j}	C_{4m}^{-}	C_{4m}
Hexoctohedral	$m3m$	C	$3R$	σ_{dp}	S_{6j}	S_{6j}^{-}	S_{4m}	S_{4m}^{-}
	$m3\underline{m}$			—			—	—

Table 3.3 (*continued*)

$\underline{m}3\underline{m}$	—	—	—	—	—	—	—
$\underline{m}3m$	—	—		—	—		
Γ_1	1	1	1	1	1	1	1
Γ_2	1	1	-1	1	1	-1	-1
Γ_3	I_2	I_2	$\Gamma_3(\sigma_{dp})$	$\Gamma_3(C_{3j}^-)$	$\Gamma_3(C_{3j})$	$\Gamma_3(C_{4m}^-)$	$\Gamma_3(C_{4m})$
Γ_4	I	$3D$	σ_{dp}	C_{3j}^-	C_{3j}	S_{4m}	S_{4m}^-
Γ_5	I	$3D$	C_{2p}	C_{3j}^-	C_{3j}	C_{4m}^-	C_{4m}
Γ_1'	-1	-1	-1	-1	-1	-1	-1
Γ_2'	-1	-1	1	-1	-1	1	1
Γ_3'	$-I_2$	$-I_2$	$-\Gamma_3(\sigma_{dp})$	$-\Gamma_3(C_{3j}^-)$	$-\Gamma_3(C_{3j})$	$-\Gamma_3(C_{4m}^-)$	$-\Gamma_3(C_{4m})$
Γ_4'	I	$3R$	C_{2p}	S_{6j}	S_{6j}^-	C_{4m}^-	C_{4m}
Γ_5'	C	$3R$	σ_{dp}	S_{6j}	S_{6j}^-	S_{4m}	S_{4m}^-

where

$$3D = D_1, D_2, D_3, \qquad S_{6j} = CM_1, R_1M_1, R_2M_1, R_3M_1,$$

$$3R = R_1, R_2, R_3, \qquad S_{6j}^- = CM_2, R_1M_2, R_2M_2, R_3M_2.$$

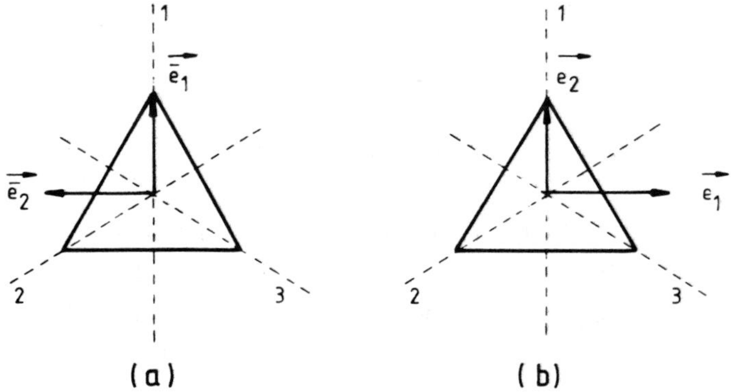

Figure 3.2. Alternative choice of basis.

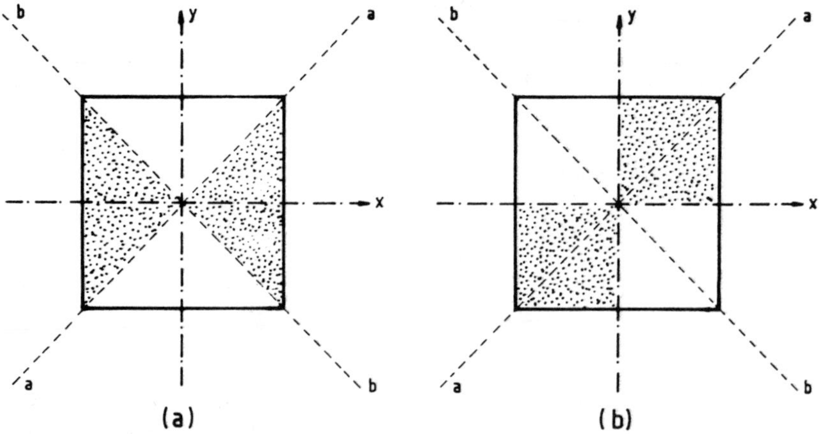

Figure 3.3. Black and white squares to illustrate (a) $\underline{4}m\underline{m}$ and (b) $\underline{4}\underline{m}m$.

The ith irreducible representation $D^{(i)}$ is designated by Γ_i. The choice of reference frame (basis) is immaterial for one-dimensional representations. Although the matrices of the representations of order 2 or 3 become different for a different choice of basis, they are all equivalent. In the construction of Table 3.3(10), the basis shown in Figure 3.2(b) is employed. The basis in Figure 3.2(a) is used in Example 14 in Appendix A, which produces a different, but equivalent, representation to Γ_3.

An Alternative Derivation of the Magnetic Point Groups

It is possible to derive the magnetic (black and white) point groups by considering the irreducible representations of the conventional point groups $\{G\}$. If the identity representation Γ_1 of $\{G\}$ is regarded as synonymous with $\{G\}$ itself, the remaining one-dimensional representations of $\{G\}$ can be related to the magnetic point groups $\{M\}$ that are generated from $\{G\}$ (Indenbom [1959]). In fact, the number of magnetic point groups is equal to the number of one-dimensional real representations that are distinct in the abstract sense (Bertaut [1968]). Two representations are said to be "distinct in the abstract sense," if they cannot be transformed into each other by changing the orientation of the axes.

We may illustrate this by again considering the point group C_{3v} and the magnetic point groups derived from it. The character table of the point group $C_{3v} = 3m$ can be found from Table 3.3(10). Γ_1 is the identity representation and corresponds to $C_{3v} = 3m$ itself. The representation Γ_3 is two-dimensional and is not relevant to the present discussion. In Γ_2, the elements $\mathbf{I}$, $\mathbf{S}_1$, and $\mathbf{S}_2$ are represented by $+1$ while $\mathbf{R}_1$, $\mathbf{R}_1\mathbf{S}_1$, and $\mathbf{R}_1\mathbf{S}_2$ are represented by -1. The elements that are represented by $+1$ are then uncolored in $\{M\}$, while the elements represented by -1 become complementary elements (colored) in $\{M\}$. Thus, Γ_2 corresponds to the magnetic point group $3\underline{m}$ that was discussed earlier.

Let us consider another example where some of the irreducible representations are not distinct in the abstract sense. Consider the group of symmetry operations of a square, $C_{4v} = 4mm$, whose character table is shown in Table 3.3(7). Γ_2 corresponds to the magnetic point groups $4\underline{m}\underline{m}$, and Γ_4 leads to the magnetic group $\underline{4}m\underline{m}$. However, $\underline{4}m\underline{m}$ derived from Γ_4, and $\underline{4}\underline{m}m$ derived from Γ_3 are not essentially different, since one of them can be turned into the other by rotating the x, y-axes by $\pi/4$ radians about z-axis (Fig. 3.3). Hence, it is possible to derive two magnetic point groups from the conventional point group $4mm$. By the application of this procedure to each of the thirty-two point groups at a time, all the fifty-eight magnetic (black and white) point groups, which were listed in Table 3.1, could be derived.

CHAPTER 4

Decomposition of Mechanical and Electromagnetic Quantities

4.1. Material Tensors and Physical Tensors

Most physical properties of crystals are defined by the relationship between two or more tensors and are therefore themselves represented by tensors. If a crystal is subjected to an influence presented by a tensor $I_{j_1 \ldots j_n}$, which produces a physical effect $E_{i_1 \ldots i_m}$, then a linear relationship between the influence and the effect is given by

$$E_{i_1 \ldots i_m} = m_{i_1 \ldots i_m j_1 \ldots j_n} I_{j_1 \ldots j_n}, \tag{4.1}$$

where the tensors **E** and **I** are called physical (field) tensors, and **m** is the material tensor. Note that, in higher-order effects, we may have more than one influence, i.e.,

$$E_{i_1 \ldots i_m} = m_{i_1 \ldots i_m, j_1 \ldots j_n, k_1 \ldots k_r} I'_{j_1 \ldots j_n} I''_{k_1 \ldots k_r}, \tag{4.2}$$

as, say, in the piezomagnetoelectrism $e_{ij} = A_{ijkr} P_k M_r$, where **e** is the strain (or stress) tensor and **P** and **M** are the electric and magnetization vectors polarization, respectively.

The invariance of (4.1), (4.2) under the spatial inversion, $\mathbf{C} = \text{diag}(-1, -1, -1)$, classifies the material tensor **m** as polar (true) or axial (pseudo), provided the physical tensors are already so classified. Note that the electric field, **E**, the electric displacement, **D**, and the polarization, **P**, are polar, but the magnetic field, **H**, the magnetic induction, **B**, and the magnetization, **M** are all axial vectors.

In three-dimensional formulations, material tensors are further classified into two types, by considering the effect of time-inversion on them. Tensors that are invariant under time-inversion are called *i*-tensors, and tensors whose components change signs under time-inversion are called *c*-tensors. The classification of physical (field) tensors, as *i*- or *c*-tensors, may be done by quantum-mechanical considerations. Note that the electric current density, **J**, the magnetic field, **H**, the flux, **B**, and the magnetic polarization, **M**, are all *c*-tensors. The remaining physical quantities which we are interested in are all time-symmetric.

4.2. Transformation Properties of Tensors

For the ordinary magnetic groups, $\{\mathbf{M}\} = \{\mathbf{G}\}$, the time reversal operator τ is not present, hence the components $t_{i_1 \ldots i_n}$ of a tensor $\mathbf{t}$ are transformed, under a symmetry operator $\mathbf{S}^\alpha$, according to

$$\bar{t}_{i_1 \ldots i_n} = S^\alpha_{i_1 j_1} \ldots S^\alpha_{i_n j_n} t_{j_1 \ldots j_n}, \qquad S^\alpha \in \{G\}, \tag{4.3}$$

for polar (true) tensors, and to

$$\bar{t}_{i_1 \ldots i_n} = |\mathbf{S}^\alpha| S^\alpha_{i_1 j_1} \ldots S^\alpha_{i_n j_n} t_{j_1 \ldots j_n}, \qquad S^\alpha \in \{G\}, \tag{4.4}$$

for axial (pseudo) tensors. These expressions are valid for both i-tensors and c-tensors.

For magnetic point groups (ferromagnetic, ferrimagnetic, and antiferromagnetic crystals) we introduced a complementary symmetry operator $\mathbf{M}^\alpha = \tau \mathbf{S}^\alpha$ where $\mathbf{S}^\alpha \in \{G\}$. For i-tensors, (4.3), (4.4) are used. For c-tensors, (4.3), (4.4) are used when the symmetry operator M^α is not a complementary operator. When, however, a complementary operator $\mathbf{M}^\alpha$ is involved, a (-1) factor is put in front of both (4.3), (4.4). This is due to the reversal of c-tensors under time inversion.

For the deformable electromagnetic solids, the relevant quantities and their properties and representations are given in Table 4.1.

4.3. Decomposition of Electromechanical Quantities

Let $\mathbf{M}^1, \mathbf{M}^2, \ldots, \mathbf{M}^g$ be the symmetry transformations (ordinary $\mathbf{S}^\alpha$ or complementary $\tau\mathbf{S}^\alpha$) which carry the material from its initial configuration into a final configuration which is indistinguishable from the original. The matrices $\{\mathbf{M}^1, \mathbf{M}^2, \ldots, \mathbf{M}^g\}$ form a matrix group $\{M\}$ which is said to be the symmetry group of the material.

The representations of $\mathbf{M}, \mathbf{J}, \ldots, \mathbf{E}$ listed in Table 4.1 are reducible. If Γ denotes any one of these representations, with the character $\chi(\mathbf{M}^\alpha)$, then we

Table 4.1. Properties of electromechanical quantities.

Type	Properties	Representations	Examples
M	Axial, c-vector	[1] a, c	Magnetization, magnetic field, and flux
J	Polar, c-vector	[1] p, c	Electric current
P	Polar, i-vector	[1] p, i	Polarization, electric field, and displacement
A	Axial, i-vector	[1] a, i	**M**, **H**, and **B** for nonmagnetic crystals
E	Polar, i-tensor second-order symmetric	[2] p, i	Stress and strain tensors

have

$$\Gamma = \sum_{i=1}^{r} n_i \Gamma_i = n_1 \Gamma_1 \dotplus \cdots \dotplus n_r \Gamma_r, \tag{4.5}$$

where $\Gamma_1 \ldots \Gamma_r$ are the irreducible representations of $\{M\}$ which are listed in Table 3.3(2–18). The number of times each irreducible representation occurs in Γ is given by

$$n_i = \frac{1}{g} \sum_{\alpha=1}^{g} \overset{*}{\chi}_i(\mathbf{M}^\alpha)\chi(\mathbf{M}^\alpha), \tag{4.6}$$

where g is the order of the group $\{M\}$, and $\overset{*}{\chi}_i(\mathbf{M}^\alpha)$ is the complex conjugate of the character of the ith irreducible representation Γ_i at the element $\mathbf{M}^\alpha$.

The characters χ_i of Γ_i are readily obtained from Table 3.3(2–18). The characters $\chi(\mathbf{M}^\alpha)$ for the reducible representations of **M**, **J**, ..., **E** are simply obtained, using transformation properties of tensors, as

$$\chi[2]_{p,i}(\mathbf{M}^\alpha) = \tfrac{1}{2}[(\operatorname{tr} \mathbf{M}^\alpha)^2 + \operatorname{tr}(\mathbf{M}^\alpha)^2], \tag{4.7}$$

$$\chi[1]_{a,i}(\mathbf{M}^\alpha) = \tfrac{1}{2}[(\operatorname{tr} \mathbf{M}^\alpha)^2 - \operatorname{tr}(\mathbf{M}^\alpha)^2], \tag{4.8}$$

$$\chi[1]_{p,i}(\mathbf{M}^\alpha) = \operatorname{tr} \mathbf{M}^\alpha, \tag{4.9}$$

where $\mathbf{M}^\alpha$ is replaced by $\mathbf{S}^\alpha \in \{G\}$. For the remaining time-asymmetric representations the characters are given by

$$\chi[1]_{p,c}(\mathbf{M}^\alpha) = (-1)^q \chi[1]_{p,i}(\mathbf{M}^\alpha), \tag{4.10}$$

$$\chi[1]_{a,c}(\mathbf{M}^\alpha) = (-1)^q \chi[1]_{a,i}(\mathbf{M}^\alpha), \tag{4.11}$$

where q is set equal to zero if $\mathbf{M}^\alpha = \mathbf{S}^\alpha$, and q is set equal to one when $\mathbf{M}^\alpha = \tau\mathbf{S}^\alpha$ (a complementary operator).

The decompositions of **M**, **J**, ..., **E** over the magnetic crystal classes are listed in Table 4.2(2–18). In fact, there is an alternative method for constructing Table 4.2(2–18) without using (4.7)–(4.11) in (4.6). Note that one of the one-dimensional irreducible representations of $\{G\}$ is obtained by associating $(+1)$ with proper rotations and (-1) with improper rotations, whose determinants are all (-1). The one-dimensional irreducible representation obtained in this manner is called the *alternating representation*, which we denote by A. There is another one-dimensional irreducible representation where ordinary transformations $\mathbf{S}^\alpha$ are associated with $(+1)$ and complementary elements by (-1). Let us call this representation T, the *time-reversal representation*. Now the decompositions of the representations of **M**, **J**, ..., **E** are readily determined, once the decomposition of $\mathbf{P} = [1]_{p,i}$ is known, with the aid of the inner direct product of the respective representations. Let

$$\mathbf{P}_i\colon\ [1]_{p,i} = \sum_{i=1}^{r} n_i \Gamma_i. \tag{4.12}$$

Table 4.2. The decompositions of **M**, **J**, ..., **E** over the magnetic crystal classes.

(2)

Type	Magnetic crystal classes		
	$\underline{\bar{1}}$	$\underline{m}$	$\underline{2}$
M	$3\Gamma_2$	$2\Gamma_1 + \Gamma_2$	$2\Gamma_1 + \Gamma_2$
J	$3\Gamma_1$	$\Gamma_1 + 2\Gamma_2$	$2\Gamma_1 + \Gamma_2$
P	$3\Gamma_2$	$2\Gamma_1 + \Gamma_2$	$\Gamma_1 + 2\Gamma_2$
A	$3\Gamma_1$	$\Gamma_1 + 2\Gamma_2$	$\Gamma_1 + 2\Gamma_2$
E	$6\Gamma_1$	$4\Gamma_1 + 2\Gamma_2$	$4\Gamma_1 + 2\Gamma_2$

M: $[1]_{a,c}$
J: $[1]_{p,c}$
P: $[1]_{p,i}$
A: $[1]_{a,i}$
E: $[2]_{p,i}$

(3)

Type	Magnetic crystal classes					
	$\underline{2}/m$	$2/\underline{m}$	$\underline{2}/\underline{m}$	$\underline{2}m\underline{m}$	$2\underline{m}\underline{m}$	$\underline{2}\underline{2}2$
M	$2\Gamma_2 + \Gamma_3$	$\Gamma_2 + 2\Gamma_3$	$2\Gamma_1 + \Gamma_4$	$\Gamma_1 + \Gamma_2 + \Gamma_3$	$\Gamma_1 + \Gamma_3 + \Gamma_4$	$\Gamma_1 + \Gamma_2 + \Gamma_3$
J	$2\Gamma_1 + \Gamma_4$	$\Gamma_1 + 2\Gamma_4$	$2\Gamma_2 + \Gamma_3$	$\Gamma_1 + \Gamma_2 + \Gamma_4$	$\Gamma_2 + \Gamma_3 + \Gamma_4$	$\Gamma_1 + \Gamma_2 + \Gamma_3$
P	$\Gamma_2 + 2\Gamma_3$	$\Gamma_2 + 2\Gamma_3$	$\Gamma_2 + 2\Gamma_3$	$\Gamma_1 + \Gamma_3 + \Gamma_4$	$\Gamma_1 + \Gamma_3 + \Gamma_4$	$\Gamma_2 + \Gamma_3 + \Gamma_4$
A	$\Gamma_1 + 2\Gamma_4$	$\Gamma_1 + 2\Gamma_4$	$\Gamma_1 + 2\Gamma_4$	$\Gamma_2 + \Gamma_3 + \Gamma_4$	$\Gamma_2 + \Gamma_3 + \Gamma_4$	$\Gamma_2 + \Gamma_3 + \Gamma_4$
E	$4\Gamma_1 + 2\Gamma_4$	$4\Gamma_1 + 2\Gamma_4$	$4\Gamma_1 + 2\Gamma_4$	$3\Gamma_1 + \Gamma_2 + \Gamma_3 + \Gamma_4$	$3\Gamma_1 + \Gamma_2 + \Gamma_3 + \Gamma_4$	$3\Gamma_1 + \Gamma_2 + \Gamma_3 + \Gamma_4$

(4)

Type	Magnetic crystal classes		
	$\underline{m}mm$	$\underline{m}\underline{m}\underline{m}$	$m\underline{m}\underline{m}$
M	$\Gamma_1' + \Gamma_2' + \Gamma_3'$	$\Gamma_2' + \Gamma_3' + \Gamma_4'$	$\Gamma_1 + \Gamma_2 + \Gamma_3$
J	$\Gamma_1 + \Gamma_2 + \Gamma_3$	$\Gamma_2 + \Gamma_3 + \Gamma_4$	$\Gamma_1' + \Gamma_2' + \Gamma_3'$
P	$\Gamma_2' + \Gamma_3' + \Gamma_4'$	$\Gamma_2' + \Gamma_3' + \Gamma_4'$	$\Gamma_2' + \Gamma_3' + \Gamma_4'$
A	$\Gamma_2 + \Gamma_3 + \Gamma_4$	$\Gamma_2 + \Gamma_3 + \Gamma_4$	$\Gamma_2 + \Gamma_3 + \Gamma_4$
E	$3\Gamma_1 + \Gamma_2 + \Gamma_3 + \Gamma_4$	$3\Gamma_1 + \Gamma_2 + \Gamma_3 + \Gamma_4$	$3\Gamma_1 + \Gamma_2 + \Gamma_3 + \Gamma_4$

(5)

Type	Magnetic crystal classes	
	$\underline{\bar{4}}$	$\underline{4}$
M	$\Gamma_2 + \Gamma_3 + \Gamma_4$	$\Gamma_2 + \Gamma_3 + \Gamma_4$
J	$\Gamma_1 + \Gamma_3 + \Gamma_4$	$\Gamma_2 + \Gamma_3 + \Gamma_4$
P	$\Gamma_2 + \Gamma_3 + \Gamma_4$	$\Gamma_1 + \Gamma_3 + \Gamma_4$
A	$\Gamma_1 + \Gamma_3 + \Gamma_4$	$\Gamma_1 + \Gamma_3 + \Gamma_4$
E	$2\Gamma_1 + 2\Gamma_2 + \Gamma_3 + \Gamma_4$	$2\Gamma_1 + 2\Gamma_2 + \Gamma_3 + \Gamma_4$

Table 4.2 (*continued*)

(6)

Type	Magnetic crystal classes		
	$\underline{4}/\underline{m}$	$4/\underline{m}$	$\underline{4}/m$
M	$\Gamma_2' + \Gamma_3' + \Gamma_4'$	$\Gamma_1' + \Gamma_3' + \Gamma_4'$	$\Gamma_2 + \Gamma_3 + \Gamma_4$
J	$\Gamma_2 + \Gamma_3 + \Gamma_4$	$\Gamma_1 + \Gamma_3 + \Gamma_4$	$\Gamma_2' + \Gamma_3' + \Gamma_4'$
P	$\Gamma_1' + \Gamma_3' + \Gamma_4'$	$\Gamma_1' + \Gamma_3' + \Gamma_4'$	$\Gamma_1' + \Gamma_3' + \Gamma_4'$
A	$\Gamma_1 + \Gamma_3 + \Gamma_4$	$\Gamma_1 + \Gamma_3 + \Gamma_4$	$\Gamma_1 + \Gamma_3 + \Gamma_4$
E	$2\Gamma_1 + 2\Gamma_2 + \Gamma_3 + \Gamma_4$	$2\Gamma_1 + 2\Gamma_2 + \Gamma_3 + \Gamma_4$	$2\Gamma_1 + 2\Gamma_2 + \Gamma_3 + \Gamma_4$

(7)

Type	Magnetic crystal classes			
	$4\underline{m}\underline{m}$	$\underline{4}m\underline{m}$	$4\underline{2}\underline{2}$	$\underline{4}2\underline{2}$
M	$\Gamma_1 + \Gamma_5$	$\Gamma_3 + \Gamma_5$	$\Gamma_1 + \Gamma_5$	$\Gamma_3 + \Gamma_5$
J	$\Gamma_2 + \Gamma_5$	$\Gamma_4 + \Gamma_5$	$\Gamma_1 + \Gamma_5$	$\Gamma_3 + \Gamma_5$
P	$\Gamma_1 + \Gamma_5$	$\Gamma_1 + \Gamma_5$	$\Gamma_2 + \Gamma_5$	$\Gamma_2 + \Gamma_5$
A	$\Gamma_2 + \Gamma_5$	$\Gamma_2 + \Gamma_5$	$\Gamma_2 + \Gamma_5$	$\Gamma_2 + \Gamma_5$
E	$2\Gamma_1 + \Gamma_3 + \Gamma_4 + \Gamma_5$	$2\Gamma_1 + \Gamma_3 + \Gamma_4 + \Gamma_5$	$2\Gamma_1 + \Gamma_3 + \Gamma_4 + \Gamma_5$	$2\Gamma_1 + \Gamma_3 + \Gamma_4 + \Gamma_5$

Type	Magnetic crystal classes		
	$\bar{\underline{4}}2\underline{m}$	$\bar{\underline{4}}\underline{2}m$	$\bar{4}\underline{2}\underline{m}$
M	$\Gamma_3 + \Gamma_5$	$\Gamma_4 + \Gamma_5$	$\Gamma_1 + \Gamma_5$
J	$\Gamma_2 + \Gamma_5$	$\Gamma_1 + \Gamma_5$	$\Gamma_4 + \Gamma_5$
P	$\Gamma_3 + \Gamma_5$	$\Gamma_3 + \Gamma_5$	$\Gamma_3 + \Gamma_5$
A	$\Gamma_2 + \Gamma_5$	$\Gamma_2 + \Gamma_5$	$\Gamma_2 + \Gamma_5$
E	$2\Gamma_1 + \Gamma_3 + \Gamma_4 + \Gamma_5$	$2\Gamma_1 + \Gamma_3 + \Gamma_4 + \Gamma_5$	$2\Gamma_1 + \Gamma_3 + \Gamma_4 + \Gamma_5$

(8)

Type	Magnetic crystal classes		
	$\underline{4}/\underline{m}m\underline{m}$	$4/\underline{m}mm$	$\underline{4}/m\underline{m}\underline{m}$
M	$\Gamma_3' + \Gamma_5'$	$\Gamma_1' + \Gamma_5'$	$\Gamma_1 + \Gamma_5$
J	$\Gamma_3 + \Gamma_5$	$\Gamma_1 + \Gamma_5$	$\Gamma_1' + \Gamma_5'$
P	$\Gamma_2' + \Gamma_5'$	$\Gamma_2' + \Gamma_5'$	$\Gamma_2' + \Gamma_5'$
A	$\Gamma_2 + \Gamma_5$	$\Gamma_2 + \Gamma_5$	$\Gamma_2 + \Gamma_5$
E	$2\Gamma_1 + \Gamma_3 + \Gamma_4 + \Gamma_5$	$2\Gamma_1 + \Gamma_3 + \Gamma_4 + \Gamma_5$	$2\Gamma_1 + \Gamma_3 + \Gamma_4 + \Gamma_5$

Type	Magnetic crystal classes	
	$4/\underline{m}\underline{m}\underline{m}$	$\underline{4}/mm\underline{m}$
M	$\Gamma_2' + \Gamma_5'$	$\Gamma_3 + \Gamma_5$
J	$\Gamma_2 + \Gamma_5$	$\Gamma_3' + \Gamma_5'$
P	$\Gamma_2' + \Gamma_5'$	$\Gamma_2' + \Gamma_5'$
A	$\Gamma_2 + \Gamma_5$	$\Gamma_2 + \Gamma_5$
E	$2\Gamma_1 + \Gamma_3 + \Gamma_4 + \Gamma_5$	$2\Gamma_1 + \Gamma_3 + \Gamma_4 + \Gamma_5$

Table 4.2 (*continued*)

(10)

Type	Magnetic crystal classes: $3\underline{m}$	$3\underline{2}$
M	$\Gamma_1 + \Gamma_3$	$\Gamma_1 + \Gamma_3$
J	$\Gamma_2 + \Gamma_3$	$\Gamma_1 + \Gamma_3$
P	$\Gamma_1 + \Gamma_3$	$\Gamma_2 + \Gamma_3$
A	$\Gamma_2 + \Gamma_3$	$\Gamma_2 + \Gamma_3$
E	$2\Gamma_1 + 2\Gamma_3$	$2\Gamma_1 + 2\Gamma_3$

(11)

Type	Magnetic crystal classes: $\underline{\bar{3}}$	$\underline{\bar{6}}$	$\underline{6}$
M	$\Gamma_4 + \Gamma_5 + \Gamma_6$	$\Gamma_2 + \Gamma_3 + \Gamma_4$	$\Gamma_2 + \Gamma_3 + \Gamma_4$
J	$\Gamma_1 + \Gamma_2 + \Gamma_3$	$\Gamma_1 + \Gamma_5 + \Gamma_6$	$\Gamma_2 + \Gamma_3 + \Gamma_4$
P	$\Gamma_4 + \Gamma_5 + \Gamma_6$	$\Gamma_2 + \Gamma_3 + \Gamma_4$	$\Gamma_1 + \Gamma_5 + \Gamma_6$
A	$\Gamma_1 + \Gamma_2 + \Gamma_3$	$\Gamma_1 + \Gamma_5 + \Gamma_6$	$\Gamma_1 + \Gamma_5 + \Gamma_6$
E	$2\Gamma_1 + 2\Gamma_2 + 2\Gamma_3$	$2\Gamma_1 + \Gamma_2 + \Gamma_3 + \Gamma_5 + \Gamma_6$	$2\Gamma_1 + \Gamma_2 + \Gamma_3 + \Gamma_5 + \Gamma_6$

(12)

Type	Magnetic crystal classes: $\bar{6}\underline{m}\underline{2}$	$\underline{\bar{6}}m\underline{2}$	$\underline{\bar{6}}\underline{m}2$	$\bar{3}\underline{m}$	$\underline{\bar{3}}m$	$\underline{\bar{3}}\underline{m}$
M	$\Gamma_1 + \Gamma_5$	$\Gamma_4 + \Gamma_6$	$\Gamma_3 + \Gamma_6$	$\Gamma_1 + \Gamma_6$	$\Gamma_3 + \Gamma_5$	$\Gamma_4 + \Gamma_5$
J	$\Gamma_4 + \Gamma_6$	$\Gamma_1 + \Gamma_5$	$\Gamma_2 + \Gamma_5$	$\Gamma_3 + \Gamma_5$	$\Gamma_1 + \Gamma_6$	$\Gamma_2 + \Gamma_6$
P	$\Gamma_3 + \Gamma_6$	$\Gamma_3 + \Gamma_6$	$\Gamma_3 + \Gamma_6$	$\Gamma_4 + \Gamma_5$	$\Gamma_4 + \Gamma_5$	$\Gamma_4 + \Gamma_5$
A	$\Gamma_2 + \Gamma_5$	$\Gamma_2 + \Gamma_5$	$\Gamma_2 + \Gamma_5$	$\Gamma_2 + \Gamma_6$	$\Gamma_2 + \Gamma_6$	$\Gamma_1 + \Gamma_6$
E	$2\Gamma_1 + \Gamma_5 + \Gamma_6$	$2\Gamma_1 + \Gamma_5 + \Gamma_6$	$2\Gamma_1 + \Gamma_5 + \Gamma_6$	$2\Gamma_1 + \Gamma_6$	$2\Gamma_1 + \Gamma_6$	$2\Gamma_1 + \Gamma_6$

Type	Magnetic crystal classes: $6/\underline{m}$	$\underline{6}/m$	$\underline{6}/\underline{m}$
M	$\Gamma_1' + \Gamma_5' + \Gamma_6'$	$\Gamma_2' + \Gamma_3' + \Gamma_4'$	$\Gamma_2 + \Gamma_3 + \Gamma_4$
J	$\Gamma_1 + \Gamma_5 + \Gamma_6$	$\Gamma_2 + \Gamma_3 + \Gamma_4$	$\Gamma_2' + \Gamma_3' + \Gamma_4'$
P	$\Gamma_1' + \Gamma_5' + \Gamma_6'$	$\Gamma_1' + \Gamma_5' + \Gamma_6'$	$\Gamma_1' + \Gamma_5' + \Gamma_6'$
A	$\Gamma_1 + \Gamma_5 + \Gamma_6$	$\Gamma_1 + \Gamma_5 + \Gamma_6$	$\Gamma_1 + \Gamma_5 + \Gamma_6$
E	$2\Gamma_1 + \Gamma_2 + \Gamma_3 + \Gamma_5 + \Gamma_6$	$2\Gamma_1 + \Gamma_2 + \Gamma_3 + \Gamma_5 + \Gamma_6$	$2\Gamma_1 + \Gamma_2 + \Gamma_3 + \Gamma_5 + \Gamma_6$

(13)

Type	Magnetic crystal classes: $6\underline{2}\underline{2}$	$\underline{6}2\underline{2}$	$6\underline{m}\underline{m}$	$\underline{6}m\underline{m}$
M	$\Gamma_1 + \Gamma_5$	$\Gamma_4 + \Gamma_6$	$\Gamma_1 + \Gamma_5$	$\Gamma_3 + \Gamma_6$
J	$\Gamma_1 + \Gamma_5$	$\Gamma_4 + \Gamma_6$	$\Gamma_2 + \Gamma_5$	$\Gamma_4 + \Gamma_6$
P	$\Gamma_2 + \Gamma_5$	$\Gamma_2 + \Gamma_5$	$\Gamma_1 + \Gamma_5$	$\Gamma_1 + \Gamma_5$
A	$\Gamma_2 + \Gamma_5$	$\Gamma_2 + \Gamma_5$	$\Gamma_2 + \Gamma_5$	$\Gamma_2 + \Gamma_5$
E	$2\Gamma_1 + \Gamma_5 + \Gamma_6$	$2\Gamma_1 + \Gamma_5 + \Gamma_6$	$2\Gamma_1 + \Gamma_5 + \Gamma_6$	$2\Gamma_1 + \Gamma_5 + \Gamma_6$

Table 4.2 (*continued*)

(14)

Type	Magnetic crystal classes				
	$6/\underline{m}mm$	$6/m\underline{mm}$	$6/\underline{mmm}$	$\underline{6}/mm\underline{m}$	$\underline{6}/\underline{m}m\underline{m}$
M	$\Gamma_1' + \Gamma_5'$	$\Gamma_1 + \Gamma_5$	$\Gamma_2' + \Gamma_5'$	$\Gamma_4' + \Gamma_6'$	$\Gamma_4 + \Gamma_6$
J	$\Gamma_1 + \Gamma_5$	$\Gamma_1' + \Gamma_5'$	$\Gamma_2 + \Gamma_5$	$\Gamma_4 + \Gamma_6$	$\Gamma_4' + \Gamma_6'$
P	$\Gamma_2' + \Gamma_5'$	$\Gamma_2' + \Gamma_5'$	$\Gamma_2' + \Gamma_5'$	$\Gamma_2' + \Gamma_5'$	$\Gamma_2' + \Gamma_5'$
A	$\Gamma_2 + \Gamma_5$	$\Gamma_2 + \Gamma_5$	$\Gamma_2 + \Gamma_5$	$\Gamma_2 + \Gamma_5$	$\Gamma_2 + \Gamma_5$
E	$2\Gamma_1 + \Gamma_5 + \Gamma_6$	$2\Gamma_1 + \Gamma_5 + \Gamma_6$	$2\Gamma_1 + \Gamma_5 + \Gamma_6$	$2\Gamma_1 + \Gamma_5 + \Gamma_6$	$2\Gamma_1 + \Gamma_5 + \Gamma_6$

(16)

Type	Magnetic crystal class
	$\underline{m}3$
M	Γ_4'
J	Γ_4
P	Γ_4'
A	Γ_4
E	$\Gamma_1 + \Gamma_2 + \Gamma_3 + \Gamma_4$

(17)

Type	Magnetic crystal classes	
	$\underline{\bar{4}}3\underline{m}$	$\underline{4}3\underline{2}$
M	Γ_4	Γ_4
J	Γ_5	Γ_4
P	Γ_4	Γ_5
A	Γ_5	Γ_5
E	$\Gamma_1 + \Gamma_3 + \Gamma_4$	$\Gamma_1 + \Gamma_3 + \Gamma_4$

(18)

Type	Magnetic crystal classes		
	$m3\underline{m}$	$\underline{m}3\underline{m}$	$\underline{m}3m$
M	Γ_4	Γ_5'	Γ_4'
J	Γ_4'	Γ_5	Γ_4
P	Γ_5'	Γ_5'	Γ_5'
A	Γ_5	Γ_5	Γ_5
E	$\Gamma_1 + \Gamma_3 + \Gamma_4$	$\Gamma_1 + \Gamma_3 + \Gamma_4$	$\Gamma_1 + \Gamma_3 + \Gamma_4$

Then, for the remaining quantities, we write

$$\begin{aligned}
\mathbf{A}:&\quad [1]_{a,i} = [1]_{p,i} \otimes A,\\
\mathbf{J}:&\quad [1]_{p,c} = [1]_{p,i} \otimes T,\\
\mathbf{M}:&\quad [1]_{a,c} = [1]_{p,i} \otimes T \otimes A,\\
\mathbf{E}:&\quad [2]_{p,i} = [1]_{p,i} \otimes [1]_{p,i} \dot{-} [1]_{a,i}.
\end{aligned} \tag{4.13}$$

This alternative technique was employed in checking the entries of Table 4.2(2–18). To illustrate, consider the magnetic crystal $\underline{2}/m = \{\mathbf{I}, \tau\mathbf{D}_3, \mathbf{R}_3, \tau\mathbf{C}\}$, whose irreducible representations are given in Table 3.3(3). Hence, we have

	$\mathbf{I}$	$\tau\mathbf{D}_3$	$\mathbf{R}_3$	$\tau\mathbf{C}$	
det $\mathbf{S}^\alpha$	1	1	-1	-1	
$A = \Gamma_2$	1	1	-1	-1	
$T = \Gamma_3$	1	-1	1	-1	
trace$[1]_{p,i}$	3	-1	1	-3	
$\Gamma_2 \otimes \Gamma_3$	1	-1	-1	1	Γ_4 and so on.

From (4.6), we get

$$\begin{aligned}
n_1 &= \tfrac{1}{4}(3 - 1 + 1 - 3) = 0,\\
n_2 &= \tfrac{1}{4}(3 - 1 - 1 + 3) = 1,\\
n_3 &= \tfrac{1}{4}(3 + 1 + 1 + 3) = 2,\\
n_4 &= \tfrac{1}{4}(3 + 1 - 1 - 3) = 0,
\end{aligned} \tag{4.14}$$

from which it follows that $[1]_{p,i} = \Gamma_2 \dot{+} 2\Gamma_3$. Now, from (4.13), we obtain

$$\begin{aligned}
\mathbf{A}:&\quad [1]_{a,i} = (\Gamma_2 \dot{+} 2\Gamma_3) \otimes \Gamma_2 = \Gamma_2 \otimes \Gamma_2 \dot{+} 2\Gamma_3 \otimes \Gamma_2 = \Gamma_1 \dot{+} 2\Gamma_4,\\
\mathbf{J}:&\quad [1]_{p,c} = (\Gamma_2 \dot{+} 2\Gamma_3) \otimes \Gamma_3 = \Gamma_2 \otimes \Gamma_3 \dot{+} 2\Gamma_3 \otimes \Gamma_3 = 2\Gamma_1 \dot{+} \Gamma_4,\\
\mathbf{M}:&\quad [1]_{a,c} = (\Gamma_2 \dot{+} 2\Gamma_3) \otimes (\Gamma_2 \otimes \Gamma_3) = (\Gamma_2 \dot{+} 2\Gamma_3) \otimes \Gamma_2) \otimes \Gamma_3\\
&\quad = (\Gamma_1 \dot{+} 2\Gamma_4) \otimes \Gamma_3 = \Gamma_1\Gamma_3 \dot{+} 2\Gamma_4 \otimes \Gamma_3 = \Gamma_3 \dot{+} 2\Gamma_2,\\
\mathbf{E}:&\quad [2]_{p,i} = (\Gamma_2 \dot{+} 2\Gamma_3) \otimes (\Gamma_2 \dot{+} 2\Gamma_3) \dot{-} (\Gamma_1 \dot{+} 2\Gamma_4) = 4\Gamma_1 \dot{+} 2\Gamma_4.
\end{aligned}$$

The Kronecker (direct) products $\Gamma_p \otimes \Gamma_q$ appearing above may be directly obtained by using Table 6.4.

4.4. Basic Quantities of Electromechanical Tensors

Having obtained the decompositions of $\mathbf{M}, \mathbf{J}, \ldots, \mathbf{E}$ over the magnetic crystallographic groups we are ready to determine the components of $\mathbf{M}, \mathbf{J}, \ldots, \mathbf{E}$ explicitly, those that form the carrier spaces of the irreducible representations Γ_i of $\{M\}$.

The transformation properties of **M**, **J**, ..., **E** yield, under an element $\mathbf{M}^\alpha$ of $\{M\}$ in the transformed configuration,

$$
\begin{aligned}
\bar{M}_i &= (-1)|\mathbf{S}^\alpha| S^\alpha_{ij} M_j && \text{when } \mathbf{M}^\alpha = \tau\mathbf{S}^\alpha, \\
&= |\mathbf{S}^\alpha| S^\alpha_{ij} M_j && \text{when } \mathbf{M}^\alpha = \mathbf{S}^\alpha, \\
\bar{J}_i &= (-1) S^\alpha_{ij} J_j && \text{when } \mathbf{M}^\alpha = \tau\mathbf{S}^\alpha, \\
&= S^\alpha_{ij} J_j && \text{when } \mathbf{M}^\alpha = \mathbf{S}^\alpha, \\
\bar{P}_i &= S^\alpha_{ij} P_j && \text{either } \mathbf{M}^\alpha = \tau\mathbf{S}^\alpha \text{ or } \mathbf{M}^\alpha = \mathbf{S}^\alpha, \\
\bar{A}_i &= |\mathbf{S}^\alpha| S^\alpha_{ij} A_j && \text{either } \mathbf{M}^\alpha = \tau\mathbf{S}^\alpha \text{ or } \mathbf{M}^\alpha = \mathbf{S}^\alpha, \\
\bar{E}_{ij} &= S^\alpha_{ik} S^\alpha_{jm} E_{km} && \text{either } \mathbf{M}^\alpha = \tau\mathbf{S}^\alpha \text{ or } \mathbf{M}^\alpha = \mathbf{S}^\alpha,
\end{aligned}
\tag{4.15}
$$

where $\mathbf{S}^\alpha$ is an element of $\{G\}$ from which $\{\mathbf{M}\}$ is generated. For instance, under the elements of the magnetic crystal class $\underline{2}/m = \{\mathbf{I}, \tau\mathbf{D}_3, \mathbf{R}_3, \tau\mathbf{C}\}$, we have for $\mathbf{M} = [1]_{a,c}$

	$\mathbf{I}$	$\tau\mathbf{D}_3$	$\mathbf{R}_3$	$\tau\mathbf{C}$
$\bar{M}_1$	M_1	M_1	$-M_1$	$-M_1$
$\bar{M}_2$	M_2	M_2	$-M_2$	$-M_2$
$\bar{M}_3$	M_3	$-M_3$	M_3	$-M_3$
T:	$\begin{pmatrix} 1 & 0 & 0 \\ 0 & 1 & 0 \\ 0 & 0 & 1 \end{pmatrix}$	$\begin{pmatrix} 1 & 0 & 0 \\ 0 & 1 & 0 \\ 0 & 0 & -1 \end{pmatrix}$	$\begin{pmatrix} -1 & 0 & 0 \\ 0 & -1 & 0 \\ 0 & 0 & 1 \end{pmatrix}$	$\begin{pmatrix} -1 & 0 & 0 \\ 0 & -1 & 0 \\ 0 & 0 & -1 \end{pmatrix}$

and for **E**, we have

	$\mathbf{I}$	$\tau\mathbf{D}_3$
$\bar{E}_{11}$	E_{11}	E_{11}
$\bar{E}_{22}$	E_{22}	E_{22}
$\bar{E}_{33}$	E_{33}	E_{33}
$\bar{E}_{23}$	E_{23}	$-E_{23}$
$\bar{E}_{31}$	E_{31}	$-E_{31}$
$\bar{E}_{12}$	E_{12}	E_{12}
T:	$\begin{pmatrix} 1 & 0 & 0 & 0 & 0 & 0 \\ 0 & 1 & 0 & 0 & 0 & 0 \\ 0 & 0 & 1 & 0 & 0 & 0 \\ 0 & 0 & 0 & 1 & 0 & 0 \\ 0 & 0 & 0 & 0 & 1 & 0 \\ 0 & 0 & 0 & 0 & 0 & 1 \end{pmatrix}$ **T(I)**	$\begin{pmatrix} 1 & 0 & 0 & 0 & 0 & 0 \\ 0 & 1 & 0 & 0 & 0 & 0 \\ 0 & 0 & 1 & 0 & 0 & 0 \\ 0 & 0 & 0 & -1 & 0 & 0 \\ 0 & 0 & 0 & 0 & -1 & 0 \\ 0 & 0 & 0 & 0 & 0 & 1 \end{pmatrix}$ $\mathbf{T}(\tau\mathbf{D}_3)$

	$\mathbf{R}_3$	$\tau\mathbf{C}$
$\bar{E}_{11}$	E_{11}	E_{11}
$\bar{E}_{22}$	E_{22}	E_{22}
$\bar{E}_{33}$	E_{33}	E_{33}
$\bar{E}_{23}$	$-E_{23}$	E_{23}
$\bar{E}_{31}$	$-E_{31}$	E_{31}
$\bar{E}_{12}$	E_{12}	E_{12}
T:	$\begin{pmatrix} 1 & 0 & 0 & 0 & 0 & 0 \\ 0 & 1 & 0 & 0 & 0 & 0 \\ 0 & 0 & 1 & 0 & 0 & 0 \\ 0 & 0 & 0 & -1 & 0 & 0 \\ 0 & 0 & 0 & 0 & -1 & 0 \\ 0 & 0 & 0 & 0 & 0 & 1 \end{pmatrix}$ $\mathbf{T}(\mathbf{R}_3)$	$\begin{pmatrix} 1 & 0 & 0 & 0 & 0 & 0 \\ 0 & 1 & 0 & 0 & 0 & 0 \\ 0 & 0 & 1 & 0 & 0 & 0 \\ 0 & 0 & 0 & 1 & 0 & 0 \\ 0 & 0 & 0 & 0 & 1 & 0 \\ 0 & 0 & 0 & 0 & 0 & 1 \end{pmatrix}$ $\mathbf{T}(\tau\mathbf{C})$

Let $z_1, z_2, \ldots, z_p$ denote the independent components of any one of the quantities $\mathbf{M}$, $\mathbf{J}$, $\ldots$, $\mathbf{E}$. From (4.15) it follows that under an element $\mathbf{M}^\alpha$, $\bar{z}_1, \bar{z}_2, \ldots, \bar{z}_p$ are given by

$$\bar{z}_i(\mathbf{M}^\alpha) = T_{ij}(\mathbf{M}^\alpha)z_j; \qquad (i, j = 1, \ldots, p;\ \alpha = 1, \ldots, g). \tag{4.16}$$

The set of g, $p \times p$ matrices

$$\Gamma = \{\mathbf{T}(\mathbf{M}^1), \mathbf{T}(\mathbf{M}^2), \ldots, \mathbf{T}(\mathbf{M}^g)\} \tag{4.17}$$

forms a matrix representation Γ of degree p of the group $\{M\}$. The independent components $z_1, \ldots, z_p$ form a carrier space for the representation Γ. We may determine the matrices (4.17) from inspection of the expressions appearing in (4.15). These matrices describe the manner in which the independent components $z_1, \ldots, z_p$ transform under an element $\mathbf{M}^\alpha$ of $\{M\}$. The representation

$$\Gamma = \{\mathbf{T}(\mathbf{M}^1), \ldots, \mathbf{T}(\mathbf{M}^g)\} \tag{4.18}$$

may be decomposed into the direct sum of irreducible representations $\Gamma_1, \ldots, \Gamma_r$ of $\{M\}$, and this has been done in the previous section. Thus, we may determine a matrix $\mathbf{Q}$ such that

$$\mathbf{Q}\mathbf{T}(\mathbf{M}^\alpha)\mathbf{Q}^{-1} = n_1\mathbf{D}^{(1)}(\mathbf{M}^\alpha) + \cdots + n_r\mathbf{D}^{(r)}(\mathbf{M}^\alpha), \tag{4.19}$$

where n_i denotes the number of times the irreducible representation $\Gamma_i = \{\mathbf{D}^{(i)}(\mathbf{M})\}$ of degree d_i appears in the decomposition of $\Gamma = \{\mathbf{T}(\mathbf{M}^\alpha)\}$ and is given by (4.6). The set of g matrices

$$\{\mathbf{Q}\mathbf{T}(\mathbf{M}^1)\mathbf{Q}^{-1}, \ldots, \mathbf{Q}\mathbf{T}(\mathbf{M}^g)\mathbf{Q}^{-1}\} = \bar{\Gamma} \tag{4.20}$$

also furnishes a matrix representation of the symmetry group $\{M\}$, which is equivalent to the representation Γ furnished by the matrices (4.18). The matrices in (4.20) describe the transformation properties of $\mathbf{Qz}$, such that

$$\mathbf{Qz} = \mathbf{U}_1^{(1)} + \cdots + \mathbf{U}_{n_1}^{(1)} + \cdots + \mathbf{U}_1^{(r)} + \cdots + \mathbf{U}_{n_r}^{(r)}, \tag{4.21}$$

where the first d_1 components of $\mathbf{Qz}$ are the d_1 components of $\mathbf{U}_1^{(1)}$, the next d_1 components of $\mathbf{Qz}$ are the d_1 components of $\mathbf{U}_2^{(1)}, \ldots,$ and the last d_r components of $\mathbf{Qz}$ are the d_r components of $\mathbf{U}_{n_r}^{(r)}$. The quantities $\mathbf{U}_1^{(1)}, \ldots, \mathbf{U}_{n_r}^{(r)}$ form carrier spaces for the irreducible representations $\Gamma_1, \ldots, \Gamma_r$, respectively. Thus, we have

$$\bar{\mathbf{U}}_j^{(i)}(\mathbf{M}^\alpha) = \mathbf{D}^{(i)}(\mathbf{M}^\alpha)\mathbf{U}_j^{(i)}; \qquad \text{any} \quad \mathbf{M}^\alpha \in \{M\}. \tag{4.22}$$

We now give a procedure for determining the quantities $\mathbf{U}_j^{(i)}$ which form the carrier spaces for the irreducible representations $\Gamma_i = \{\mathbf{D}^{(i)}(\mathbf{M}^\alpha)\}$ appearing in the decomposition (4.19) of $\Gamma = \{T(M^\alpha)\}$. Suppose that the irreducible representation Γ_i of degree d_i appears n_i times in the decomposition of $\Gamma = \{\mathbf{T}(\mathbf{M}^\alpha)\}$. Then the quantity $\boldsymbol{\phi}^{(sqi)}$, whose d_i independent components are given by

$$\phi_k^{(sqi)} = \sum_{\alpha=1}^{g} \overset{*}{D}{}_{sk}^{(i)}(\mathbf{M}^\alpha)\, T_{qr}(\mathbf{M}^\alpha) z_r \qquad (k = 1, \ldots, d_i), \tag{4.23}$$

forms a carrier space for the irreducible representation Γ_i, provided that $\phi_k^{(sqi)}$ are nonzero. We vary s and q ($s = 1, \ldots, d_i$; $q = 1, \ldots, p$) so as to obtain n_i linearly independent quantities which we denote by

$$\mathbf{U}_j^{(i)} = \boldsymbol{\phi}^{(s_j q_j i)} \qquad (j = 1, \ldots, n_i), \tag{4.24}$$

where the components of $\mathbf{U}_j^{(i)}$ will be linear combinations (see (4.23)), of $z_1, \ldots, z_p$ and are called the *basic quantities* of $\mathbf{z}$ associated with the group $\{M\}$.

We may readily show that the components $\phi_1^{(sqi)}, \ldots, \phi_{d_i}^{(sqi)}$, of the quantity $\boldsymbol{\phi}^{(sqi)}$ obtained from (4.23), form a carrier space for the irreducible representation Γ_i. To begin with, since the irreducible representations of concern here are unitary, we have

$$\overset{*}{D}{}_{sk}^{(i)}(\mathbf{M}^\alpha) = ([D^{(i)}(\mathbf{M}^\alpha)]^{-1})_{ks} = D_{ks}^{(i)}[(\mathbf{M}^\alpha)^{-1}],$$

hence, (4.23) may be rewritten as

$$\phi_k^{(sqi)} = \sum_{\alpha=1}^{g} D_{ks}^{(i)}[(\mathbf{M}^\alpha)^{-1}]\, T_{qr}(\mathbf{M}^\alpha) z_r. \tag{4.25}$$

Under an element $\mathbf{M}^\beta$, this becomes

$$\bar{\phi}_k^{(sqi)}(\mathbf{M}^\beta) = \sum_{\alpha=1}^{g} D_{ks}^{(i)}[(M^\alpha)^{-1}]\, T_{qr}(\mathbf{M}^\alpha) \bar{z}_r(\mathbf{M}^\beta) \tag{4.26}$$

$$= \sum_{\alpha=1}^{g} D_{ks}^{(i)}[(M^\alpha)^{-1}]\, T_{qr}(\mathbf{M}^\alpha)\, T_{ru}(\mathbf{M}^\beta) z_u \tag{4.27}$$

$$= \sum_{\alpha=1}^{g} D_{ks}^{(i)}[(M^\alpha)^{-1}]\, T_{qu}(\mathbf{M}^\alpha\mathbf{M}^\beta) z_u, \tag{4.28}$$

since the set $\{\mathbf{T}(\mathbf{M}^\alpha)\}$ ($\alpha = 1, \ldots, g$) forms a representation of the group $\{M\}$, and hence $\mathbf{T}(\mathbf{M}^\alpha)\mathbf{T}(\mathbf{M}^\beta) = \mathbf{T}(\mathbf{M}^\alpha\mathbf{M}^\beta)$. We introduce $\mathbf{M}^\alpha$ as

$$\mathbf{M}^\alpha\mathbf{M}^\beta = \mathbf{M}^\gamma; \qquad \mathbf{M}^\alpha = \mathbf{M}^\gamma(\mathbf{M}^\beta)^{-1}; \qquad (\mathbf{M}^\alpha)^{-1} = \mathbf{M}^\beta(\mathbf{M}^\gamma)^{-1}. \tag{4.29}$$

Substituting from (4.29) into (4.28) yields

$$\bar{\phi}_k^{(sqi)}(\mathbf{M}^\beta) = \sum_{\gamma=1}^{g} D_{ks}^{(i)}[\mathbf{M}^\beta(\mathbf{M}^\gamma)^{-1}]T_{qu}(\mathbf{M}^\gamma)z_u \tag{4.30}$$

$$= \sum_{\gamma=1}^{g} D_{km}^{(i)}(\mathbf{M}^\beta)D_{ms}^{(i)}(\mathbf{M}^\gamma)^{-1}]T_{qu}(\mathbf{M}^\gamma)z_u. \tag{4.31}$$

Thus with (4.25), we have

$$\bar{\phi}_k^{(sqi)}(\mathbf{M}^\beta) = D_{km}^{(i)}(\mathbf{M}^\beta)\phi_m^{(sqi)}; \qquad k, m = 1, \ldots, d_i, \tag{4.32}$$

which verifies that $\phi_k^{(sqi)}$ forms a carrier space for the representation Γ_i.

We give an example of the application of the formula (4.23) to the problem of splitting the vector $\mathbf{M}$ (axial, c-tensor) into the sum of the basic quantities associated with the irreducible representations of $\underline{2}/m = \{\mathbf{I}, \tau\mathbf{D}_3, \mathbf{R}_3, \tau\mathbf{C}\}$. At the beginning of this section we obtained the matrices $\mathbf{T}(\mathbf{M}^\alpha)$ in

$$\overline{M}_i(\mathbf{M}^\alpha) = T_{ij}(\mathbf{M}^\alpha)M_j, \tag{4.33}$$

where

$$\operatorname{tr}\mathbf{T}(\mathbf{I}) = 3, \quad \operatorname{tr}\mathbf{T}(\tau\mathbf{D}_3) = 1, \quad \operatorname{tr}\mathbf{T}(\mathbf{R}_3) = -1, \quad \operatorname{tr}\mathbf{T}(\tau\mathbf{C}) = -3, \tag{4.34}$$

from this, and with Table 3.3(3) and equation (4.6), we get

$$\begin{aligned} n_1 &= \tfrac{1}{4}(3 + 1 - 1 - 3) = 0, \\ n_2 &= \tfrac{1}{4}(3 + 1 + 1 + 3) = 2, \\ n_3 &= \tfrac{1}{4}(3 - 1 - 1 + 3) = 1, \\ n_4 &= \tfrac{1}{4}(3 - 1 + 1 - 3) = 0. \end{aligned} \tag{4.35}$$

Hence, the decomposition of the representation

$$\Gamma = \{\mathbf{T}(\mathbf{I}), \mathbf{T}(\tau\mathbf{D}_3), \mathbf{T}(\mathbf{R}_3), \mathbf{T}(\tau\mathbf{C})\}$$

is given by

$$\Gamma = 2\Gamma_2 \dotplus \Gamma_3. \tag{4.36}$$

Since the irreducible representations of $\underline{2}/m$ are all one-dimensional, so are the basic quantities. Furthermore, (4.36) indicates that there are two basic quantities associated with Γ_2 and one with Γ_3, and there are no basic quantities forming carrier spaces of Γ_1 and Γ_4. For Γ_2, we have

$$\phi_k^{(sq2)} = \sum_{\alpha=1}^{4} \overset{*}{D}{}_{sk}^{(2)}(\mathbf{M}^\alpha)\bar{z}_q(\mathbf{M}^\alpha) \qquad (s, k = 1; q = 1, 2, 3), \tag{4.37}$$

from which there follows

$$\phi_1^{(112)} = 4M_1, \qquad \phi_1^{(122)} = 4M_2, \qquad \phi_1^{(132)} = 0, \tag{4.38}$$

hence we set $U_{11}^{(2)} = M_1$ and $U_{21}^{(2)} = M_2$. For Γ_3, we have

$$\phi_k^{(sq3)} = \sum_{\alpha=1}^{4} \overset{*}{D}{}_{sk}^{(3)}(\mathbf{M}^\alpha)\bar{z}_q(\mathbf{M}^\alpha) \qquad (s, k = 1; q = 1, 2, 3), \tag{4.39}$$

from which we get

$$\phi_1^{(113)} = 0, \qquad \phi_1^{(123)} = 0, \qquad \phi_1^{(123)} = 4M_3,$$

hence we set $U_{11}^{(3)} = M_3$. Note that if $(U_1, U_2, \ldots, U_p)$ are the components of a basic quantity $\mathbf{U}$, we could equally well take $(aU_1, aU_2, \ldots, aU_p)$ to be the components of $\mathbf{U}$ where a is any constant.

We list, in Table 4.3(1–58), those linear combinations of the components of **M**, **J**, **P**, **A**, and **E**, which yield the basic quantities appearing in the decomposition, for each of the magnetic point groups.

Table 4.3. Basic quantities for **M**, **J**, **P**, **A**, and **E**.

(1): $\underline{\bar{1}}$

Γ_1		J_1, J_2, J_3		$E_{11}, E_{22}, E_{33}, E_{23}, E_{31}, E_{12}$	A_1, A_2, A_3
Γ_2	M_1, M_2, M_3		P_1, P_2, P_3		

(2): $\underline{m}$

Γ_1	M_1, M_2	J_3	P_1, P_2	$E_{11}, E_{22}, E_{33}, E_{12}$	A_3
Γ_2	M_3	J_1, J_2	P_3	E_{23}, E_{31}	A_1, A_2

(3): $\underline{2}$

Γ_1	M_1, M_2	J_1, J_2	P_3	$E_{11}, E_{22}, E_{33}, E_{12}$	A_3
Γ_2	M_3	J_3	P_1, P_2	E_{23}, E_{31}	A_1, A_2

(4): $\underline{2}/m$

Γ_1		J_1, J_2		$E_{11}, E_{22}, E_{33}, E_{12}$	A_3
Γ_2	M_1, M_2		P_3		
Γ_3	M_3		P_1, P_2		
Γ_4		J_3		E_{32}, E_{31}	A_1, A_2

(5): $2/\underline{m}$

Γ_1		J_3		$E_{11}, E_{22}, E_{33}, E_{12}$	A_3
Γ_2	M_3		P_3		
Γ_3	M_1, M_2		P_1, P_2		
Γ_4		J_1, J_2		E_{32}, E_{31}	A_1, A_2

(6): $\underline{2}/\underline{m}$

Γ_1	M_1, M_2			$E_{11}, E_{22}, E_{33}, E_{12}$	A_3
Γ_2		J_1, J_2	P_3		
Γ_3		J_3	P_1, P_2		
Γ_4	M_3			E_{32}, E_{31}	A_1, A_2

(7): $\underline{2}/m\underline{m}$

Γ_1	M_2	J_1	P_3	E_{11}, E_{22}, E_{33}	
Γ_2	M_1	J_2		E_{12}	A_3
Γ_3	M_3		P_2	E_{23}	A_1
Γ_4		J_3	P_1	E_{31}	A_2

Table 4.3 (*continued*)

(8): $2\underline{mm}$

Γ_1	M_3		P_3	E_{11}, E_{22}, E_{33}	
Γ_2		J_3		E_{12}	A_3
Γ_3	M_2	J_1	P_2	E_{23}	A_1
Γ_4	M_1	J_2	P_1	E_{31}	A_2

(9): $\underline{22}2$

Γ_1	M_3	J_3		E_{11}, E_{22}, E_{33}	
Γ_2	M_2	J_2	P_1	E_{23}	A_1
Γ_3	M_1	J_1	P_2	E_{31}	A_2
Γ_4			P_3	E_{12}	A_3

(10): $\underline{m}mm$

Γ_1		J_3		E_{11}, E_{22}, E_{33}	
Γ_2		J_2		E_{23}	A_1
Γ_3		J_1		E_{31}	A_2
Γ_4				E_{21}	A_3
Γ_1'	M_3				
Γ_2'	M_2		P_1		
Γ_3'	M_1		P_2		
Γ_4'			P_3		

(11): $\underline{mmm}$

Γ_1				E_{11}, E_{22}, E_{33}	
Γ_2		J_1		E_{23}	A_1
Γ_3		J_2		E_{31}	A_2
Γ_4		J_3		E_{21}	A_3
Γ_1'					
Γ_2'	M_1		P_1		
Γ_3'	M_2		P_2		
Γ_4'	M_3		P_3		

(12): $m\underline{mm}$

Γ_1	M_3			E_{11}, E_{22}, E_{33}	
Γ_2	M_2			E_{23}	A_1
Γ_3	M_1			E_{31}	A_2
Γ_4				E_{21}	A_3
Γ_1'		J_3			
Γ_2'		J_2	P_1		
Γ_3'		J_1	P_2		
Γ_4'			P_3		

(13): $\underline{\bar{4}}$

Γ_1		J_3		$E_{11} + E_{22}, E_{33}$	A_3
Γ_2	M_3		P_3	$E_{11} - E_{22}, E_{12}$	
Γ_3	$M_1 - iM_2$	$J_1 + iJ_2$	$P_1 - iP_2$	$E_{13} + iE_{23}$	$A_1 + iA_2$
Γ_4	$M_1 + iM_2$	$J_1 - iJ_2$	$P_1 + iP_2$	$E_{13} - iE_{23}$	$A_1 - iA_2$

Table 4.3 (*continued*)

(14): $\underline{4}$					
Γ_1			P_3	$E_{11}+E_{22}, E_{33}$	A_3
Γ_2	M_3	J_3		$E_{11}-E_{22}, E_{12}$	
Γ_3	M_1-iM_2	J_1-iJ_2	P_1+iP_2	$E_{13}+iE_{23}$	A_1+iA_2
Γ_4	M_1+iM_2	J_1+iJ_2	P_1-iP_2	$E_{13}-iE_{23}$	A_1-iA_2
(15): $\underline{4}/\underline{m}$					
Γ_1				$E_{11}+E_{22}, E_{33}$	A_3
Γ_2		J_3		$E_{11}-E_{22}, E_{12}$	
Γ_3		J_1-iJ_2		$E_{31}+iE_{32}$	A_1+iA_2
Γ_4		J_1+iJ_2		$E_{31}-iE_{32}$	A_1-iA_2
Γ_1'			P_3		
Γ_2'	M_3				
Γ_3'	M_1-iM_2		P_1+iP_2		
Γ_4'	M_1+iM_2		P_1-iP_2		
(16): $4/\underline{m}$					
Γ_1		J_3		$E_{11}+E_{22}, E_{33}$	A_3
Γ_2				$E_{11}-E_{22}, E_{12}$	
Γ_3		J_1+iJ_2		$E_{31}+iE_{23}$	A_1+iA_2
Γ_4		J_1-iJ_2		$E_{31}-iE_{23}$	A_1-iA_2
Γ_1'	M_3		P_3		
Γ_2'					
Γ_3'	M_1+iM_2		P_1+iP_2		
Γ_4'	M_1-iM_2		P_1-iP_2		
(17): $\underline{4}/m$					
Γ_1				$E_{11}+E_{22}, E_{33}$	A_3
Γ_2	M_3			$E_{11}-E_{22}, E_{12}$	
Γ_3	M_1-iM_2			$E_{31}+iE_{23}$	A_1+iA_2
Γ_4	M_1+iM_2			$E_{31}-iE_{23}$	A_1-iA_2
Γ_1'			P_3		
Γ_2'		J_3			
Γ_3'		J_1-iJ_2	P_1+iP_2		
Γ_4'		J_1+iJ_2	P_1-iP_2		
(18): $4\underline{mm}$					
Γ_1	M_3		P_3	$E_{11}+E_{22}, E_{33}$	
Γ_2		J_3			A_3
Γ_3				E_{12}	
Γ_4				$E_{11}-E_{22}$	
Γ_5	(M_1, M_2)	$(J_2, -J_1)$	(P_1, P_2)	(E_{31}, E_{32})	$(A_2, -A_1)$
(19): $\underline{4}m\underline{m}$					
Γ_1			P_3	$E_{11}+E_{22}, E_{33}$	
Γ_2					A_3
Γ_3	M_3			E_{12}	
Γ_4		J_3		$E_{11}-E_{22}$	
Γ_5	(M_2, M_1)	$(J_1, -J_2)$	(P_1, P_2)	(E_{31}, E_{32})	$(A_2, -A_1)$

Table 4.3 (*continued*)

(20): $4\underline{2}\underline{2}$					
Γ_1	M_3	J_3		$E_{11} + E_{22}, E_{33}$	
Γ_2			P_3		A_3
Γ_3				E_{12}	
Γ_4				$E_{11} - E_{22}$	
Γ_5	$(M_2, -M_1)$	$(J_2, -J_1)$	(P_1, P_2)	$(E_{23}, -E_{31})$	(A_1, A_2)
(21): $\underline{4}2\underline{2}$					
Γ_1				$E_{11} + E_{22}, E_{33}$	
Γ_2			P_3		A_3
Γ_3	M_3	J_3		E_{12}	
Γ_4				$E_{11} - E_{22}$	
Γ_5	$(M_1, -M_2)$	$(J_1, -J_2)$	(P_1, P_2)	$(E_{32}, -E_{31})$	(A_1, A_2)
(22): $\underline{\bar{4}}2\underline{m}$					
Γ_1				$E_{11} + E_{22}, E_{33}$	
Γ_2		J_3			A_3
Γ_3	M_3		P_3	E_{12}	
Γ_4				$E_{11} - E_{22}$	
Γ_5	(M_1, M_2)	$(J_1, -J_2)$	(P_1, P_2)	(E_{23}, E_{31})	$(A_1, -A_2)$
(23): $\underline{\bar{4}}\underline{2}m$					
Γ_1		J_3		$E_{11} + E_{22}, E_{33}$	
Γ_2					A_3
Γ_3			P_3	E_{12}	
Γ_4	M_3			$E_{11} - E_{22}$	
Γ_5	$(M_2, -M_1)$	(J_2, J_1)	(P_1, P_2)	(E_{23}, E_{31})	$(A_1, -A_2)$
(24): $\bar{4}\underline{2}\underline{m}$					
Γ_1	M_3			$E_{11} + E_{22}, E_{33}$	
Γ_2					A_3
Γ_3			P_3	E_{12}	
Γ_4		J_3		$E_{11} - E_{22}$	
Γ_5	(M_2, M_1)	$(J_2, -J_1)$	(P_1, P_2)	(E_{23}, E_{31})	$(A_1, -A_2)$
(25): $\underline{4}/\underline{m}m\underline{m}$					
Γ_1				$E_{11} + E_{22}, E_{33}$	
Γ_2					A_3
Γ_3		J_3		E_{12}	
Γ_4				$E_{11} - E_{22}$	
Γ_5		$(J_1, -J_2)$		$(E_{23}, -E_{31})$	(A_1, A_2)
Γ_1'					
Γ_2'			P_3		
Γ_3'	M_3				
Γ_4'					
Γ_5'	$(M_1, -M_2)$		(P_1, P_2)		

Table 4.3 (*continued*)

(26): $4/\underline{m}mm$

Γ_1		J_3		$E_{11} + E_{22}, E_{33}$	
Γ_2					A_3
Γ_3				E_{12}	
Γ_4				$E_{11} - E_{22}$	
Γ_5		$(J_2, -J_1)$		$(E_{23}, -E_{31})$	(A_1, A_2)
Γ_1'	M_3				
Γ_2'			P_3		
Γ_3'					
Γ_4'					
Γ_5'	$(M_2, -M_1)$		(P_1, P_2)		

(27): $\underline{4}/m\underline{mm}$

Γ_1	M_3			$E_{11} + E_{22}, E_{33}$	
Γ_2					A_3
Γ_3				E_{12}	
Γ_4				$E_{11} - E_{22}$	
Γ_5	$(M_2, -M_1)$			$(E_{23}, -E_{31})$	(A_1, A_2)
Γ_1'		J_3			
Γ_2'			P_3		
Γ_3'					
Γ_4'					
Γ_5'		$(J_2, -J_1)$	(P_1, P_2)		

(28): $4/\underline{mmm}$

Γ_1				$E_{11} + E_{22}, E_{33}$	
Γ_2		J_3			A_3
Γ_3				E_{12}	
Γ_4				$E_{11} - E_{22}$	
Γ_5		(J_1, J_2)		$(E_{32}, -E_{31})$	(A_1, A_2)
Γ_1'					
Γ_2'	M_3		P_3		
Γ_3'					
Γ_4'					
Γ_5'	(M_1, M_2)		(P_1, P_2)		

(29): $\underline{4}/mm\underline{m}$

Γ_1				$E_{11} + E_{22}, E_{33}$	
Γ_2					A_3
Γ_3	M_3			E_{12}	
Γ_4				$E_{11} - E_{22}$	
Γ_5	$(M_1, -M_2)$			$(E_{23}, -E_{31})$	(A_1, A_2)
Γ_1'					
Γ_2'			P_3		
Γ_3'		J_3			
Γ_4'					
Γ_5'		$(J_1, -J_2)$	(P_1, P_2)		

Table 4.3 (*continued*)

(30): $3\underline{m}$					
Γ_1	M_3		P_3	$E_{11}+E_{22}, E_{33}$	
Γ_2		J_3			A_3
Γ_3	(M_1, M_2)	$(J_2, -J_1)$	(P_1, P_2)	$(E_{13}, E_{23}), (2E_{12}, E_{11}-E_{22})$	$(A_2, -A_1)$
(31): $3\underline{2}$					
Γ_1	M_3	J_3		$E_{11}+E_{22}, E_{33}$	
Γ_2			P_3		A_3
Γ_3	(M_1, M_2)	(J_1, J_2)	$(P_2, -P_1)$	$(E_{13}, E_{23}), (2E_{12}, E_{11}-E_{22})$	$(A_2, -A_1)$
(32): $\underline{\bar{3}}$					
Γ_1		J_3		$E_{11}+E_{22}, E_{33}$	A_3
Γ_2		$J_1 - iJ_2$		$E_{13}-iE_{23}, E_{11}-E_{22}+2iE_{12}$	$A_1 - iA_2$
Γ_3		$J_1 + iJ_2$		$E_{13}+iE_{23}, E_{11}-E_{22}-2iE_{12}$	$A_1 + iA_2$
Γ_4	M_3		P_3		
Γ_5	$M_1 - iM_2$		$P_1 - iP_2$		
Γ_6	$M_1 + iM_2$		$P_1 + iP_2$		
(33): $\underline{\bar{6}}$					
Γ_1		J_3		$E_{11}+E_{22}, E_{33}$	A_3
Γ_2	$M_1 - iM_2$		$P_1 - iP_2$	$E_{11}-E_{22}+2iE_{12}$	
Γ_3	$M_1 + iM_2$		$P_1 + iP_2$	$E_{11}-E_{22}-2iE_{12}$	
Γ_4	M_3		P_3		
Γ_5		$J_1 - iJ_2$		$E_{13}-iE_{23}$	$A_1 - iA_2$
Γ_6		$J_1 + iJ_2$		$E_{13}+iE_{23}$	$A_1 + iA_2$
(34): $\underline{6}$					
Γ_1			P_3	$E_{11}+E_{22}, E_{33}$	A_3
Γ_2	$M_1 - iM_2$	$J_1 - iJ_2$		$E_{11}-E_{22}+2iE_{12}$	
Γ_3	$M_1 + iM_2$	$J_1 + iJ_2$		$E_{11}-E_{22}-2iE_{12}$	
Γ_4	M_3	J_3			
Γ_5			$P_1 - iP_2$	$E_{13}-iE_{23}$	$A_1 - iA_2$
Γ_6			$P_1 + iP_2$	$E_{13}+iE_{23}$	$A_1 + iA_2$
(35): $\bar{6}\underline{m}\underline{2}$					
Γ_1	M_3			$E_{11}+E_{22}, E_{33}$	
Γ_2					A_3
Γ_3			P_3		
Γ_4		J_3			
Γ_5	$(M_2, -M_1)$			$(E_{23}, -E_{31})$	(A_1, A_2)
Γ_6		$(J_2, -J_1)$	(P_1, P_2)	$(2E_{12}, E_{11}-E_{22})$	
(36): $\underline{\bar{6}}m\underline{2}$					
Γ_1		J_3		$E_{11}+E_{22}, E_{33}$	
Γ_2					A_3
Γ_3			P_3		
Γ_4	M_3				
Γ_5		$(J_2, -J_1)$		$(E_{23}, -E_{31})$	(A_1, A_2)
Γ_6	$(M_2, -M_1)$		(P_1, P_2)	$(2E_{12}, E_{11}-E_{22})$	

Table 4.3 (*continued*)

(37): $\underline{\bar{6}m}2$

Γ_1				$E_{11}+E_{22}, E_{33}$	
Γ_2		J_3			A_3
Γ_3	M_3		P_3		
Γ_4					
Γ_5		(J_1, J_2)		$(E_{23}, -E_{31})$	(A_1, A_2)
Γ_6	(M_1, M_2)		(P_1, P_2)	$(2E_{12}, E_{11}-E_{22})$	

(38): $\bar{3}\underline{m}$

Γ_1	M_3			$E_{11}+E_{22}, E_{33}$	
Γ_2					A_3
Γ_3		J_3			
Γ_4			P_3		
Γ_5		$(J_2, -J_1)$	(P_1, P_2)		
Γ_6	(M_1, M_2)			$(E_{13}, E_{23}), (2E_{12}, E_{11}-E_{22})$	$(A_2, -A_1)$

(39): $\underline{\bar{3}}m$

Γ_1		J_3		$E_{11}+E_{22}, E_{33}$	
Γ_2					A_3
Γ_3	M_3				
Γ_4			P_3		
Γ_5	$(M_2, -M_1)$		(P_1, P_2)		
Γ_6		(J_1, J_2)		$(E_{13}, E_{23}), (2E_{12}, E_{11}-E_{22})$	$(A_2, -A_1)$

(40): $\underline{\bar{3}m}$

Γ_1				$E_{11}+E_{22}, E_{33}$	
Γ_2		J_3			A_3
Γ_3					
Γ_4	M_3		P_3		
Γ_5	(M_1, M_2)		(P_1, P_2)		
Γ_6		$(J_2, -J_1)$		$(E_{13}, E_{23}), (2E_{12}, E_{11}-E_{22})$	$(A_2, -A_1)$

(41): $6\underline{22}$

Γ_1	M_3	J_3		$E_{11}+E_{22}, E_{33}$	
Γ_2			P_3		A_3
Γ_3					
Γ_4					
Γ_5	$(M_2, -M_1)$	$(J_2, -J_1)$	(P_1, P_2)	$(E_{23}, -E_{13})$	(A_1, A_2)
Γ_6				$(2E_{12}, E_{11}-E_{22})$	

(42): $\underline{62}2$

Γ_1				$E_{11}+E_{22}, E_{33}$	
Γ_2			P_3		A_3
Γ_3					
Γ_4	M_3	J_3			
Γ_5			(P_1, P_2)	$(E_{23}, -E_{13})$	(A_1, A_2)
Γ_6	$(M_2, -M_1)$	$(J_2, -J_1)$		$(2E_{12}, E_{11}-E_{22})$	

Table 4.3 (*continued*)

(43): $6\underline{mm}$

Γ_1	M_3		P_3	$E_{11}+E_{22}, E_{33}$	
Γ_2		J_3			A_3
Γ_3					
Γ_4					
Γ_5	(M_1, M_2)	$(J_2, -J_1)$	(P_1, P_2)	(E_{13}, E_{23})	$(A_2, -A_1)$
Γ_6				$(2E_{12}, E_{11}-E_{22})$	

(44): $\underline{6}m\underline{m}$

Γ_1			P_3	$E_{11}+E_{22}, E_{33}$	
Γ_2					A_3
Γ_3	M_3				
Γ_4		J_3			
Γ_5			(P_1, P_2)	(E_{13}, E_{23})	$(A_2, -A_1)$
Γ_6	$(M_2, -M_1)$	(J_1, J_2)		$(2E_{12}, E_{11}-E_{22})$	

(45): $6/\underline{m}$

Γ_1		J_3		$E_{11}+E_{22}, E_{33}$	A_3
Γ_2				$E_{11}-E_{22}+2iE_{12}$	
Γ_3				$E_{11}-E_{22}-2iE_{12}$	
Γ_4					
Γ_5		J_1-iJ_2		$E_{13}-iE_{23}$	A_1-iA_2
Γ_6		J_1+iJ_2		$E_{13}+iE_{23}$	A_1+iA_2
Γ_1'	M_3		P_3		
Γ_2'					
Γ_3'					
Γ_4'					
Γ_5'	M_1-iM_2		P_1-iP_2		
Γ_6'	M_1+iM_2		P_1+iP_2		

(46): $\underline{6}/m$

Γ_1				$E_{11}+E_{22}, E_{33}$	A_3
Γ_2		J_1-iJ_2		$E_{11}-E_{22}+2iE_{12}$	
Γ_3		J_1+iJ_2		$E_{11}-E_{22}-2iE_{12}$	
Γ_4		J_3			
Γ_5				$E_{13}-iE_{23}$	A_1-iA_2
Γ_6				$E_{13}+iE_{23}$	A_1+iA_2
Γ_1'			P_3		
Γ_2'	M_1-iM_2				
Γ_3'	M_1+iM_2				
Γ_4'	M_3				
Γ_5'			P_1-iP_2		
Γ_6'			P_1+iP_2		

Table 4.3 (*continued*)

(47): $\underline{6}/\underline{m}$

Γ_1				$E_{11}+E_{22}, E_{33}$	A_3
Γ_2	M_1-iM_2			$E_{11}-E_{22}+2iE_{12}$	
Γ_3	M_1+iM_2			$E_{11}-E_{22}-2iE_{12}$	
Γ_4	M_3				
Γ_5				$E_{13}-iE_{23}$	A_1-iA_2
Γ_6				$E_{13}+iE_{23}$	A_1+iA_2
Γ_1'			P_3		
Γ_2'		J_1-iJ_2			
Γ_3'		J_1+iJ_2			
Γ_4'		J_3			
Γ_5'			P_1-iP_2		
Γ_6'			P_1+iP_2		

(48): $6/\underline{m}mm$

Γ_1		J_3		$E_{11}+E_{22}, E_{33}$	
Γ_2					A_3
Γ_3					
Γ_4					
Γ_5		$(J_2, -J_1)$		$(E_{23}, -E_{31})$	(A_1, A_2)
Γ_6				$(2E_{12}, E_{11}-E_{22})$	
Γ_1'	M_3				
Γ_2'			P_3		
Γ_3'					
Γ_4'					
Γ_5'	$(M_2, -M_1)$		(P_1, P_2)		
Γ_6'					

(49): $6/\underline{m}\underline{m}\underline{m}$

Γ_1				$E_{11}+E_{22}, E_{33}$	
Γ_2		J_3			A_3
Γ_3					
Γ_4					
Γ_5		(J_1, J_2)		$(E_{23}, -E_{31})$	(A_1, A_2)
Γ_6				$(2E_{12}, E_{11}-E_{22})$	
Γ_1'					
Γ_2'	M_3		P_3		
Γ_3'					
Γ_4'					
Γ_5'	(M_1, M_2)		(P_1, P_2)		
Γ_6'					

Table 4.3 (*continued*)

(50): $6/m\underline{m}\underline{m}$

Γ_1	M_3			$E_{11} + E_{22}, E_{33}$	
Γ_2					A_3
Γ_3					
Γ_4					
Γ_5	$(M_2, -M_1)$			$(E_{23}, -E_{31})$	(A_1, A_2)
Γ_6				$(2E_{12}, E_{11} - E_{22})$	
Γ_1'		J_3			
Γ_2'			P_3		
Γ_3'					
Γ_4'					
Γ_5'		$(J_2, -J_1)$	(P_1, P_2)		
Γ_6'					

(51): $\underline{6}/mm\underline{m}$

Γ_1				$E_{11} + E_{22}, E_{33}$	
Γ_2					A_3
Γ_3					
Γ_4		J_3			
Γ_5				$(E_{23}, -E_{31})$	(A_1, A_2)
Γ_6		$(J_2, -J_1)$		$(2E_{12}, E_{11} - E_{22})$	
Γ_1'					
Γ_2'			P_3		
Γ_3'					
Γ_4'	M_3				
Γ_5'			(P_1, P_2)		
Γ_6'	$(M_2, -M_1)$				

(52): $\underline{6}/\underline{m}m\underline{m}$

Γ_1				$E_{11} + E_{22}, E_{33}$	
Γ_2					A_3
Γ_3					
Γ_4	M_3				
Γ_5				$(E_{23}, -E_{31})$	(A_1, A_2)
Γ_6	$(M_2, -M_1)$			$(2E_{12}, E_{11} - E_{22})$	
Γ_1'					
Γ_2'			P_3		
Γ_3'					
Γ_4'		J_3			
Γ_5'			(P_1, P_2)		
Γ_6'		$(J_2, -J_1)$			

(53): $\underline{m}3$

Γ_1				$E_{11} + E_{22} + E_{33}$	
Γ_2				$E_{11} + w^2E_{22} + wE_{33}$	
Γ_3				$E_{11} + wE_{22} + w^2E_{33}$	
Γ_4		(J_1, J_2, J_3)		(E_{23}, E_{31}, E_{12})	(A_1, A_2, A_3)
Γ_1'					
Γ_2'					
Γ_3'					
Γ_4'	(M_1, M_2, M_3)		(P_1, P_2, P_3)		

Table 4.3 (*continued*)

(54): $\underline{\bar{4}}3\underline{m}$					
Γ_1				$E_{11} + E_{22} + E_{33}$	
Γ_2					
Γ_3				$[(2E_{11} - E_{22} - E_{33}), \sqrt{3}(E_{22} - E_{33})]$	
Γ_4	(M_1, M_2, M_3)		(P_1, P_2, P_3)	(E_{23}, E_{31}, E_{12})	
Γ_5		(J_1, J_2, J_3)			(A_1, A_2, A_3)
(55): $\underline{4}3\underline{2}$					
Γ_1				$E_{11} + E_{22} + E_{33}$	
Γ_2					
Γ_3				$[(2E_{11} - E_{22} - E_{33}), \sqrt{3}(E_{22} - E_{33})]$	
Γ_4	(M_1, M_2, M_3)	(J_1, J_2, J_3)		(E_{23}, E_{31}, E_{12})	
Γ_5			(P_1, P_2, P_3)		(A_1, A_2, A_3)
(56): $\underline{m}3\underline{m}$					
Γ_1				$E_{11} + E_{22} + E_{33}$	
Γ_2					
Γ_3				$[(2E_{11} - E_{22} - E_{33}), \sqrt{3}(E_{22} - E_{33})]$	
Γ_4				(E_{23}, E_{31}, E_{12})	
Γ_5		(J_1, J_2, J_3)			(A_1, A_2, A_3)
Γ_1'					
Γ_2'					
Γ_3'					
Γ_4'					
Γ_5'	(M_1, M_2, M_3)		(P_1, P_2, P_3)		
(57): $m3\underline{m}$					
Γ_1				$E_{11} + E_{22} + E_{33}$	
Γ_2					
Γ_3				$[(2E_{11} - E_{22} - E_{33}), \sqrt{3}(E_{22} - E_{33})]$	
Γ_4	(M_1, M_2, M_3)			(E_{23}, E_{31}, E_{12})	
Γ_5					(A_1, A_2, A_3)
Γ_1'					
Γ_2'					
Γ_3'					
Γ_4'		(J_1, J_2, J_3)			
Γ_5'			(P_1, P_2, P_3)		
(58): $\underline{m}3m$					
Γ_1				$E_{11} + E_{22} + E_{33}$	
Γ_2					
Γ_3				$[(2E_{11} - E_{22} - E_{33}), \sqrt{3}(E_{22} - E_{33})]$	
Γ_4		(J_1, J_2, J_3)		(E_{23}, E_{31}, E_{12})	
Γ_5					(A_1, A_2, A_3)
Γ_1'					
Γ_2'					
Γ_3'					
Γ_4'	(M_1, M_2, M_3)				
Γ_5'			(P_1, P_2, P_3)		

CHAPTER 5

Material Symmetry Restrictions

Let the matrices $\{\mathbf{M}^1, \mathbf{M}^2, \ldots, \mathbf{M}^g\}$ be the symmetry group $\{M\}$ of the material under consideration. Since $\{M\}$ is obtained from $\{G\}$ by $\{M\} = \{H\} + \tau\{G - H\}$, $\mathbf{M}^\alpha$ is either an ordinary symmetry element $\mathbf{S}^\alpha$, or it is a complementary element in the form $\tau\mathbf{S}^\alpha$ ($\alpha = 1, \ldots, g$).

Consider a general nonlinear constitutive relation given by

$$E_{i_1\cdots i_n} = E_{i_1\cdots i_n}(A_{i_1\cdots i_r}, B_{i_1\cdots i_p}, \ldots, C_{i_1\cdots i_q}), \tag{5.1}$$

where $\mathbf{E}, \mathbf{A}, \mathbf{B}, \ldots, \mathbf{C}$ are the argument tensors of orders indicated by their indices. The components of $\mathbf{E}, \mathbf{A}, \mathbf{B}, \ldots, \mathbf{C}$ under the symmetry element $\mathbf{M}^\alpha$ are given by

$$\begin{aligned} \bar{E}_{i_1\cdots i_n}(\mathbf{M}^\alpha) &= (-1)^\lambda |\mathbf{M}^\alpha|^\mu M^\alpha_{i_1j_1} \cdots M^\alpha_{i_nj_n} E_{j_1\cdots j_n}, \\ \bar{C}_{i_1\cdots i_q}(\mathbf{M}^\alpha) &= (-1)^\lambda |M^\alpha|^\mu M_{i_1j_1} \cdots M^\alpha_{i_qj_q} C_{j_1\cdots j_q}, \end{aligned} \tag{5.2}$$

where λ, μ is in the set $\{0, 1\}$ according to the nature of the tensors $\mathbf{E}, \ldots, \mathbf{C}$. In accordance with the principal of material invariance, the relation (5.1) must have the same form under each of the symmetry elements $\mathbf{M}^\alpha$ associated with the material. Therefore, we write

$$\bar{E}_{i_1\cdots i_n}(\mathbf{M}^\alpha) = E_{i_1\cdots i_n}[\bar{A}_{i_1\cdots i_r}(\mathbf{M}^\alpha), \bar{B}_{i_1\cdots i_p}(\mathbf{M}^\alpha), \ldots] \tag{5.3}$$

for each $\mathbf{M}^\alpha \in \{M\}$. Using (5.2), it is seen that the functions $E_{i_1\cdots i_n}(\mathbf{A}, \mathbf{B}, \ldots, \mathbf{C})$ must satisfy the equations

$$\begin{aligned} &(-1)^\lambda |\mathbf{M}^\alpha|^\mu M^\alpha_{i_1j_1} \cdots M^\alpha_{i_nj_n} E_{j_1\cdots j_n}(\mathbf{A}, \mathbf{B}, \ldots, \mathbf{C}) \\ &\qquad = E_{i_1\cdots i_n}[\bar{\mathbf{A}}(\mathbf{M}^\alpha), \bar{\mathbf{B}}(\mathbf{M}^\alpha), \ldots, \bar{\mathbf{C}}(\mathbf{M}^\alpha)] \qquad \text{for all} \quad \mathbf{M}^\alpha \in \{M\}. \end{aligned} \tag{5.4}$$

It is then said that the tensor-valued function $\mathbf{E}(\mathbf{A}, \mathbf{B}, \ldots, \mathbf{C})$ is invariant under $\{M\}$. In the case of a scalar-valued polynomial function $W(\mathbf{A}, \mathbf{B}, \ldots, \mathbf{C})$, the invariance requirement is given by

$$W(\mathbf{A}, \mathbf{B}, \ldots, \mathbf{C}) = W[\bar{\mathbf{A}}(\mathbf{M}^\alpha), \bar{\mathbf{B}}(\mathbf{M}^\alpha), \ldots, \bar{\mathbf{C}}(\mathbf{M}^\alpha)] \tag{5.5}$$

for all $\mathbf{M}^\alpha \in \{M\}$, which is a special case of (5.4). In this case, a solution to (5.5) is furnished by giving a set of scalar-valued invariants $I_p(\mathbf{A}, \mathbf{B}, \ldots, \mathbf{C})$

($p = 1, 2, \ldots$), such that any polynomial function invariant under $\mathbf{M}^\alpha$ is expressible as a polynomial in $I_1, I_2, \ldots$, i.e., $W = W(I_1, I_2, \ldots)$. The invariants I_1, $I_2, \ldots$ are said to form an integrity basis for the polynomial functions of $\mathbf{A}, \mathbf{B}, \ldots, \mathbf{C}$, which are invariant under $\{M\}$. We may omit from such an integrity basis any elements which can be expressed as polynomials in the remaining ones. The set of invariants so obtained is called an *irreducible integrity basis.*

Note that the system (5.4) may be converted into a scalar relation by introducing an arbitrary tensor $\boldsymbol{\psi}$, which has the same order and symmetry properties as $\mathbf{E}$. Now the scalar quantity V, defined by

$$V = \psi_{i_1 \cdots i_n} E_{i_1 \cdots i_n}, \tag{5.6}$$

is invariant under $\{M\}$, and it is a function of $\mathbf{A}, \mathbf{B}, \ldots, \mathbf{C}$ and $\boldsymbol{\psi}$, being linear in $\boldsymbol{\psi}$. Thus, the restriction imposed by the material symmetry upon a tensor-valued constitutive equation is reduced to the problem of determining a set of integrity bases for the scalar invariants of an appropriate set of tensors, at the expense that an extra argument $\boldsymbol{\psi}$ is introduced into the argument list.

Suppose that the irreducible integrity basis for $\mathbf{A}, \mathbf{B}, \ldots, \mathbf{C}$, and $\boldsymbol{\psi}$ under $\{M\}$, consists of the elements $I_1, \ldots, I_r$ which are independent of $\boldsymbol{\psi}$, of the elements $L_1, \ldots, L_m$ which are linear in $\boldsymbol{\psi}$, and of the further elements which are of higher degree than the first in $\boldsymbol{\psi}$. Since V is linear in $\boldsymbol{\psi}$, it may be represented in the form

$$V = \sum_{p=1}^{m} \alpha_p(I_1, \ldots, I_r) L_p, \tag{5.7}$$

where the α_p's are scalar polynomials in the indicated arguments. From (5.6), (5.7) it follows that

$$E_{i_1 \cdots i_n}(\mathbf{A}, \mathbf{B}, \ldots, \mathbf{C}) = \frac{\partial V}{\partial \psi_{i_1 \cdots i_n}} = \sum_{p=1}^{m} \alpha_p(I_1, \ldots, I_r) \frac{\partial L_p}{\partial \psi_{i_1 \cdots i_n}}. \tag{5.8}$$

Letting $\mathbf{T}^{(p)} = \partial L_p / \partial \boldsymbol{\psi}$, (5.8) becomes

$$E_{i_1 \cdots i_n} = \sum_{p=1}^{m} \alpha_p(I_1, \ldots, I_r) \mathbf{T}^{(p)}_{i_1 \cdots i_n}(\mathbf{A}, \ldots, \mathbf{C}), \tag{5.9}$$

which indicates that the form invariance of a constitutive equation (5.1), relating one tensor to any finite number of tensors of any order, can be expressed by writing the dependent tensor $\mathbf{E}$ as the sum of a finite number of tensor polynomials of the argument tensors $\mathbf{A}, \mathbf{B}, \ldots, \mathbf{C}$, each of which is form-invariant under the material symmetry group $\{M\}$.

The physical characterization of a particular material is then provided by specifying the scalar invariant coefficients $\alpha_p(I_1, \ldots, I_r)$. Note that the tensor polynomials depend on the symmetry group of the material and on the nature of the argument tensors $\mathbf{A}, \mathbf{B}, \ldots, \mathbf{C}$, however, they do not depend on the particular material under consideration.

Let us introduce a single quantity $\mathbf{Z}$ by

$$\mathbf{Z} = \mathbf{A} \dotplus \mathbf{B} \dotplus \cdots \dotplus \mathbf{C} \tag{5.10}$$

with $N = r + p + \cdots + q$, where the independent components Z_i $(i = 1, \ldots, N)$ of $\mathbf{Z}$ are the direct sum of the independent components of $\mathbf{A}, \mathbf{B}, \ldots, \mathbf{C}$. Hence, without loss of generality, we may consider $\mathbf{E}$ to be a function of a single quantity $\mathbf{Z}$. Let E_p and $\bar{E}_p$, and Z_i and $\bar{Z}_i$ be the components of $\mathbf{Z}$ and $\mathbf{E}$ in the original and transformed configuration under $\mathbf{M}^\alpha$. We write

$$\bar{Z}_i(\mathbf{M}^\alpha) = T_{ij}(\mathbf{M}^\alpha) Z_j \qquad (i, j = 1, \ldots, N), \tag{5.11}$$

$$\bar{E}_p(\mathbf{M}^\alpha) = U_{pq}(\mathbf{M}^\alpha) E_q \qquad (p, q = 1, \ldots, Q), \tag{5.12}$$

where $\mathbf{T}(\mathbf{M}^\alpha)$ and $\mathbf{U}(\mathbf{M}^\alpha)$ are the matrix-valued functions of the transformation matrix $\mathbf{M}^\alpha$, and Q is the number of independent components of $E_{i_1 \cdots i_n}$. The requirement, that the form of $E_{i_1 \cdots i_n}(\mathbf{Z})$ be the same under the symmetry operations $\mathbf{M}^\alpha$ $(\alpha = 1, \ldots, g)$, is that

$$\bar{E}_p(\mathbf{M}^\alpha) = E_p[\bar{Z}_i(\mathbf{M}^\alpha)] \qquad \text{for all} \quad \mathbf{M}^\alpha \in \{M\}, \tag{5.13}$$

or, using (5.11), (5.12), we write (5.13) explicitly as

$$U_{pq}(\mathbf{M}^\alpha) E_q(Z_i) = E_p[T_{ij}(\mathbf{M}^\alpha) Z_j], \tag{5.14}$$

for all $\mathbf{M}^\alpha \in \{M\}$. The set of matrices

$$\{T\} = \{\mathbf{T}(\mathbf{M}^1), \ldots, \mathbf{T}(\mathbf{M}^g)\} \tag{5.15}$$

and

$$\{U\} = \{\mathbf{U}(\mathbf{M}^1), \ldots, \mathbf{U}(\mathbf{M}^g)\} \tag{5.16}$$

form matrix representations of the symmetry group $\{M\}$. The representations $\{T\}$ and $\{U\}$ may be decomposed into the direct sum of irreducible representations Γ_i of $\{M\}$. We may find constant nonsingular matrices $\mathbf{Q}$ and $\mathbf{P}$ such that

$$\mathbf{Q}\mathbf{T}(\mathbf{M}^\alpha)\mathbf{Q}^{-1} = n_1 \mathbf{D}^{(1)}(\mathbf{M}^\alpha) \dotplus \cdots \dotplus n_r \mathbf{D}^{(r)}(\mathbf{M}^\alpha) \tag{5.17}$$

and

$$\mathbf{P}\mathbf{U}(\mathbf{M}^\alpha)\mathbf{P}^{-1} = n_1' \mathbf{D}^{(1)}(\mathbf{M}^\alpha) \dotplus \cdots \dotplus n_r' \mathbf{D}^{(r)}(\mathbf{M}^\alpha), \tag{5.18}$$

where n_i and n_i' denote the number of times the irreducible representation Γ_i of degree d_i appears in the decomposition of $\{T\}$ and $\{U\}$, respectively. They are given by

$$n_i = \frac{1}{g} \sum_{\alpha=1}^{g} \overset{*}{\chi}{}^{(i)}(\mathbf{M}^\alpha) \operatorname{tr} \mathbf{T}(\mathbf{M}^\alpha) \tag{5.19}$$

and

$$n_i' = \frac{1}{g} \sum_{\alpha=1}^{g} \overset{*}{\chi}{}^{(i)}(\mathbf{M}^\alpha) \operatorname{tr} \mathbf{U}(\mathbf{M}^\alpha). \tag{5.20}$$

The matrices $\mathbf{Q}\mathbf{T}(\mathbf{M}^\alpha)\mathbf{Q}^{-1}$ and $\mathbf{P}\mathbf{U}(\mathbf{M}^\alpha)\mathbf{P}^{-1}$ $(\alpha = 1, \ldots, g)$ describe the transformation properties of

$$\mathbf{QZ} = \mathbf{U}_1^{(1)} \dotplus \cdots \dotplus \mathbf{U}_{n_1}^{(1)} \dotplus \cdots \dotplus \mathbf{U}_1^{(r)} \dotplus \cdots \dotplus \mathbf{U}_{n_r}^{(r)}, \tag{5.21}$$

and

$$\mathbf{PE} = \mathbf{v}_1^{(1)} + \cdots + \mathbf{v}_{n_1'}^{(1)} + \cdots + \mathbf{v}_1^{(r)} + \cdots + \mathbf{v}_{n_r'}^{(r)}, \tag{5.22}$$

respectively. In (5.21), the first d_1 components of $\mathbf{QZ}$ are the d_1 components of $\mathbf{U}_1^{(1)}$, and the last d_{n_r} components of $\mathbf{QZ}$ are the d_{n_r} components of $\mathbf{U}_{n_r}^{(r)}$. Similarly, in (5.22), the first d_1 components of $\mathbf{PE}$ are the d_1 components of $\mathbf{v}_1^{(1)}$, and so on.

The basic quantities $\mathbf{U}_1^{(1)}, \ldots, \mathbf{U}_{n_r}^{(r)}$ and $\mathbf{v}_1^{(1)}, \ldots, \mathbf{v}_{n_r'}^{(r)}$ form carrier spaces for the irreducible representations $\Gamma_i = \{\mathbf{D}^{(i)}(\mathbf{M}^\alpha)\}$ ($\alpha = 1, \ldots, g$; $i = 1, \ldots, r$), that is,

$$\bar{\mathbf{U}}_j^{(i)}(\mathbf{M}^\alpha) = \mathbf{D}^{(i)}(\mathbf{M}^\alpha)\mathbf{U}_j^{(i)}, \tag{5.23}$$

and

$$\bar{\mathbf{v}}_k^{(i)}(\mathbf{M}^\alpha) = \mathbf{D}^{(i)}(\mathbf{M}^\alpha)\mathbf{v}_k^{(i)}, \tag{5.24}$$

$$\begin{cases} (\alpha = 1, \ldots, g), \\ (i = 1, \ldots, r), \\ (j = 1, \ldots, n_r), \\ (k = 1, \ldots, n_r'), \end{cases}$$

where the summation on i is suspended. The basic quantities $\mathbf{U}_1^{(1)}, \ldots, \mathbf{U}_{n_r}^{(r)}$ and $\mathbf{v}_1^{(1)}, \ldots, \mathbf{v}_{n_r'}^{(r)}$ are obtained from

$$\mathbf{U}_j^{(i)} = \boldsymbol{\phi}^{(p_j q_j i)} \qquad \text{and} \qquad \mathbf{v}_k^{(i)} = \boldsymbol{\phi}^{(p_k q_k' i)}, \tag{5.25}$$

where

$$\phi_m^{(pqi)} = \sum_{\alpha=1}^{g} \overset{*}{D}{}_{pm}^{(i)}(\mathbf{M}^\alpha)\bar{Z}_q(\mathbf{M}^\alpha), \tag{5.26}$$

and

$$\phi_m^{(pq'i)} = \sum_{\alpha=1}^{g} \overset{*}{D}{}_{pm}^{(i)}(\mathbf{M}^\alpha)\bar{E}_{q'}(\mathbf{M}^\alpha), \tag{5.27}$$

$$\begin{cases} (p, m = 1, \ldots, d_i), \\ (q = 1, \ldots, N), \\ (q' = 1, \ldots, Q). \end{cases}$$

The problem of determining the general form of $\mathbf{E}(\mathbf{Z})$ with the invariance requirement (5.13) or (5.14) may now be replaced by that of determining the general form of the functions $\mathbf{v}_1^{(1)}, \ldots, \mathbf{v}_{n_r'}^{(r)}$ which satisfy

$$\bar{\mathbf{v}}_k^{(i)}(\underbrace{\mathbf{U}_j^{(1)}, \ldots, \mathbf{U}_m^{(r)}}) = \mathbf{v}_k^{(i)}(\bar{\mathbf{U}}_j^{(1)}, \ldots, \bar{\mathbf{U}}_m^{(r)}) \tag{5.28}$$

$$(j = 1, \ldots, n_1; \ldots; m = 1, \ldots, n_r),$$
$$(k = 1, \ldots, n_1'; \ldots; \ldots; 1, \ldots, n_r'),$$
$$(i = 1, \ldots, r),$$

for each $\mathbf{M}^\alpha$ in $\{M\}$. In other words, instead of working with the explicit components $E_{i_1 \cdots i_n}$ of $\mathbf{E}$ and $A_{i_1 \cdots i_n}, B_{i_1 \cdots i_p}, C_{i_1 \cdots i_q}$ of $\mathbf{A}, \mathbf{B}, \ldots, \mathbf{C}$, respectively, we deal with the linear combinations $\mathbf{v}_k^{(i)}$ of $\mathbf{E}$ that form carrier spaces of the irreducible representations Γ_i of $\{M\}$, so that the transformation properties of $\mathbf{v}_k^{(i)}$ and $\mathbf{U}_j^{(i)}$ are much more simplified, as observed from a comparison of (5.11),

(5.12) and (5.23), (5.24). The solution of (5.28), for $i = 1, \ldots, r$ for the case in which $n_1, \ldots, n_r$ are arbitrary, will furnish the solution to the problem of determining the form of $\mathbf{E}(\mathbf{Z})$ which is invariant under $\{M\}$ for any $\mathbf{E}$ and any $\mathbf{Z}$, in which case $\mathbf{Z}$ is the direct sum of any number of tensors of any order and symmetry type. For example, consider the isomorphic magnetic crystal classes given in Table 3.3(3). We have $\{M\} = \{\mathbf{M}^1, \mathbf{M}^2, \mathbf{M}^3, \mathbf{M}^4\}$, and the irreducible representations $\Gamma_1, \ldots, \Gamma_4$ are all of degree one and are listed below for ease of reference:

	$\mathbf{M}^1$	$\mathbf{M}^2$	$\mathbf{M}^3$	$\mathbf{M}^4$	Basic quantities of $\mathbf{Z}$
Γ_1	1	1	1	1	$\alpha_1, \alpha_2, \ldots, \alpha_p$
Γ_2	1	1	-1	-1	$\beta_1, \beta_2, \ldots, \beta_q$
Γ_3	1	-1	1	-1	$\gamma_1, \gamma_2, \ldots, \gamma_r$
Γ_4	1	-1	-1	1	$\delta_1, \delta_2, \ldots, \delta_s$

The quantities forming the carrier spaces for the irreducible representations $\Gamma_1, \ldots, \Gamma_4$ are denoted by α, β, γ, δ, respectively.

We readily find that every scalar-valued polynomial function $W(\alpha_i, \beta_j, \gamma_k, \delta_m)$ of the quantities α_i, β_j, γ_k, and δ_m is expressible as a polynomial in

$$\begin{array}{lll} \text{(i)} & \alpha_i \quad (i = 1, \ldots, p) & \\ \text{(ii)} & \beta_{j_1}\beta_{j_2}, \gamma_{k_1}\gamma_{k_2}, \delta_{m_1}\delta_{m_2}, & \\ \text{(iii)} & \beta_j\gamma_k\delta_m, & \end{array} \quad \left\{ \begin{array}{l} (j, j_1, j_2 = 1, \ldots, q), \\ (k, k_1, k_2 = 1, \ldots, r), \\ (m, m_1, m_2 = 1, \ldots, s). \end{array} \right. \tag{5.29}$$

Every polynomial function $v^{(1)}(\alpha_i, \beta_j, \gamma_k, \delta_m)$ whose transformation properties under $\{M\}$ are defined by Γ_1 is a scalar invariant and it is expressible in terms of the elements (5.29).

Every polynomial function $v^{(2)}(\alpha_i, \beta_j, \gamma_k, \delta_m)$, whose transformation properties under $\{M\}$ are defined by $\Gamma_2 = \{\mathbf{D}^{(2)}(\mathbf{M}^\alpha)\}$, is expressible as a linear combination of the quantities

$$\begin{array}{lll} \text{(i)} & \beta_j & (j = 1, \ldots, q), \\ \text{(ii)} & \gamma_k\delta_m & (k = 1, \ldots, r;\ m = 1, \ldots, s), \end{array} \tag{5.30}$$

with scalar coefficients which are polynomials in the quantities (5.29).

Every polynomial function $v^{(3)}(\delta_i, \beta_j, \gamma_k, \delta_m)$, whose transformation properties under $\{M\}$ are defined by Γ_3, is expressible as a linear combination of the quantities

$$\begin{array}{lll} \text{(i)} & \gamma_k & (k = 1, \ldots, q), \\ \text{(ii)} & \beta_j\delta_m & (j = 1, \ldots, q;\ m = 1, \ldots, s), \end{array} \tag{5.31}$$

with scalar coefficients which are polynomials in the quantities (5.29).

Every polynomial function $v^{(4)}(\alpha_i, \beta_j, \gamma_k, \delta_m)$, whose transformation properties under $\{M\}$ are defined by Γ_4, is expressible as a linear combination of the quantities

$$\begin{aligned} &\text{(i)}\quad \delta_m \qquad (m = 1, \ldots, s), \\ &\text{(ii)}\quad \beta_j\gamma_k \qquad (j = 1, \ldots, q;\ k = 1, \ldots, r), \end{aligned} \tag{5.32}$$

with scalar coefficients which are polynomials in the quantities (5.29).

In particular, let us determine the form of an axial, *c*-vector-valued function $\mathbf{M}(\mathbf{E})$ of a symmetric second-order polar *i*-tensor, $\mathbf{E}$, which is invariant under $\underline{2}/m = \{\mathbf{I}, \tau\mathbf{D}_3, \mathbf{R}_3, \tau\mathbf{C}\}$. We employ the notation

$$\begin{aligned} (Z_1, \ldots, Z_6) &= (E_{11}, E_{22}, E_{33}, E_{23}, E_{31}, E_{12}), \\ (E_1, E_2, E_3) &= (M_1, M_2, M_3), \end{aligned} \tag{5.33}$$

we then write

$$\bar{Z}_i(M^\alpha) = T_{ij}(\mathbf{M}^\alpha)Z_j, \qquad \begin{cases} (\alpha = 1, \ldots, 4), \\ (i, j = 1, \ldots, 6), \\ (p, q = 1, 2, 3), \end{cases} \tag{5.34}$$

and

$$\bar{E}_p(M^\alpha) = U_{pq}(\mathbf{M}^\alpha)E_q, \tag{5.35}$$

where the matrices $\mathbf{T}(\mathbf{M}^\alpha)$ and $\mathbf{U}(\mathbf{M}^\alpha)$ are, from the example at the very beginning of Section 4.4,

$$\begin{aligned} \mathbf{T}(\mathbf{I}) &= \text{diag}(1, 1, 1, 1, 1, 1), & \mathbf{T}(\tau\mathbf{D}_3) &= \text{diag}(1, 1, 1, -1, -1, 1), \\ \mathbf{T}(\mathbf{R}_3) &= \text{diag}(1, 1, 1, -1, -1, 1), & \mathbf{T}(\tau\mathbf{C}) &= \text{diag}(1, 1, 1, 1, 1, 1), \\ \mathbf{U}(\mathbf{I}) &= \text{diag}(1, 1, 1), & \mathbf{U}(\tau\mathbf{D}_3) &= \text{diag}(1, 1, -1), \\ \mathbf{U}(\mathbf{R}^3) &= \text{diag}(-1, -1, 1), & \mathbf{U}(\tau\mathbf{C}) &= \text{diag}(-1, -1, -1). \end{aligned} \tag{5.36}$$

The character table for the representations $\{T\}$ and $\{U\}$ are

	$\mathbf{I}$	$\tau\mathbf{D}_3$	$\mathbf{R}_3$	$\tau\mathbf{C}$
$\text{tr}\{T\}$	6	2	2	6
$\text{tr}\{U\}$	3	1	-1	-3

(5.37)

With (5.19) and (5.20) we find

$$\{T\} = 4\Gamma_1 \dotplus 2\Gamma_4, \qquad \{U\} = 2\Gamma_2 \dotplus \Gamma_3. \tag{5.38}$$

From (5.26) and (5.27), or by inspection (or see Table 4.3(3)), the basic quantities, which form carrier spaces for the irreducible representations $\Gamma_1, \ldots, \Gamma_4$ of $\underline{2}/m$, are obtained and are listed below

	M_i	E_{ij}	
Γ_1		$E_{11}, E_{22}, E_{33}, E_{12}$	$\alpha_1, \alpha_2, \alpha_3, \alpha_4$
Γ_2	M_1, M_2		0
Γ_3	M_3		0
Γ_4		E_{32}, E_{31}	δ_1, δ_2

(5.39)

With (5.29)–(5.32) and (5.39), we see that

$$M_1 = M_2 = M_3 = 0,$$

that is, for the magnetic crystal $\underline{2}/m$, the piezomagnetic effect is forbidden, even for the higher-order effects!

However, if we consider the isomorphic magnetic class $\underline{2}/\underline{m} = \{\mathbf{I}, \tau\mathbf{D}_3, \tau\mathbf{R}_3, \mathbf{C}\}$, the basic quantities associated with M_i and E_{ij} are obtained in similar manner and are listed below (see Table 4.3(6)):

	M_i	E_{ij}	
Γ_1	M_1, M_2	$E_{11}, E_{22}, E_{33}, E_{12}$	$\alpha_1, \alpha_2, \alpha_3, \alpha_4$
Γ_2			0
Γ_3			0
Γ_4	M_3	E_{32}, E_{31}	δ_1, δ_2

(5.40)

With (5.29)–(5.32) and (5.40), we see that

$$\left.\begin{aligned} M_1 &= \phi_1, \\ M_2 &= \phi_2, \\ M_3 &= \phi_3 E_{32} + \phi_4 E_{31}, \end{aligned}\right\} \quad M_k = \phi_1 \delta_{1k} + \phi_2 \delta_{2k} + \phi_3 \delta_{3k} E_{32} + \phi_4 \delta_{3k} E_{31}, \tag{5.41}$$

where ϕ_i are the scalar polynomials in the invariant quantities obtained from (5.29) as

$$E_{11}, E_{22}, E_{33}, E_{12}, E_{32}^2, E_{31}^2, E_{32}, E_{31}.$$

If we assume that the dependence of M_i on E_{ij} is a linear one, we set

$$\begin{aligned} \phi_1 &= J_{11}E_{11} + J_{12}E_{22} + J_{13}E_{33} + J_{16}E_{12}, \\ \phi_2 &= J_{21}E_{11} + J_{22}E_{22} + J_{23}E_{33} + J_{26}E_{12}, \\ \phi_3 &= \text{constant} = J_{34}, \\ \phi_4 &= \text{constant} = J_{35}, \end{aligned} \tag{5.42}$$

hence, (5.41) takes the form

$$\begin{aligned} M_1 &= J_{11}E_{11} + J_{12}E_{22} + J_{13}E_{33} + J_{16}E_{12}, \\ M_2 &= J_{21}E_{11} + J_{22}E_{22} + J_{23}E_{33} + J_{26}E_{12}, \\ M_3 &= J_{34}E_{32} + J_{35}E_{31}, \end{aligned} \tag{5.43}$$

which is associated with the linear piezomagnetism and the scheme for the material tensors J_{ij} ($i = 1, 2, 3; j = 1, \ldots, 6$) is in agreement with, e.g., Birss [1964, p. 141] and Bhagavantam [1966, p. 173].

Of course, if we are interested in linear relationships alone, more immediate methods are available (see, e.g., Birss [1964, p. 102]). One interestingly simple method follows from this general information, and is the topic of the next chapter.

CHAPTER 6

Linear Constitutive Equations

In general, a linear relationship between the influence **I** and the effect **E** is given by (4.1), i.e.,

$$E_{i_1 \ldots i_r} = m_{i_1 \ldots i_m} I_{i_{r+1} \ldots i_m}. \tag{6.1}$$

The material property tensor $m_{i_1 \ldots i_m}$ is required to be invariant under the crystallographic group $\{M\}$, that is,

$$m_{i_1 \ldots i_m} = (-1)^{\lambda} |\mathbf{M}^{\alpha}|^{\mu} M^{\alpha}_{i_1 j_1} \ldots M^{\alpha}_{i_m j_m} m_{j_1 \ldots j_m} \tag{6.2}$$

for all $\mathbf{M}^{\alpha}$ in $\{M\}$, where λ and μ are in $\{0, 1\}$ depending on the properties of **m**. Equation (6.2) is a special case of (5.4), and it leads to as many linear equations as there are independent components of **m**, between the transformed components and the original components. Exhausting all the symmetry transformations in the crystallographic group $\{M\}$ (in fact, using the generating elements is enough), will result in certain components surviving. This process of finding nonvanishing independent components of the material tensor **m** becomes tedious when it is of an order higher than 2, and when trigonal and hexagonal crystals are involved.

We now give a method based on the basic quantities of the physical (field) tensors **E** and **I**, and on the considerations given at the end of Chapter 5, which completely eliminates the algebraic calculations mentioned above.

Suppose that the field tensors **E** and **I** (effect and influence) are decomposed into the basic quantities associated with the appropriate irreducible representations of the symmetry group $\{M\}$ of the material considered. We found that the basic quantities $\mathbf{v}^{(i)}$ of **E**, whose transformation properties under $\{M\}$ are defined by $\Gamma_i = \{D^{(i)}(\mathbf{M}^{\alpha})\}$, were expressed as linear combinations of the basic quantities (and their products) of the influence tensor **I**, such that they are invariant under the elements of Γ_i (they form carrier spaces of Γ_i). Since we are interested in the linear relationship between **E** and **I**, only the linear terms forming carrier spaces of Γ_i will be considered, for instance, part (i) of (5.29)–(5.32) only will come into play. Hence we have proved

Theorem. *In the case of the linear constitutive equations, the part of the basic quantities of the influence tensor belonging to a given irreducible representation produces an effect belonging to the same representation.*

To illustrate, let us consider the piezomagnetism described by

$$M_i = \lambda_{ijk} E_{jk} \tag{6.3}$$

for the magnetic crystal $\underline{2}/\underline{m} = \{\mathbf{I}, \tau\mathbf{D}_3, \tau\mathbf{R}_3, \mathbf{C}\}$. Referring to (5.40), since M_1 and M_2 form the carrier space of Γ_1, they depend on (E_{11}, E_{22}, E_{33}, and E_{12}) and not on (E_{32}, E_{11}) which are in Γ_4. Similarly, M_3 forms the carrier space of Γ_4 and thus it depends on (E_{32}, E_{31}), but not on ($E_{11}, E_{22}, E_{33}, E_{12}$) which are in Γ_1. Therefore, we immediately arrive at (5.43) that was obtained as a special case of the nonlinear representation of $M_i = M_i(E_{jk})$.

Let us now consider the same effect defined by (6.3) for the cubic magnetic crystal $\underline{\bar{4}}3\underline{m}$. We have, from Table 4.2(17),

$$\mathbf{M}\colon\ [1]_{a,c} \doteq \Gamma_4 \qquad \text{and} \qquad \mathbf{E}\colon\ [2]_{p,i} \doteq \Gamma_1 \dotplus \Gamma_3 \dotplus \Gamma_4, \tag{6.4}$$

and the respective basic quantities are listed as follows:

Γ	Effect $\mathbf{M}$	Influence $\mathbf{E}$
Γ_1	0	$E_{11} + E_{22} + E_{33}$
Γ_2	0	0
Γ_3	0	$\begin{pmatrix} 2E_{11} - E_{22} - E_{33} \\ \sqrt{3}(E_{22} - E_{33}) \end{pmatrix}$
Γ_4	$\begin{pmatrix} M_1 \\ M_2 \\ M_3 \end{pmatrix}$	$\begin{pmatrix} E_{23} \\ E_{31} \\ E_{12} \end{pmatrix}$

(6.5)

whose entries are reproduced from Table 4.3(54). With (6.5) we write

$$\begin{pmatrix} M_1 \\ M_2 \\ M_3 \end{pmatrix} = \lambda_{123} \begin{pmatrix} E_{23} \\ E_{31} \\ E_{12} \end{pmatrix} \tag{6.6}$$

which furnishes only one piezomagnetic modula, as is given by Birss [1964, p. 142].

To illustrate the case where the basic quantities are complex, consider again the same effect described by (6.3) for the tetragonal magnetic crystal $\underline{\bar{4}} = \{\mathbf{I}, \mathbf{D}_3, \tau\mathbf{D}_1\mathbf{T}_3, \tau\mathbf{D}_2\mathbf{T}_3\}$. From Table 4.2(5) we write

$$M\colon\ [1]_{a,c} \doteq \Gamma_2 \dotplus \Gamma_3 \dotplus \Gamma_4 \qquad \text{and} \qquad E\colon\ [2]_{p,i} \doteq 2\Gamma_1 \dotplus 2\Gamma_2 \dotplus \Gamma_3 \dotplus \Gamma_4, \tag{6.7}$$

and the corresponding basic quantities are taken from Table 4.3(13) as

Γ	Effect $\mathbf{M}$	Influence $\mathbf{E}$
Γ_1	0	$E_{11} + E_{22}, E_{33}$
Γ_2	M_3	$E_{11} - E_{22}, E_{12}$
Γ_3	$M_1 - iM_2$	$E_{13} + iE_{23}$
Γ_4	$M_1 + iM_2$	$E_{13} - iE_{23}$

(6.8)

from which we obtain

$$\begin{aligned} M_1 + iM_2 &= \alpha(E_{13} - iE_{23}) \\ &= (\alpha_1 + i\alpha_2)(E_{13} - iE_{23}) \\ &= (\alpha_1 E_{13} + \alpha_2 E_{23}) + i(-\alpha_1 E_{23} + \alpha_2 E_{13}), \end{aligned} \tag{6.9}$$

where $i = \sqrt{-1}$. With (6.9) we get

$$\begin{aligned} M_1 &= \alpha_2 E_{23} + \alpha_1 E_{13}, \\ M_2 &= -\alpha_1 E_{23} + \alpha_2 E_{13}, \end{aligned} \tag{6.10}$$

and for M_3, from (6.8), it is expressed as

$$M_3 = \beta(E_{11} - E_{22}) + \gamma E_{12}. \tag{6.11}$$

Hence, from (6.10), (6.11), piezomagnetism, for the crystal class $\underline{\bar{4}}$, is governed by the four coefficients $\alpha_1, \alpha_2, \beta$, and γ. The scheme in matrix form is given by

$$\begin{pmatrix} 0 & 0 & 0 & \alpha_2 & \alpha_1 & 0 \\ 0 & 0 & 0 & -\alpha_1 & \alpha_2 & 0 \\ \beta & -\beta & 0 & 0 & 0 & \gamma \end{pmatrix} \tag{6.12}$$

which is in agreement with Birss [1964, p. 141].

The piezomagnetic effect was first observed experimentally by Borovik-Romanov [1959, 1960] in measurements on single crystals of CoF_2 and MnF_2. Since these antiferromagnetic difluorides belong to the magnetic symmetry group $\underline{4}/mm\underline{m}$, from Table 4.3(29) we have

Γ	Effect $\mathbf{M}$	Influence $\mathbf{E}$
Γ_3	M_3	E_{12}
Γ_5	$\begin{pmatrix} M_1 \\ -M_2 \end{pmatrix}$	$\begin{pmatrix} E_{23} \\ -E_{31} \end{pmatrix}$

(6.13)

from which there follows

$$\begin{aligned} M_1 &= \alpha E_{23}, \\ M_2 &= \alpha E_{31}, \\ M_3 &= \beta E_{12}. \end{aligned} \tag{6.14}$$

The values of α and β, in units of gauss kg^{-1} cm^2 and at 20.4 K, are

Coefficients	CoF_2	MnF_2
α	2.1×10^{-3}	$\simeq 10^{-5}$
β	0.8×10^{-3}	$\simeq 0$

(6.15)

The magnetoelectric effect, described by $M_i = \lambda_{ij} P_j$, was first observed experimentally by Astrov [1960] on the single crystal Cr_2O_3. Since Cr_2O_3 has the magnetic symmetry $\bar{3}\underline{m}$, we have, from Table 4.3(40),

Γ	Effect **M**	Influence **P**
Γ_4	M_3	P_3
Γ_5	$\begin{pmatrix} M_1 \\ M_2 \end{pmatrix}$	$\begin{pmatrix} P_1 \\ P_1 \end{pmatrix}$

(6.16)

from which there follows

$$\begin{aligned} M_1 &= \alpha P_1, \\ M_2 &= \alpha P_2, \\ M_3 &= \beta P_3. \end{aligned} \tag{6.17}$$

The first observations of the inverse magnetoelectric effect, in which an electric polarization is produced by an applied magnetic field, were made by Rado and Folen [1962] with the same crystal Cr_2O_3.

Let us now consider the pyromagnetic effect which is described by

$$M_i = \lambda_i \Delta\theta, \tag{6.18}$$

where $\Delta\theta$ is the difference between the temperature of the crystal and some reference temperature. Since $\Delta\theta$ is an invariant scalar quantity, only those crystals displaying a spontaneous magnetization (at least one component of M_i forms the carrier space of the identity representation Γ_1) can exhibit pyromagnetism. Observation of Table 4.3(1–58) reveals that only the following sixteen magnetic crystals can be pyromagnetic:

Crystal	Components in Γ_1	Matrix representation of λ_i
$\underline{m}$, $\underline{2}$, $\underline{2}/\underline{m}$	M_1, M_2	$(\lambda_1, \lambda_2, 0)$
$\underline{2}m\underline{m}$	M_2	$(0, \lambda_2, 0)$
$2\underline{m}\underline{m}$, $\underline{2}\underline{2}2$, $m\underline{m}\underline{m}$, $4\underline{m}\underline{m}$, $4\underline{2}\underline{2}$, $\bar{4}\underline{2}\underline{m}$, $\underline{4}/m\underline{m}\underline{m}$, $3\underline{m}$ $3\underline{2}$, $\bar{6}\underline{m}\underline{2}$, $\bar{3}\underline{m}$, $6\underline{2}\underline{2}$, $6\underline{m}\underline{m}$, $6/m\underline{m}\underline{m}$,	M_3	$(0, 0, \lambda_3)$

Table 4.3(1–58) was primarily prepared for the magnetic crystals $\{M\} = \{H\} + \tau\{G - H\}$ originated from a conventional crystal $\{G\}$. However, it also furnishes information for the nonmagnetic crystals $\{G\}$, if $\mathbf{M} = [1]_{a,c}$ is replaced by $\mathbf{A} = [1]_{a,i}$ and $\mathbf{J} = [1]_{p,c}$ is replaced by $\mathbf{P} = [1]_{p,i}$. The time-symmetric tensors **P**, **A** and **E** remain the same.

The coefficients λ_{ij} of the magnetoelectric polarizability in $M_i = \lambda_{ij} P_j$ (and β_{ij} in $P_i = \beta_{ij} M_j$) are available in the literature (e.g., Birss [1964, p. 137]). Under the conventional operation $S^\alpha \in \{G\}$, λ_{ij} transforms as an axial tensor of second order, and hence it follows that it vanishes in those twenty-one classes where the central inversion operator **C** occurs as a conventional symmetry operator. These classes are, from Tables 2.5 and 4.1:

$$\begin{gathered}\bar{1},\ 2/m,\ mmm,\ 4/m,\ 4/mmm,\ \bar{3},\ \bar{3}m,\ 6/m,\ 6/mmm,\ m3,\ m3m,\ \underline{2}/\underline{m},\\ m\underline{mm},\ \underline{4}/m,\ \underline{4}/m\underline{mm},\ \underline{4}/mm\underline{m},\ \bar{3}\underline{m},\ \underline{6}/\underline{m},\ 6/m\underline{mm},\ \underline{6}/\underline{m}m\underline{m},\ m3\underline{m}.\end{gathered} \tag{6.19}$$

Furthermore, observation of Table 4.3(1–58) reveals that the material tensor λ_{ij} vanishes in the following eleven classes as well:

$$\bar{6},\ \underline{6},\ \underline{6}/m,\ \bar{6}m2,\ \bar{6}\underline{m}\underline{2},\ \underline{6}m\underline{m},\ \underline{6}2\underline{2},\ \underline{6}/mm\underline{m},\ \bar{4}3m,\ \underline{4}3\underline{2},\ \underline{m}3m, \tag{6.20}$$

due to the fact that the representations of **M** and **P** are disjoint. Thus, we are left with fifty-eight classes out of ninety; the detailed schemes of the coefficients λ_{ij} for each of these classes are listed in Table 6.1 for ease of reference.

The material tensor λ_{ijk} in piezomagnetism, described by (6.3), is an axial, c-tensor of order 3. It is apparent that any axial, c-tensor of odd order vanishes in those classes where the complementary symmetry operation $\tau\mathbf{C}$ is present. From Table 3.1 (or, more conveniently from Table 3.3) it is seen that there are twenty-one such classes:

$$\begin{gathered}\underline{\bar{1}},\ \underline{2}/m,\ 2/\underline{m},\ \underline{m}mm,\ \underline{mmm},\ \underline{4}/\underline{m},\ 4/\underline{m},\ \underline{4}/\underline{m}m\underline{m},\ 4/\underline{m}mm,\ 4/\underline{mmm},\\ \underline{\bar{3}},\ 6/\underline{m},\ \underline{6}/m,\ \underline{\bar{3}}m,\ \underline{\bar{3}}\underline{m},\ 6/\underline{m}mm,\ 6/\underline{mmm},\ \underline{6}/mm\underline{m},\ \underline{m}3,\ \underline{m}3\underline{m},\ \underline{m}3m.\end{gathered} \tag{6.21}$$

Moreover, from Table 4.3(1–58) it is seen that λ_{ijk} vanishes in three more conventional classes, namely, $\bar{4}3m$, 432, and $m3m$, in which the representations of A_i and E_{ij} are disjoint. Detailed schemes of the nonvanishing coefficients of λ_{ijk} are given for the remaining sixty-six classes in Table 6.2 for ease of reference.

Note that converse piezomagnetism may be regarded as the physical property which relates a magnetic field $\mathbf{M}_i$ as the influence tensor, and a strain E_{ij} in the crystal as the effect tensor. This relation is represented by, in the linear case,

$$E_{ij} = \Lambda_{ijk} M_k. \tag{6.22}$$

In determining the components of Λ_{ijk}, say, for the magnetic crystal class $\underline{\bar{4}} = \{\mathbf{I}, \mathbf{D}_3, \tau\mathbf{D}_1\mathbf{T}_3, \tau\mathbf{D}_2\mathbf{T}_3\}$, from (6.8) we write

$$\begin{gathered}E_{11} - E_{22} = \alpha M_3, \qquad E_{11} + E_{22} = 0, \qquad E_{33} = 0,\\ E_{12} = \beta M_3,\\ E_{13} + iE_{23} = (\gamma_1 + i\gamma_2)(M_1 - iM_2),\end{gathered} \tag{6.23}$$

from which we obtain

$$\begin{aligned} E_{11} &= (\alpha/2)M_3,\\ E_{22} &= (-\alpha/2)M_3,\\ E_{33} &= 0,\\ E_{23} &= \gamma_2 M_1 - \gamma_1 M_2,\\ E_{13} &= \gamma_1 M_1 + \gamma_2 M_2,\\ E_{12} &= \beta M_3,\end{aligned} \quad \text{or} \quad \begin{bmatrix} E_{11}\\ E_{22}\\ E_{33}\\ E_{23}\\ E_{31}\\ E_{12}\end{bmatrix} = \begin{bmatrix} 0 & 0 & \alpha/2\\ 0 & 0 & -\alpha/2\\ 0 & 0 & 0\\ \gamma_2 & -\gamma_1 & 0\\ \gamma_1 & \gamma_2 & 0\\ 0 & 0 & \beta\end{bmatrix} \begin{bmatrix} M_1\\ M_2\\ M_3\end{bmatrix}. \tag{6.24}$$

Table 6.1. Magnetoelectric coefficients.

Classes	Magnetoelectric coefficients	Classes	Magnetoelectric coefficients
$1, \underline{\bar{1}}$	$\begin{bmatrix} \lambda_{11} & \lambda_{12} & \lambda_{13} \\ \lambda_{21} & \lambda_{22} & \lambda_{23} \\ \lambda_{31} & \lambda_{32} & \lambda_{33} \end{bmatrix}$	$\underline{4}, \bar{4}, \underline{4}/\underline{m}$	$\begin{bmatrix} \lambda_{11} & \lambda_{12} & 0 \\ \lambda_{21} & -\lambda_{11} & 0 \\ 0 & 0 & 0 \end{bmatrix}$
$2, \underline{m}, 2/\underline{m}$	$\begin{bmatrix} \lambda_{11} & \lambda_{12} & 0 \\ \lambda_{21} & \lambda_{22} & 0 \\ 0 & 0 & \lambda_{33} \end{bmatrix}$	$422, 4\underline{mm}, \bar{4}2\underline{m}, 4/\underline{mmm}, 32, 3\underline{m}, 6/\underline{mmm}, \underline{\bar{3}}\underline{m}, 622, 6\underline{mm}, \underline{\bar{6}}\underline{m}2$	$\begin{bmatrix} \lambda_{11} & 0 & 0 \\ 0 & \lambda_{11} & 0 \\ 0 & 0 & \lambda_{33} \end{bmatrix}$
$\underline{2}, m, \underline{2}/m$	$\begin{bmatrix} 0 & 0 & \lambda_{13} \\ 0 & 0 & \lambda_{23} \\ \lambda_{31} & \lambda_{32} & 0 \end{bmatrix}$	$\underline{4}2\underline{2}, \underline{4}m\underline{m}, \bar{4}2m, \underline{4}/\underline{m}m\underline{m}$	$\begin{bmatrix} \lambda_{11} & 0 & 0 \\ 0 & -\lambda_{11} & 0 \\ 0 & 0 & 0 \end{bmatrix}$
$222, 2\underline{mm}, \underline{mmm}$	$\begin{bmatrix} \lambda_{11} & 0 & 0 \\ 0 & \lambda_{22} & 0 \\ 0 & 0 & \lambda_{33} \end{bmatrix}$	$4\underline{22}, 4mm, \underline{\bar{4}}\underline{2}m, 3\underline{2}, 3m, 4/\underline{m}mm, \underline{\bar{3}}m, 6\underline{22}, 6mm, \underline{\bar{6}}m\underline{2}, 6/\underline{m}mm$	$\begin{bmatrix} 0 & \lambda_{12} & 0 \\ -\lambda_{12} & 0 & 0 \\ 0 & 0 & 0 \end{bmatrix}$
$2\underline{22}, 2mm, 2\underline{mm}, \underline{m}mm$	$\begin{bmatrix} 0 & \lambda_{12} & 0 \\ \lambda_{21} & 0 & 0 \\ 0 & 0 & 0 \end{bmatrix}$	$\bar{4}\underline{2}\underline{m}, \underline{4}m\underline{m}$	$\begin{bmatrix} 0 & \lambda_{12} & 0 \\ \lambda_{12} & 0 & 0 \\ 0 & 0 & 0 \end{bmatrix}$
$4, \underline{\bar{4}}, 4/\underline{m}, 3, \underline{\bar{3}}, 6, \underline{\bar{6}}, 6/\underline{m}$	$\begin{bmatrix} \lambda_{11} & \lambda_{12} & 0 \\ -\lambda_{21} & \lambda_{11} & 0 \\ 0 & 0 & \lambda_{33} \end{bmatrix}$	$23, \underline{m}3, 432, \underline{\bar{4}}3\underline{m}, \underline{m}3\underline{m}$	$\begin{bmatrix} \lambda_{11} & 0 & 0 \\ 0 & \lambda_{11} & 0 \\ 0 & 0 & \lambda_{11} \end{bmatrix}$

Table 6.2. Piezomagnetism coefficients.

Classes	Piezomagnetism coefficients
$1, \bar{1}$	$\begin{bmatrix} \lambda_{111} & \lambda_{122} & \lambda_{133} & \lambda_{123} & \lambda_{131} & \lambda_{121} \\ \lambda_{211} & \lambda_{222} & \lambda_{233} & \lambda_{223} & \lambda_{231} & \lambda_{221} \\ \lambda_{311} & \lambda_{322} & \lambda_{333} & \lambda_{323} & \lambda_{331} & \lambda_{321} \end{bmatrix}$
$2, m, 2/m$	$\begin{bmatrix} 0 & 0 & 0 & \lambda_{123} & \lambda_{131} & 0 \\ 0 & 0 & 0 & \lambda_{223} & \lambda_{231} & 0 \\ \lambda_{311} & \lambda_{322} & \lambda_{333} & 0 & 0 & \lambda_{321} \end{bmatrix}$
$\underline{2}, \underline{m}, \underline{2}/\underline{m}$	$\begin{bmatrix} \lambda_{111} & \lambda_{122} & \lambda_{133} & 0 & 0 & \lambda_{121} \\ \lambda_{211} & \lambda_{222} & \lambda_{233} & 0 & 0 & \lambda_{221} \\ 0 & 0 & 0 & \lambda_{323} & \lambda_{331} & 0 \end{bmatrix}$
$222, 2mm, mmm$	$\begin{bmatrix} 0 & 0 & 0 & \lambda_{123} & 0 & 0 \\ 0 & 0 & 0 & 0 & \lambda_{231} & 0 \\ 0 & 0 & 0 & 0 & 0 & \lambda_{321} \end{bmatrix}$
$\underline{22}2, 2\underline{mm}, \underline{2}m\underline{m}, m\underline{mm}$	$\begin{bmatrix} 0 & 0 & 0 & 0 & \lambda_{131} & 0 \\ 0 & 0 & 0 & \lambda_{232} & 0 & 0 \\ \lambda_{311} & \lambda_{322} & \lambda_{333} & 0 & 0 & 0 \end{bmatrix}$
$4, \bar{4}, 4/m, 6, \bar{6}, 6/m$	$\begin{bmatrix} 0 & 0 & 0 & \lambda_{123} & \lambda_{131} & 0 \\ 0 & 0 & 0 & \lambda_{131} & -\lambda_{123} & 0 \\ \lambda_{311} & \lambda_{311} & \lambda_{333} & 0 & 0 & 0 \end{bmatrix}$

Table 6.2 (*continued*)

Classes	Piezomagnetism coefficients
$4, \underline{\bar{4}}, \underline{4}/m$	$\begin{bmatrix} 0 & 0 & 0 & \lambda_{123} & \lambda_{131} & 0 \\ 0 & 0 & 0 & -\lambda_{131} & \lambda_{123} & 0 \\ \lambda_{311} & -\lambda_{311} & 0 & 0 & 0 & \lambda_{321} \end{bmatrix}$
$422, 4mm, \bar{4}2m, 4/mmm, 622, 6mm, \bar{6}m2, 6/mmm$	$\begin{bmatrix} 0 & 0 & 0 & \lambda_{123} & 0 & 0 \\ 0 & 0 & 0 & 0 & -\lambda_{123} & 0 \\ 0 & 0 & 0 & 0 & 0 & 0 \end{bmatrix}$
$\underline{4}2\underline{2}, \underline{4}m\underline{m}, \underline{\bar{4}}2\underline{m}, \underline{4}/mm\underline{m}$	$\begin{bmatrix} 0 & 0 & 0 & \lambda_{123} & 0 & 0 \\ 0 & 0 & 0 & 0 & \lambda_{123} & 0 \\ 0 & 0 & 0 & 0 & 0 & \lambda_{321} \end{bmatrix}$
$\underline{\bar{4}}\underline{2}m$	$\begin{bmatrix} 0 & 0 & 0 & 0 & -\lambda_{131} & 0 \\ 0 & 0 & 0 & \lambda_{131} & 0 & 0 \\ \lambda_{311} & -\lambda_{311} & 0 & 0 & 0 & 0 \end{bmatrix}$
$4\underline{22}, 4\underline{mm}, \bar{4}\underline{2m}, \underline{4}/m\underline{mm}, 6\underline{22}, 6\underline{mm}, \bar{6}\underline{m2}, 6/m\underline{mm}$	$\begin{bmatrix} 0 & 0 & 0 & 0 & \lambda_{131} & 0 \\ 0 & 0 & 0 & \lambda_{131} & 0 & 0 \\ \lambda_{311} & \lambda_{311} & \lambda_{333} & 0 & 0 & 0 \end{bmatrix}$
$3, \bar{3}$	$\begin{bmatrix} \lambda_{111} & -\lambda_{111} & 0 & \lambda_{123} & \lambda_{131} & -2\lambda_{222} \\ -\lambda_{222} & \lambda_{222} & 0 & \lambda_{131} & -\lambda_{123} & -2\lambda_{111} \\ \lambda_{311} & \lambda_{311} & \lambda_{333} & 0 & 0 & 0 \end{bmatrix}$
$32, 3m, \bar{3}m$	$\begin{bmatrix} \lambda_{111} & -\lambda_{111} & 0 & \lambda_{123} & 0 & 0 \\ 0 & 0 & 0 & 0 & -\lambda_{123} & -2\lambda_{111} \\ 0 & 0 & 0 & 0 & 0 & 0 \end{bmatrix}$
$3\underline{2}, 3\underline{m}, \bar{3}\underline{m}$	$\begin{bmatrix} 0 & 0 & 0 & 0 & \lambda_{131} & -2\lambda_{222} \\ -\lambda_{222} & \lambda_{222} & 0 & \lambda_{231} & 0 & 0 \\ \lambda_{311} & \lambda_{311} & \lambda_{333} & 0 & 0 & 0 \end{bmatrix}$
$\underline{6}, \underline{\bar{6}}, \underline{6}/\underline{m}$	$\begin{bmatrix} \lambda_{111} & -\lambda_{111} & 0 & 0 & 0 & -2\lambda_{222} \\ -\lambda_{222} & \lambda_{222} & 0 & 0 & 0 & -2\lambda_{111} \\ 0 & 0 & 0 & 0 & 0 & 0 \end{bmatrix}$
$\underline{6}2\underline{2}, \underline{6}m\underline{m}, \underline{\bar{6}}m\underline{2}, \underline{6}/\underline{m}m\underline{m}$	$\begin{bmatrix} \lambda_{111} & -\lambda_{111} & 0 & 0 & 0 & 0 \\ 0 & 0 & 0 & 0 & 0 & -2\lambda_{111} \\ 0 & 0 & 0 & 0 & 0 & 0 \end{bmatrix}$
$\underline{\bar{6}}\underline{m}2$	$\begin{bmatrix} 0 & 0 & 0 & 0 & 0 & -2\lambda_{222} \\ -\lambda_{222} & \lambda_{222} & 0 & 0 & 0 & 0 \\ 0 & 0 & 0 & 0 & 0 & 0 \end{bmatrix}$
$23, m3, \underline{4}3\underline{2}, \underline{\bar{4}}3\underline{m}, m3\underline{m}$	$\begin{bmatrix} 0 & 0 & 0 & \lambda_{123} & 0 & 0 \\ 0 & 0 & 0 & 0 & \lambda_{123} & 0 \\ 0 & 0 & 0 & 0 & 0 & \lambda_{123} \end{bmatrix}$

Note that, for the magnetic class $\underline{4}$, the alternating and time-reversal representations A and T, respectively, are identical, namely,

	$\mathbf{I}$	$\mathbf{D}_3$	$\tau\mathbf{D}_1\mathbf{T}_3$	$\tau\mathbf{D}_2\mathbf{T}_3$
A	1	1	-1	-1
T	1	1	-1	-1
$A \otimes T$	1	1	1	1

from this and (4.13) it follows that

$$[1]_{a,c} = [1]_{p,i} \otimes A \otimes T = [1]_{p,i}.$$

Hence, for this crystal symmetry, converse piezomagnetism and piezoelectricity ($E_{ij} = \bar{\Lambda}_{ijk} P_k$) become the same. The scheme (6.24) is in agreement with the one given by Nye [1957, p. 297].

6.1. Higher-Order Effects

Magnetostriction

We may assume that the strain tensor E_{ij} may also be made up of terms of second order in M_i in addition to the first order, (6.22) will then be written as

$$E_{ij} = \Lambda_{ijk} M_k + \Lambda_{ijkl} M_k M_l. \tag{6.25}$$

The second term on the right-hand side is a second-order property, and which is called magnetostriction. It is to be noted that in practice this effect becomes measurable only when there is a strong magnetic moment involved, such as in the ferromagnetic materials.

Note also that the quantity $M_k M_l$ transforms as if it is a second-order symmetric i-tensor, since it appears in its even power. We do not, therefore, have to consider the magnetic classes separately, since E_{ij} is also an i-tensor. Λ_{ijkl} has the intrinsic symmetry

$$\Lambda_{ijkl} = \Lambda_{jikl} = \Lambda_{ijlk}. \tag{6.26}$$

The nonvanishing components Λ_{ijkl}, with respect of the thirty-two conventional classes, will be identical to those given for the *elasticity* tensor, if we assume that Λ_{ijkl} is further symmetric with respect to an interchange of (ij) with (kl). On the other hand, if we do not make such an assumption, the components will be identical with those given for the *photoelasticity* tensor (see Nye [1957, pp. 140–141 and pp. 250–251]). We may reproduce these results readily, in an independent manner, by making use of Table 4.3(1–58). For instance, for the conventional classes $2mm$, 222, mmm, and for the magnetic classes originating from them, we have

	Effect **E**	Influence **MM**
Γ_1	E_{11}, E_{22}, E_{33}	M_1M_1, M_2M_2, M_3M_3
Γ_2	E_{12}	M_1M_2
Γ_3	E_{23}	M_2M_3
Γ_4	E_{31}	M_3M_1

(6.27)

from which it follows that

$$\left.\begin{aligned} E_{11} &= \Lambda_{1111}M_1M_1 + \Lambda_{1122}M_2M_2 + \Lambda_{1133}M_3M_3, \\ E_{22} &= \Lambda_{2211}M_1M_1 + \Lambda_{2222}M_2M_2 + \Lambda_{2233}M_3M_3, \\ E_{33} &= \Lambda_{3311}M_1M_1 + \Lambda_{3322}M_2M_2 + \Lambda_{3333}M_3M_3, \\ E_{23} &= \Lambda_{2323}M_2M_3, \\ E_{31} &= \Lambda_{3131}M_3M_1, \\ E_{12} &= \Lambda_{1212}M_1M_2. \end{aligned}\right\} \tag{6.28}$$

Electrostriction

We note that another second-order property, which is called electrostriction, is described by

$$E_{ij} = \alpha_{ijkl}P_kP_l, \tag{6.29}$$

where the coefficients α_{ijkl} will be identical to Λ_{ijkl}.

Piezomagnetoelectricity

Another important effect is the appearance of a deformation E_{ij} under the simultaneous existence of the polarization, **P** (or electric field E_i), and the magnetization, **M** (or magnetic field H_i), which may be described by

$$E_{ij} = \beta_{ijkl}M_kP_l \tag{6.30}$$

and which has been called (by Rado [1962]) the piezomagnetoelectric effect. Under the conventional operator $\mathbf{S}^\alpha \in \{G\}$, β_{ijkl} transforms as an axial tensor of fourth order, and hence it follows that it vanishes in those twenty-one classes where the central inversion operator **C** occurs as a conventional symmetry operator. These classes have already been listed in (6.19).

The determination of the nonzero independent components of β_{ijkl} is made possible if we introduce an asymmetric second-order tensor **V** by $\mathbf{V}_{ij} = M_iP_j$. We thus write

$$E_{ij} = \beta_{ijkl}V_{kl}, \tag{6.31}$$

which now indicates a linear relationship between E_{ij} and V_{kl}. Once the basic quantities of V_{kl} are known for a given crystal the enumeration of β_{ijkl} is obtained immediately.

To illustrate, consider the magnetic crystal $m\underline{m}m$ under whose symmetry elements the vectors **M** and **P** are transformed as

	I	$\tau\mathbf{D}_1$	$\tau\mathbf{D}_2$	$\mathbf{D}_3$	$\tau\mathbf{C}$	$\mathbf{R}_1$	$\mathbf{R}_2$	$\tau\mathbf{R}_3$
$\bar{M}_1$	M_1	$-M_1$	M_1	$-M_1$	$-M_1$	M_1	$-M_1$	$-M_1$
$\bar{M}_2$	M_2	M_2	$-M_2$	$-M_2$	$-M_2$	$-M_2$	M_2	M_2
$\bar{M}_3$	M_3	M_3	M_3	M_3	$-M_3$	$-M_3$	$-M_3$	$-M_3$
$\bar{P}_1$	P_1	P_1	$-P_1$	$-P_1$	$-P_1$	$-P_1$	P_1	P_1
$\bar{P}_2$	P_2	$-P_2$	P_2	$-P_2$	$-P_2$	P_2	$-P_2$	P_2
$\bar{P}_3$	P_3	$-P_3$	$-P_3$	P_3	$-P_3$	P_3	P_3	$-P_3$
$\chi(\mathbf{M})$	3	1	1	-1	-3	-1	-1	1
$\chi(\mathbf{P})$	3	-1	-1	-1	-3	1	1	1

(6.32)

From (6.32), transformations of the elements $V_{ij} = M_i P_j$ are obtained as

	I	$\tau\mathbf{D}_1$	$\tau\mathbf{D}_2$	$\mathbf{D}_3$	$\tau\mathbf{C}$	$\mathbf{R}_1$	$\mathbf{R}_2$	$\tau\mathbf{R}_3$	Γ
$\bar{V}_{11}$	V_{11}	$-V_{11}$	$-V_{11}$	V_{11}	V	$-V_{11}$	$-V_{11}$	V_{11}	Γ_4
$\bar{V}_{22}$	V_{22}	$-V_{22}$	$-V_{22}$	V_{22}	V_{22}	$-V_{22}$	$-V_{22}$	V_{22}	Γ_4
$\bar{V}_{33}$	V_{33}	$-V_{33}$	$-V_{33}$	V_{33}	V_{33}	$-V_{33}$	$-V_{33}$	V_{33}	Γ_4
$\bar{V}_{23}$	V_{23}	$-V_{23}$	V_{23}	$-V_{23}$	V_{23}	$-V_{23}$	V_{23}	$-V_{23}$	Γ_3
$\bar{V}_{32}$	V_{32}	$-V_{32}$	V_{32}	$-V_{32}$	V_{32}	$-V_{32}$	V_{32}	$-V_{32}$	Γ_3
$\bar{V}_{31}$	V_{31}	V_{31}	$-V_{31}$	$-V_{31}$	V_{31}	V_{31}	$-V_{31}$	$-V_{31}$	Γ_2
$\bar{V}_{13}$	V_{13}	V_{13}	$-V_{13}$	$-V_{13}$	V_{13}	V_{13}	$-V_{13}$	$-V_{13}$	Γ_2
$\bar{V}_{12}$	V_{12}	V_{12}	V_{12}	V_{12}	V_{12}	V_{12}	V_{12}	V_{12}	Γ_1
$\bar{V}_{21}$	V_{21}	V_{21}	V_{21}	V_{21}	V_{21}	V_{21}	V_{21}	V_{21}	Γ_1
$\chi(\mathbf{V})$	9	-1	-1	1	9	-1	-1	1	

(6.33)

Hence, basic quantities are simply obtained by inspection of (6.33) without referring to (6.23), (6.24). They are listed below

Γ	Effect E_{ij}	Influence $M_i P_j = V_{ij}$
Γ_1	E_{11}, E_{22}, E_{33}	$M_1 P_2, M_2 P_1$
Γ_2	E_{23}	$M_1 P_3, M_3 P_1$
Γ_3	E_{31}	$M_2 P_3, M_3 P_2$
Γ_4	E_{12}	$M_1 P_1, M_2 P_2, M_3 P_3$

(6.34)

In (6.34) the basic quantities for the strain tensor E_{ij} are reproduced from Table

4.3(10). From (6.34) we have

$$\left.\begin{aligned} E_{11} &= \beta_{1112}M_1P_2 + \beta_{1121}M_2P_1, \\ E_{22} &= \beta_{2212}M_1P_2 + \beta_{2221}M_2P_1, \\ E_{33} &= \beta_{3312}M_1P_2 + \beta_{3321}M_2P_1, \\ E_{23} &= \beta_{2313}M_1P_3 + \beta_{2331}M_3P_1, \\ E_{31} &= \beta_{3123}M_2P_3 + \beta_{3132}M_3P_2, \\ E_{12} &= \beta_{1211}M_1P_1 + \beta_{1222}M_2P_2 + \beta_{1233}M_3P_3. \end{aligned}\right\} \tag{6.35}$$

Thus, piezomagnetoelectricity is governed by thirteen nonzero distinct components of β_{ijkl}. This number may be obtained in an alternative manner, which also provides a valuable check.

6.2. The Number of Independent Components

Consider the material property tensor **m** in (6.1). If the physical (field) tensors **E** and **I** are of order p and q, respectively, the material tensor **m** is of order $(p + q)$, i.e.,

$$E_{i_1 \cdots i_p} = m_{i_1 \cdots i_p j_1 \cdots i_q} I_{j_1 \cdots j_q}. \tag{6.36}$$

The tensor $m_{i_1 \cdots i_p j_1 \cdots j_q}$ acquires the intrinsic symmetry of the tensor $E_{i_1 \cdots i_p}$ with respect to its $i_1 \cdots i_p$ suffixes, and also the intrinsic symmetry of the tensor $I_{j_1 \cdots j_q}$ with respect to its indices $j_1 \cdots j_q$. In the case where the tensor **m** does not have any extra intrinsic symmetry, besides that acquired from **E** and **I**, the representation formed by it is the same as the Kronecker (direct) product of the two representations over $\{M\}$ formed by **E** and **I** (see Appendix A (A.44)). The characters of the product representation are known to be equal to the product of the characters of the factor (constituent) representations. Thus, if we know the characters of the representations formed by **E** and **I**, we can find the characters of the representation formed by **m**.

On the other hand, from (6.2), we write

$$\bar{m}_A = m_A = T_{AB}(\mathbf{M}^\alpha) m_B \qquad \text{for all} \quad \mathbf{M}^\alpha \text{ in } \{M\} \qquad (A, B = 1, \ldots, p + q), \tag{6.37}$$

where m_A are the independent components of $m_{i_1 \cdots i_p j_1 \cdots j_q}$. Equation (6.37) is the mathematical statement of the invariance of the material tensor **m** under $\{M\}$.

Let us introduce other material tensors U_A which are linear combinations of m_A as

$$U_A = R_{AB} m_B \qquad (A, B = 1, \ldots, p + q), \tag{6.38}$$

where **R** is a constant nonsingular matrix. Under a symmetry element $\mathbf{M}^\alpha$, U_A

becomes

$$\bar{\mathbf{U}}(\mathbf{M}^\alpha) = \mathbf{R}\bar{\mathbf{m}}(\mathbf{M}^\alpha) = \mathbf{R}\mathbf{T}(\mathbf{M}^\alpha)\mathbf{m} = \underbrace{\mathbf{R}\mathbf{T}(\mathbf{M}^\alpha)\mathbf{R}^{-1}}\mathbf{U}. \tag{6.39}$$

Hence the representation $\bar{\Gamma}$ defined by

$$\bar{\Gamma} = \{\mathbf{R}\mathbf{T}(\mathbf{M}^1)\mathbf{R}^{-1}, \mathbf{R}\mathbf{T}(\mathbf{M}^2)\mathbf{R}^{-1}, \ldots, \mathbf{R}\mathbf{T}(\mathbf{M}^g)\mathbf{R}^{-1}\}$$

describes the transformation properties of $\mathbf{U} = \mathbf{R}\mathbf{m}$. The invariance requirement (6.37) is then replaced by

$$\mathbf{U}(\mathbf{M}^\alpha) = \underbrace{\mathbf{R}\mathbf{T}(\mathbf{M}^\alpha)\mathbf{R}^{-1}}\mathbf{U} \qquad \text{for all} \quad \mathbf{M}^\alpha \text{ in } \{M\}, \tag{6.40}$$

which simply implies that either the corresponding tensor component vanishes at $U_A = 0$, or is invariant under all elements $\mathbf{M}^\alpha$ in $\{M\}$. Thus the only nonvanishing tensor components are those which belong to the representation Γ_1 in $\bar{\Gamma}$. If the identity occurs n_1 times, the material tensor is therefore determined by just n_1 numbers.

The occurence n_1 of the identity representation Γ_1 in $\bar{\Gamma}$ is readily obtained from the character system of $\bar{\Gamma}$ which is identical to Γ (they are equivalent). Hence from (4.6) we have

$$n_1 = \frac{1}{g}\sum_{\alpha=1}^{g} \operatorname{tr} \mathbf{T}(\mathbf{M}^\alpha), \tag{6.41}$$

since $\overset{*}{\chi}{}^{(1)}(\mathbf{M}^\alpha) = 1$ for any $\mathbf{M}^\alpha$.

To illustrate, consider the material tensor β_{ijk} in piezomagnetoelectricity, (6.30), for the magnetic crystal $m\underline{m}m$. The representation of $\boldsymbol{\beta}$ is the direct product of those of E_{ij} and $V_{ij} = M_i P_j$. The character system of the representations of $\mathbf{E}$ and $\mathbf{V}$ are, from (6.32), (6.33),

	$\mathbf{I}$	$\tau\mathbf{D}_1$	$\tau\mathbf{D}_2$	$\mathbf{D}_3$	$\tau\mathbf{C}$	$\mathbf{R}_1$	$\mathbf{R}_2$	$\tau\mathbf{R}_3$
$\chi(E_{ij})$	6	2	2	2	6	2	2	2
$\chi(V_{ij})$	9	-1	-1	1	9	-1	-1	1
$\chi(\beta_{ijkl})$	54	-2	-2	2	54	-2	-2	2

(6.42)

With (6.41) and noting that $g = 8$, we get $n_1 = 13$. The characters $\chi(E_{ij}) = \chi[2]_{p,i}$ are obtained from (4.7). For ease of reference they are listed in Table 6.3 for each conventional symmetry element $\mathbf{S}^\alpha$ in $\{G\}$.

Referring to (4.13) we also find that

$$\begin{aligned}
\mathbf{A}:&\quad \chi[1]_{a,i} = \chi[1]_{p,i}\cdot\chi(A),\\
\mathbf{J}:&\quad \chi[1]_{p,c} = \chi[1]_{p,i}\cdot\chi(T),\\
\mathbf{M}:&\quad \chi[1]_{a,c} = \chi[1]_{p,i}\cdot\chi(T)\cdot\chi(A),\\
\mathbf{E}:&\quad \chi[2]_{p,i} = (\chi[1]_{p,i})^2 - \chi[1]_{a,i},
\end{aligned} \tag{6.43}$$

Table 6.3. Traces of the representations of a vector, and second-order symmetric and skew-symmetric tensors.

Transformation $\mathbf{S}$	$\mathrm{tr}\,\mathbf{S}$	$\mathrm{tr}\,\mathbf{S}^2$	$[(\mathrm{tr}\,\mathbf{S})^2 + \mathrm{tr}\,\mathbf{S}^2]/2$	$[(\mathrm{tr}\,\mathbf{S})^2 - \mathrm{tr}\,\mathbf{S}^2]/2$
$\mathbf{I}$	3	3	6	3
$\mathbf{C}$	-3	3	6	3
$\mathbf{R}_1, \mathbf{R}_2, \mathbf{R}_3,$ $\mathbf{T}_1, \mathbf{T}_2, \mathbf{T}_3$ $\mathbf{D}_1\mathbf{T}_1, \mathbf{D}_2\mathbf{T}_2, \mathbf{D}_3\mathbf{T}_3$	1	3	2	-1
$\mathbf{D}_1\mathbf{D}_2\mathbf{D}_3,$ $\mathbf{C}\mathbf{T}_1, \mathbf{C}\mathbf{T}_2, \mathbf{C}\mathbf{T}_3$ $\mathbf{R}_1\mathbf{T}_1, \mathbf{R}_2\mathbf{T}_3, \mathbf{R}_3\mathbf{T}_3$	-1	3	2	-1
$\mathbf{R}_2\mathbf{T}_1, \mathbf{R}_3\mathbf{T}_1, \mathbf{R}_1\mathbf{T}_2,$ $\mathbf{R}_3\mathbf{T}_2, \mathbf{R}_1\mathbf{T}_3, \mathbf{R}_2\mathbf{T}_3$	1	-1	0	1
$\mathbf{D}_2\mathbf{T}_1, \mathbf{D}_3\mathbf{T}_1, \mathbf{D}_1\mathbf{T}_2$ $\mathbf{D}_3\mathbf{T}_2, \mathbf{D}_1\mathbf{T}_3, \mathbf{D}_2\mathbf{T}_3$	-1	-1	0	1
$\mathbf{M}_1, \mathbf{D}_1\mathbf{M}_1, \mathbf{D}_2\mathbf{M}_1, \mathbf{D}_3\mathbf{M}_1,$ $\mathbf{M}_2, \mathbf{D}_1\mathbf{M}_2, \mathbf{D}_2\mathbf{M}_3, \mathbf{D}_3\mathbf{M}_2$	0	0	0	0
$\mathbf{C}\mathbf{M}_1, \mathbf{R}_1\mathbf{M}_1, \mathbf{R}_2\mathbf{M}_1, \mathbf{R}_3\mathbf{M}_1,$ $\mathbf{C}\mathbf{M}_2, \mathbf{R}_1\mathbf{M}_2, \mathbf{R}_2\mathbf{M}_3, \mathbf{R}_3\mathbf{M}_2$	0	0	0	0

where A is the alternating representation and T is the time-reversal representation associated with the crystal class considered. For instance, for the magnetic crystal $\underline{m}mm$, they are

	$\mathbf{I}$	$\tau\mathbf{D}_1$	$\tau\mathbf{D}_2$	$\mathbf{D}_3$	$\tau\mathbf{C}$	$\mathbf{R}_1$	$\mathbf{R}_2$	$\tau\mathbf{R}_3$
A	1	1	1	1	-1	-1	-1	-1
T	1	-1	-1	1	-1	1	1	-1

(6.44)

The importance of (6.43) is that the relevant character systems for the representations associated with higher-order effects are readily obtained, once the character system for the representation $[1]_{p,i}$ (polar, i-vector) is known.

As a final illustration, consider the converse piezomagnetism, (6.22), for the magnetic class $\underline{\bar{4}} = \{\mathbf{I}, \mathbf{D}_3, \tau\mathbf{D}_1\mathbf{T}_3, \tau\mathbf{D}_2\mathbf{T}_3\}$. We need the character system for the representation Λ_{ijk} which is expressed by

$$\chi(\Lambda_{ijk}) = \mathrm{tr}([2]_{p,i} \otimes [1]_{a,c}). \tag{6.45}$$

We form the following table:

	$\mathbf{I}$	$\mathbf{D}_3$	$\tau\mathbf{D}_1\mathbf{T}_3$	$\tau\mathbf{D}_2\mathbf{T}_3$
T	1	1	-1	-1
A	1	1	-1	-1
$\chi[1]_{p,i}$	3	-1	-1	-1
$\chi[1]_{a,i}$	3	-1	1	1
$\chi[1]_{a,c}$	3	-1	-1	-1
$\chi[2]_{p,i}$	6	2	0	0
$\chi[1]_{a,c}\chi[2]_{p,i}$	18	-2	0	0

(6.46)

With (6.41) and $g = 4$, we get $n_1 = 16/4 = 4$, which is in agreement with (6.24).

6.3. An Alternative Procedure for Finding Independent Components

Finding the number of independent components of a material tensor associated with a given crystal $\{M\}$ can be achieved in an alternative manner, and which may be easier than the previous procedure.

Note that the inner Kronecker (direct) product of two irreducible representations Γ_i and Γ_j of $\{M\}$ is, in general, reducible and given by (see Appendix A(A.49))

$$\Gamma_i \otimes \Gamma_j = \sum_{i=1}^{r} \gamma_{k,ij} \Gamma_k, \tag{6.47}$$

where the coefficients $\gamma_{k,ij}$ are determined from

$$\gamma_{k,ij} = \frac{1}{g} \sum_{\alpha=1}^{g} \overset{*}{\chi}^{(k)}(\mathbf{M}^\alpha) \chi^{(i)}(\mathbf{M}^\alpha) \chi^{(j)}(\mathbf{M}^\alpha). \tag{6.48}$$

With the aid of Table 3.3(2–18) and equations (6.47), (6.48), decomposition of the inner direct product representations are obtained and listed in Table 6.4(2–18) for each distinct family of magnetic crystallographic groups.

On the other hand, the decomposition of the electromechanical quantities

$$\mathbf{M} = [1]_{a,c}, \qquad \mathbf{J} = [1]_{p,c}, \qquad \mathbf{P} = [1]_{p,i}, \qquad \mathbf{A} = [1]_{a,i}, \qquad \mathbf{E} = [2]_{p,i}, \tag{6.49}$$

are obtained in Chapter 4 and listed in Table 4.2(2–18).

Now, if the representation formed by the material tensor $\mathbf{m}$ in (6.36) is obtained as the Kronecker (direct) product of the two representations of $\{M\}$ formed by the effect, $\mathbf{E}$, and the influence, $\mathbf{I}$, we have

$$\text{Rep}(\mathbf{m}) = \text{Rep}(\mathbf{E}) \otimes \text{Rep}(\mathbf{I}), \tag{6.50}$$

where Rep($\mathbf{m}$) stands for the representation formed by $\mathbf{m}$. Note that Rep($\mathbf{E}$) and Rep($\mathbf{I}$) are either equal to one of the representations appearing in (6.49) or they may be equal to the Kronecker (direct) product of some of those in (6.49). Hence, the number of the identity representations Γ_1 occurring in Rep($\mathbf{m}$) gives the number of the surviving nonzero components of the material tensor $\mathbf{m}$.

As an example, consider the converse piezomagnetism defined by (6.22), i.e., $E_{ij} = \Lambda_{ijk} M_k$. We have

$$\text{Rep}(\Lambda_{ijk}) = \text{Rep}(E_{ij}) \otimes \text{Rep}(M_k). \tag{6.51}$$

In particular, let us find the number of nonzero components of Λ_{ijk} for the magnetic crystal $\underline{\bar{4}} = \{\mathbf{I}, \mathbf{D}_3, \tau\mathbf{D}_1\mathbf{T}_3, \tau\mathbf{D}_2\mathbf{T}_3\}$. Referring to Table 4.2(5) and

Table 6.4(5) we have

$$\text{Rep}\,\mathbf{E} = [2]_{p,i} = 2\Gamma_1 \dotplus 2\Gamma_2 \dotplus \Gamma_3 \dotplus \Gamma_4,$$
$$\text{Rep}\,\mathbf{M} = [\mathbf{1}]_{q,i} = \Gamma_2 \dotplus \Gamma_3 \dotplus \Gamma_4,$$

and

	Γ_1	Γ_2	Γ_3	Γ_4
Γ_1	Γ_1			
Γ_2	Γ_2	Γ_1		
Γ_3	Γ_3	Γ_4	Γ_2	
Γ_4	Γ_4	Γ_3	Γ_1	Γ_2

(6.52)

Hence

$$\begin{aligned}\text{Rep}(\Lambda_{ijk}) &= (2\Gamma_1 \dotplus 2\Gamma_2 \dotplus \Gamma_3 \dotplus \Gamma_4) \otimes (\Gamma_2 \dotplus \Gamma_3 \dotplus \Gamma_4)\\ &= 2\Gamma_1 \otimes \Gamma_2 \dotplus 2\Gamma_1 \otimes \Gamma_3 \dotplus 2\Gamma_1 \otimes \Gamma_4 \dotplus 2\Gamma_2 \otimes \Gamma_2\\ &\quad \dotplus 3\Gamma_2 \otimes \Gamma_3 \dotplus 3\Gamma_2 \otimes \Gamma_4 \dotplus \Gamma_3 \otimes \Gamma_3 \dotplus 2\Gamma_3 \otimes \Gamma_4 \dotplus \Gamma_4 \otimes \Gamma_4\\ &= 4\Gamma_1 \dotplus 4\Gamma_2 \dotplus 5\Gamma_3 \dotplus 5\Gamma_4. \end{aligned} \tag{6.53}$$

Since Γ_1 appears four times, four components of Λ_{ijk} will survive. This result is in agreement with (6.24).

If we wish to consider piezomagnetoelectricity for the same crystal, its representation is given by

$$\text{Rep}(\Lambda_{ijkl}) = \text{Rep}(E_{ij}) \otimes \text{Rep}(M_k) \otimes \text{Rep}(P_l) \tag{6.54}$$

and from Table 4.2(5), we get $\text{Rep}(P_l) = [1]_{p,i} = \Gamma_2 \dotplus \Gamma_3 \dotplus \Gamma_4$. Hence, from (6.53) we have

$$\begin{aligned}\text{Rep}(\Lambda_{ijkl}) &= (4\Gamma_1 \dotplus 4\Gamma_2 \dotplus 5\Gamma_3 \dotplus 5\Gamma_4) \otimes (\Gamma_2 \dotplus \Gamma_3 \dotplus \Gamma_4)\\ &= 14\Gamma_1 \dotplus 14\Gamma_2 \dotplus 13\Gamma_3 \dotplus 13\Gamma_4 \end{aligned} \tag{6.55}$$

from which we find $n_1 = 14$, which is in agreement with the results given by Lyubimov [1966].

Table 6.4. Multiplication tables of the irreducible representations

$$\left(\Gamma_p \otimes \Gamma_q = \sum_{i=1}^{r} \gamma_{i,pq}\Gamma_i = \Gamma_q \otimes \Gamma_p\right).$$

$\bar{1}; \underline{\bar{1}}$ $\quad m; \underline{m}$ $\quad 2; \underline{2}$

	Γ_1	Γ_2
Γ_1	Γ_1	
Γ_2	Γ_2	Γ_1

(2)

Table 6.4 (*continued*)

$2/m$; $\underline{2}/m$, $2/\underline{m}$, $\underline{2}/\underline{m}$
$2mm$; $\underline{2}m\underline{m}$, $2\underline{m}\underline{m}$
222; $\underline{2}\underline{2}2$

	Γ_1	Γ_2	Γ_3	Γ_4
Γ_1	Γ_1			
Γ_2	Γ_2	Γ_1		
Γ_3	Γ_3	Γ_4	Γ_1	
Γ_4	Γ_4	Γ_3	Γ_2	Γ_1

(3)

mmm; $\underline{m}mm$, $\underline{m}\underline{m}\underline{m}$, $m\underline{m}\underline{m}$

	Γ_1	Γ_2	Γ_3	Γ_4	Γ_1'	Γ_2'	Γ_3'	Γ_4'
Γ_1	Γ_1							
Γ_2	Γ_2	Γ_1						
Γ_3	Γ_3	Γ_4	Γ_1					
Γ_4	Γ_4	Γ_3	Γ_2	Γ_1				
Γ_1'	Γ_1'	Γ_2'	Γ_3'	Γ_4'	Γ_1			
Γ_2'	Γ_2'	Γ_1'	Γ_4'	Γ_3'	Γ_2	Γ_1		
Γ_3'	Γ_3'	Γ_4'	Γ_1'	Γ_2'	Γ_3	Γ_4	Γ_1	
Γ_4'	Γ_4'	Γ_3'	Γ_2'	Γ_1'	Γ_4	Γ_3	Γ_2	Γ_1

(4)

$\bar{4}$; $\underline{\bar{4}}$ 4; $\underline{4}$

	Γ_1	Γ_2	Γ_3	Γ_4
Γ_1	Γ_1			
Γ_2	Γ_2	Γ_1		
Γ_3	Γ_3	Γ_4	Γ_2	
Γ_4	Γ_4	Γ_3	Γ_1	Γ_2

(5)

$4/m$; $\underline{4}/\underline{m}$, $4/\underline{m}$, $\underline{4}/m$

	Γ_1	Γ_2	Γ_3	Γ_4	Γ_1'	Γ_2'	Γ_3'	Γ_4'
Γ_1	Γ_1							
Γ_2	Γ_2	Γ_1						
Γ_3	Γ_3	Γ_4	Γ_2					
Γ_4	Γ_4	Γ_3	Γ_1	Γ_2				
Γ_1'	Γ_1'	Γ_2'	Γ_3'	Γ_4'	Γ_1			
Γ_2'	Γ_2'	Γ_1'	Γ_4'	Γ_3'	Γ_2	Γ_1		
Γ_3'	Γ_3'	Γ_4'	Γ_2'	Γ_1'	Γ_3	Γ_4	Γ_2	
Γ_4'	Γ_4'	Γ_3'	Γ_1'	Γ_2'	Γ_4	Γ_3	Γ_1	Γ_2

(6)

Table 6.4 (*continued*)

$4mm$; $4\underline{m}\underline{m}$, $\underline{4}m\underline{m}$ 422; $4\underline{2}\underline{2}$, $\underline{4}2\underline{2}$ $\bar{4}2m$; $\underline{\bar{4}}2\underline{m}$, $\underline{\bar{4}}\underline{2}m$, $\bar{4}\underline{2}\underline{m}$

	Γ_1	Γ_2	Γ_3	Γ_4	Γ_5
Γ_1	Γ_1				
Γ_2	Γ_2	Γ_1			
Γ_3	Γ_3	Γ_4	Γ_1		
Γ_4	Γ_4	Γ_3	Γ_2	Γ_1	
Γ_5	Γ_5	Γ_5	Γ_5	Γ_5	a

(7)

$a = \Gamma_1 + \Gamma_2 + \Gamma_3 + \Gamma_4$.

$4/mmm$ $\underline{4}/\underline{m}m\underline{m}$ $4/\underline{m}mm$ $\underline{4}/m\underline{m}\underline{m}$ $4/\underline{m}\underline{m}\underline{m}$ $\underline{4}/mm\underline{m}$

	Γ_1	Γ_2	Γ_3	Γ_4	Γ_5	Γ_1'	Γ_2'	Γ_3'	Γ_4'	Γ_5'
Γ_1	Γ_1									
Γ_2	Γ_2	Γ_1								
Γ_3	Γ_3	Γ_4	Γ_1							
Γ_4	Γ_4	Γ_3	Γ_2	Γ_1						
Γ_5	Γ_5	Γ_5	Γ_5	Γ_5	a					
Γ_1'	Γ_1'	Γ_2'	Γ_3'	Γ_4'	Γ_5'	Γ_1				
Γ_2'	Γ_2'	Γ_1'	Γ_4'	Γ_3'	Γ_5'	Γ_2	Γ_1			
Γ_3'	Γ_3'	Γ_4'	Γ_1'	Γ_2'	Γ_5'	Γ_3	Γ_4	Γ_1		
Γ_4'	Γ_4'	Γ_3'	Γ_2'	Γ_1'	Γ_5'	Γ_4	Γ_3	Γ_2	Γ_1	
Γ_5'	Γ_5'	Γ_5'	Γ_5'	Γ_5'	b	Γ_5	Γ_5	Γ_5	Γ_5	a

(8)

$a = \Gamma_1 + \Gamma_2 + \Gamma_3 + \Gamma_4$; $b = \Gamma_1' + \Gamma_2' + \Gamma_3' + \Gamma_4'$.

$3m$; $3\underline{m}$ 32; $3\underline{2}$ $a = \Gamma_1 + \Gamma_2 + \Gamma_3$.

	Γ_1	Γ_2	Γ_3
Γ_1	Γ_1		
Γ_2	Γ_2	Γ_1	
Γ_3	Γ_3	Γ_3	a

(10)

$\bar{3}$; $\underline{\bar{3}}$ $\bar{6}$; $\underline{\bar{6}}$ 6; $\underline{6}$

	Γ_1	Γ_2	Γ_3	Γ_4	Γ_5	Γ_6
Γ_1	Γ_1					
Γ_2	Γ_2	Γ_3				
Γ_3	Γ_3	Γ_1	Γ_2			
Γ_4	Γ_4	Γ_5	Γ_6	Γ_1		
Γ_5	Γ_5	Γ_6	Γ_4	Γ_2	Γ_3	
Γ_6	Γ_6	Γ_4	Γ_5	Γ_3	Γ_1	Γ_2

(11)

Table 6.4 (*continued*)

$6/m$; $6/\underline{m}$, $\underline{6}/m$, $\underline{6}/\underline{m}$

	Γ_1	Γ_2	Γ_3	Γ_4	Γ_5	Γ_6	Γ_1'	Γ_2'	Γ_3'	Γ_4'	Γ_5'	Γ_6'
Γ_1	Γ_1											
Γ_2	Γ_2	Γ_3										
Γ_3	Γ_3	Γ_1	Γ_2									
Γ_4	Γ_4	Γ_5	Γ_6	Γ_1								
Γ_5	Γ_5	Γ_6	Γ_4	Γ_2	Γ_3							
Γ_6	Γ_6	Γ_4	Γ_5	Γ_3	Γ_1	Γ_2						
Γ_1'	Γ_1'	Γ_2'	Γ_3'	Γ_4'	Γ_5'	Γ_6'	Γ_1					
Γ_2'	Γ_2'	Γ_3'	Γ_1'	Γ_5'	Γ_6'	Γ_4'	Γ_2	Γ_3				
Γ_3'	Γ_3'	Γ_1'	Γ_2'	Γ_6'	Γ_4'	Γ_5'	Γ_3	Γ_1	Γ_2			
Γ_4'	Γ_4'	Γ_5'	Γ_6'	Γ_1'	Γ_2'	Γ_3'	Γ_4	Γ_5	Γ_6	Γ_1		
Γ_5'	Γ_5'	Γ_6'	Γ_4'	Γ_2'	Γ_3'	Γ_1'	Γ_5	Γ_6	Γ_4	Γ_2	Γ_3	
Γ_6'	Γ_6'	Γ_4'	Γ_5'	Γ_3'	Γ_1'	Γ_2'	Γ_6	Γ_4	Γ_5	Γ_3	Γ_1	Γ_2

(12)

$\bar{6}m2$; $\bar{6}\underline{m}\underline{2}$, $\underline{\bar{6}}m\underline{2}$, $\underline{\bar{6}}\underline{m}2$ $\bar{3}m$; $\bar{3}\underline{m}$, $\underline{\bar{3}}m$, $\underline{\bar{3}}\underline{m}$ 622; $6\underline{2}\underline{2}$, $\underline{6}2\underline{2}$ $6mm$; $6\underline{m}\underline{m}$, $\underline{6}m\underline{m}$

	Γ_1	Γ_2	Γ_3	Γ_4	Γ_5	Γ_6
Γ_1	Γ_1					
Γ_2	Γ_2	Γ_1				
Γ_3	Γ_3	Γ_4	Γ_1			
Γ_4	Γ_4	Γ_3	Γ_2	Γ_1		
Γ_5	Γ_5	Γ_5	Γ_6	Γ_6	a	
Γ_6	Γ_6	Γ_6	Γ_5	Γ_5	b	a

(13)

$a = \Gamma_1 + \Gamma_2 + \Gamma_6$; $b = \Gamma_3 + \Gamma_4 + \Gamma_5$.

$6/mmm$; $6/\underline{m}mm$, $6/m\underline{m}\underline{m}$, $6/\underline{m}\underline{m}\underline{m}$, $\underline{6}/mm\underline{m}$, $\underline{6}/\underline{m}m\underline{m}$

	Γ_1	Γ_2	Γ_3	Γ_4	Γ_5	Γ_6	Γ_1	Γ_2	Γ_3	Γ_4	Γ_5	Γ_6
Γ_1	Γ_1											
Γ_2	Γ_2	Γ_1										
Γ_3	Γ_3	Γ_4	Γ_1									
Γ_4	Γ_4	Γ_3	Γ_2	Γ_1								
Γ_5	Γ_5	Γ_5	Γ_6	Γ_6	a							
Γ_6	Γ_6	Γ_6	Γ_5	Γ_5	b	a						
Γ_1'	Γ_1'	Γ_2'	Γ_3'	Γ_4'	Γ_5'	Γ_6'	Γ_1					
Γ_2'	Γ_2'	Γ_1'	Γ_4'	Γ_3'	Γ_5'	Γ_6'	Γ_2	Γ_1				
Γ_3'	Γ_3'	Γ_4'	Γ_1'	Γ_2'	Γ_6'	Γ_5'	Γ_3	Γ_4	Γ_1			
Γ_4'	Γ_4'	Γ_3'	Γ_2'	Γ_1'	Γ_6'	Γ_5'	Γ_4	Γ_3	Γ_2	Γ_1		
Γ_5'	Γ_5'	Γ_5'	Γ_6'	Γ_6'	a'	b'	Γ_5	Γ_5	Γ_6	Γ_6	a	
Γ_6'	Γ_6'	Γ_6'	Γ_5'	Γ_5'	b'	a'	Γ_6	Γ_6	Γ_5	Γ_5	b	a

(14)

$a = \Gamma_1 + \Gamma_2 + \Gamma_6$; $a' = \Gamma_1' + \Gamma_2' + \Gamma_6'$;
$b = \Gamma_3 + \Gamma_4 + \Gamma_5$; $b' = \Gamma_3' + \Gamma_4' + \Gamma_5'$.

Table 6.4 (*continued*)

$m3$: $\underline{m}3$

	Γ_1	Γ_2	Γ_3	Γ_4	Γ_1'	Γ_2'	Γ_3'	Γ_4'
Γ_1	Γ_1							
Γ_2	Γ_2	Γ_3						
Γ_3	Γ_3	Γ_1	Γ_2					
Γ_4	Γ_4	Γ_4	Γ_4	a				
Γ_1'	Γ_1'	Γ_2'	Γ_3'	Γ_4	Γ_1			
Γ_2'	Γ_2'	Γ_3'	Γ_1'	Γ_4'	Γ_2	Γ_3		
Γ_3'	Γ_3'	Γ_1'	Γ_2'	Γ_4'	Γ_3	Γ_1	Γ_2	
Γ_4'	Γ_4'	Γ_4'	Γ_4'	a'	Γ_4	Γ_4	Γ_4	a

(16)

$a = \Gamma_1 + \Gamma_2 + \Gamma_3 + 2\Gamma_4$; $\quad a' = \Gamma_1' + \Gamma_2' + \Gamma_3' + 2\Gamma_4'$.

$\bar{4}3m$: $\bar{4}3\underline{m}$ $\quad 432$: $\underline{4}3\underline{2}$

	Γ_1	Γ_2	Γ_3	Γ_4	Γ_5
Γ_1	Γ_1				
Γ_2	Γ_2	Γ_1			
Γ_3	Γ_3	Γ_3	a		
Γ_4	Γ_4	Γ_5	b	c	
Γ_5	Γ_5	Γ_4	b	d	c

(17)

$a = \Gamma_1 + \Gamma_2 + \Gamma_3$; $\quad b = \Gamma_4 + \Gamma_5$;
$c = \Gamma_1 + \Gamma_3 + \Gamma_4 + \Gamma_5$; $\quad d = \Gamma_2 + \Gamma_3 + \Gamma_4 + \Gamma_5$.

$m3m$: $m3\underline{m}$, $\underline{m}3\underline{m}$, $\underline{m}3m$

	Γ_1	Γ_2	Γ_3	Γ_4	Γ_5	Γ_1'	Γ_2'	Γ_3'	Γ_4'	Γ_5'
Γ_1	Γ_1									
Γ_2	Γ_2	Γ_1								
Γ_3	Γ_3	Γ_3	a							
Γ_4	Γ_4	Γ_5	b	c						
Γ_5	Γ_5	Γ_4	b	d	c					
Γ_1'	Γ_1'	Γ_2'	Γ_3'	Γ_4'	Γ_5'	Γ_1				
Γ_2'	Γ_2'	Γ_1'	Γ_3'	Γ_5'	Γ_4'	Γ_2	Γ_1			
Γ_3'	Γ_3'	Γ_3'	a'	b'	b'	Γ_3	Γ_3	a		
Γ_4'	Γ_4'	Γ_5'	b'	c'	d'	Γ_4	Γ_5	b	c	
Γ_5'	Γ_5'	Γ_4'	b'	d'	c'	Γ_5	Γ_4	b	d	c

(18)

$a = \Gamma_1 + \Gamma_2 + \Gamma_3$; $\quad b = \Gamma_4 + \Gamma_5$; $\quad c = \Gamma_1 + \Gamma_3 + \Gamma_4 + \Gamma_5$;
$d = \Gamma_2 + \Gamma_3 + \Gamma_4 + \Gamma_5$; $\quad a' = \Gamma_1' + \Gamma_2' + \Gamma_3'$; $\quad b' = \Gamma_4' + \Gamma_5'$;
$c' = \Gamma_1' + \Gamma_3' + \Gamma_4' + \Gamma_5'$; $\quad d' = \Gamma_2' + \Gamma_3' + \Gamma_4' + \Gamma_5'$.

CHAPTER 7

Nonlinear Constitutive Equations for Electromagnetic Crystalline Solids

In this chapter we tabulate the elements of the integrity basis which are invariant under the magnetic symmetry group of the electromagnetic crystalline solids. For the sake of simplicity, the electromagnetic solids considered here are assumed to have the constitutive equations

$$W = W(E_{KL}, P_K, M_K, \theta) \tag{7.1}$$

for the internal energy. We have similar expressions for the entropy, the electric field, the magnetic induction, the heat flux, and the stress tensor. The arguments E_{KL}, P_K, M_K, and θ appearing in (7.1) are, respectively, the material description of the strain tensor (a true, symmetric, and *i*-tensor), the polarization (a polar and time-symmetric vector), the magnetization (an axial, time-asymmetric vector), and the temperature (a true time-symmetric scalar). The dependent quantities such as the entropy, the electric field, the magnetic induction, and the stress tensor are derived from the internal energy, W, since these materials are conservative (cf. eq. (1.40)).

Kiral and Smith [1974] considered constitutive relations of the form $W = W(\mathbf{B}, \ldots, \mathbf{C})$, where W is a scalar polynomial function of the components of $\mathbf{B}, \ldots, \mathbf{C}$ which is invariant under the classical point groups. They made no restriction as to the number or symmetry type of the quantities $\mathbf{B}, \ldots, \mathbf{C}$ appearing as arguments of W. The elements of the integrity basis listed by Kiral and Smith are given in terms of *all possible* basic quantities associated with the crystal class considered. The results given by them are complete, due to the fact that any tensorial quantity will form the carrier spaces of some of the irreducible representations of a given crystallographic group.

The cases covered by Kiral and Smith [1974] are the conventional triclinic, monoclinic, orthorhombic, tetragonal, and hexagonal (except for the crystal D_{6h}) crystal systems. The crystal class D_{6h} is considered separately by Kiral et al. [1980].

The fact that a conventional crystallographic group, and the family of magnetic groups derived from it, are isomorphic is shown by Indenbom [1960]. For an alternative derivation for the point groups the reader is referred to Mert [1975]. This result not only allows us the use of the irreducible representations of the conventional point groups as those of the corresponding magnetic groups, but it also enables us to determine the integrity basis for the magnetic classes *directly* from the result obtained for the conventional ones. This again follows from the fact that the basic quantities associated with a magnetic group happen to be some of the basic quantities under the conventional point group, from which the magnetic group under consideration is generated—just because of the isomorphism between them.

We note that the results given by Kiral and Smith are in typical multilinear form. The nonlinear elements of the integrity basis are readily obtained once the multilinear elements are given. We also note that the actual elements of the integrity basis, in terms of the components of the arguments E, P, M, are determined using Tables B.1–B.13 of Appendix B, copied from Kiral and Smith [1974]. In the case of magnetic point groups this association is provided in Table 4.3(1–58) of the present work.

The following example should illustrate the use of these tables and the scheme for obtaining the integrity basis.

Let us determine the polynomial integrity basis of a true scalar $W = W(E_{kL}, P_k, M_k, \theta)$ for the magnetic class $4\underline{mm} = \{I, \mathbf{R}_2, \mathbf{R}_1, \mathbf{D}_3, \tau\mathbf{T}_3, \tau\mathbf{R}_2\mathbf{T}_3, \tau\mathbf{R}_1\mathbf{T}_3, \tau\mathbf{D}_3\mathbf{T}_3\}$.

The irreducible representations of $4\underline{mm}$ and the basic quantities associated with **E**, **P**, and **M** are given in Table 3.3(7) and Table 4.3(19), respectively. We write

$$\begin{aligned} \phi, \phi', \phi'' &= E_{11} + E_{22}, E_{33}, P_3 &&: \Gamma_1, \\ \nu, \nu' &= M_3, E_{12} &&: \Gamma_3, \\ \tau &= E_{11} - E_{22} &&: \Gamma_4, \\ \mathbf{a}, \mathbf{b}, \mathbf{c} &= \begin{pmatrix} M_2 \\ M_1 \end{pmatrix}, \begin{pmatrix} E_{31} \\ E_{32} \end{pmatrix}, \begin{pmatrix} P_1 \\ P_2 \end{pmatrix} &&: \Gamma_5. \end{aligned} \tag{7.2}$$

The typical multilinear elements of the integrity basis are given by (B.5) (in Appendix B), in terms of the basic quantities ($\phi, \phi', \ldots; \psi, \psi', \ldots; \nu, \nu', \ldots; \tau, \tau', \ldots; a, b, \ldots$). Noting that $\psi = 0$, we list the elements as:

Degree 1: ϕ, ϕ', ϕ''.

Degree 2: $a_1^2 + a_2^2, b_1^2 + b_2^2, c_1^2 + c_2^2, a_1 b_1 + a_2 b_2, a_1 c_1 + a_2 c_2, b_1 c_1 + b_2 c_2$, and $\nu^2, \nu\nu', \nu'^2, \tau^2$.

Degree 3: $\nu a_1 a_2, \nu(a_1 b_2 + a_2 b_1)$,
$\nu b_1 b_2, \nu(a_1 c_2 + a_2 c_1)$,
$\nu c_1 c_2, \nu(b_1 c_2 + b_2 c_1)$,
$\nu' a_1 a_2, \nu'(a_1 b_2 + a_2 b_1)$,
$\nu' b_1 b_2, \nu'(a_1 c_2 + a_2 c_1)$,
$\nu' c_1 c_2, \nu'(b_1 c_2 + b_2 c_1)$,

and

$$\tau(a_1^2 - a_2^2),\ \tau(a_1 b_1 - a_2 b_1),$$
$$\tau(b_1^2 - b_2^2),\ \tau(a_1 c_1 - a_2 c_2),$$
$$\tau(c_1^2 - c_2^2),\ \tau(b_1 c_1 - b_2 c_2).$$

Degree 4: $a_1^4 + a_2^4,\ a_1^3 b_1 + a_2^3 b_2,\ b_1^3 c_1 + b_2^3 c_2,$
$b_1^4 + b_2^4,\ a_1^3 c_1 + a_2^3 c_2,\ c_1^3 a_1 + c_2^3 a_2,$
$c_1^4 + c_2^4,\ b_1^3 a_1 + b_2^3 a_2,\ c_1^3 b_1 + c_2^3 b_2,$
$a_1^2 b_1^2 + a_2^2 b_2^2,\ a_1^2 b_1 c_1 + a_2^2 b_2 c_2,$
$a_1^2 c_1^2 + a_2^2 c_2^2,\ b_1^2 a_1 c_1 + b_2^2 a_2 c_2,$
$b_1^2 c_1^2 + b_2^2 c_2^2,\ c_1^2 a_1 b_1 + c_2^2 a_2 b_2,$

and

$$\nu\tau(a_1 b_2 - a_2 b_1),\ \nu\tau(a_1 c_2 - a_2 c_1),\ \nu\tau(b_1 c_2 - b_2 c_1),$$
$$\nu'\tau(a_1 b_2 - a_2 b_1),\ \nu'\tau(a_1 c_2 - a_2 c_1),\ \nu'\tau(b_1 c_2 - b_2 c_1).$$

In the final step we substitute the basic quantities ϕ, ϕ', ϕ'', ν, ν', τ, $\mathbf{a}$, $\mathbf{b}$, and $\mathbf{c}$ by the explicit components of **E**, **P**, and **M** given by (7.2), and rearrange the elements in the seven groups (E), (P), (M), (E, P), (E, M), (P, M), and (E, P, M).

Alternatively, we may proceed by introducing the three symmetric tensors **A**, **B**, and **C** by

$$A_{ij} = E_{ij} \qquad (i, j = 1, 2, 3),$$
$$B_{11} = B_{22} = 0, \qquad B_{33} = P_3, \qquad B_{12} = 0, \qquad B_{31} = P_1, \qquad B_{32} = P_2,$$
$$C_{11} = C_{22} = C_{33} = 0, \qquad C_{12} = M_3, \qquad C_{31} = M_2, \qquad C_{32} = M_1, \quad (7.3)$$

and use the results given by Smith and Kiral [1969, p. 11], where arbitrary numbers of the symmetric tensors are considered.

7.1. Magnetic Crystal Class $\underline{\bar{1}} = \{I, \tau C\}$

From Table 4.3(1) and Table 1.B (in Appendix B) we see that

$$a, a', a'', a''', a^{\mathrm{IV}}, a^{\mathrm{V}} = E_{11}, E_{22}, E_{33}, E_{23}, E_{31}, E_{12}$$

are

$$b, b', b'', b''', b^{\mathrm{IV}}, b^{\mathrm{V}} = P_1, P_2, P_3, M_1, M_2, M_3.$$

The typical multilinear elements (TMEs) of the integrity basis are given by

$$1.\ \ a, \qquad 2.\ \ bb'. \qquad (7.4)$$

Elements in E_{ij} only: (6)
Degree 1: $E_{11}, E_{22}, E_{33}, E_{23}, E_{31}, E_{12}$.

Elements in P_i only: (6)
Degree 1: None.
Degree 2: $P_1^2, P_2^2, P_3^2, P_1P_2, P_1P_3, P_2P_3$.

Elements in M_i only: (6)
Degree 1: None.
Degree 2: $M_1^2, M_2^2, M_3^2, M_1M_2, M_1M_3, M_2M_3$.

Elements in E_{ij} and P_i only: None.

Elements in E_{ij} and M_i only: None.

Elements in P_i and M_i only: (9)
Degree 2: $M_1P_1, M_1P_2, M_1P_3, M_2P_1, M_2P_3, M_3P_1, M_3P_2, M_3P_3, M_2P_2$.

Elements in E_{ij}, P_i and M_i: None.

7.2. Magnetic Crystal Class $\underline{m} = \{I, \tau R_3\}$

From Table 4.3(2) and Table B.2 (in Appendix B), we have

$$a, a', a'', a''', a^{\mathrm{IV}}, a^{\mathrm{V}}, a^{\mathrm{VI}}, a^{\mathrm{VII}} = M_1, M_2, E_{11}, E_{22}, E_{33}, E_{12}, P_1, P_2,$$

and

$$b, b', b'', b''' = M_3, E_{23}, E_{31}, P_3.$$

The TMEs of the integrity basis are given by (7.4). The actual elements of the integrity basis are:

Elements in E_{ij} only: (7)
Degree 1: $E_{11}, E_{22}, E_{33}, E_{12}$.
Degree 2: $E_{23}^2, E_{31}^2, E_{23}E_{31}$.

Elements in P_i only: (3)
Degree 1: P_1, P_2.
Degree 2: P_3^2.

Elements in M_i only: (3)
Degree 1: M_1, M_2.
Degree 2: M_3^2.

Elements in E_{ij} and P_i only: (2)
Degree 2: $E_{23}P_3, E_{31}P_3$.

Elements in E_{ij} and M_i only: (2)
Degree 2: $E_{23}M_3, E_{31}M_3$.

Elements in P_i and M_i only: (1)
Degree 2: P_3M_3.

Elements in E_{ij}, P_i, and M_i: None.

7.3. Magnetic Crystal Class $\underline{2} = \{I, \tau D_3\}$

From Table 4.3(3) and Table B.2 (in Appendix B), we have

$$a, a', a'', a''', a^{\mathrm{IV}}, a^{\mathrm{V}}, a^{\mathrm{VI}} = M_1, M_2, E_{11}, E_{22}, E_{33}, E_{12}, P_3,$$

and

$$b, b', b'', b''', b^{\mathrm{IV}} = M_3, E_{23}, E_{31}, P_1, P_2.$$

The TMEs of the integrity basis are given by (7.4). The actual elements are

Elements in E_{ij} only: (7)
Degree 1: $E_{11}, E_{22}, E_{33}, E_{12}$.
Degree 2: $E_{23}^2, E_{31}^2, E_{23}E_{31}$.

Elements in P_i only: (4)
Degree 1: P_3.
Degree 2: P_1^2, P_2^2, P_1P_2.

Elements in M_i only: (3)
Degree 1: M_1, M_2.
Degree 2: M_3^2.

Elements in E_{ij} and P_i only: (4)
Degree 2: $E_{23}P_1, E_{23}P_2, E_{31}P_1, E_{31}P_2$.

Elements in E_{ij} and M_i only: (2)
Degree 2: $E_{23}M_3, E_{31}M_3$.

Elements in P_i and M_i only: (2)
Degree 2: P_1M_3, P_2M_3.

Elements in E_{ij}, P_i and M_i: None.

7.4. Magnetic Crystal Class $\underline{2}/m = \{I, \tau D_3, R_3, \tau C\}$

From Table 4.3(4) and Table B.3 (in Appendix B), we have

$$a, a', a'', a''' = E_{11}, E_{22}, E_{33}, E_{12},$$

$$b, b', b'' = M_1, M_2, P_3,$$

$$c, c', c'' = M_3, P_1, P_2,$$

$$d, d' = E_{23}, E_{31}.$$

The TMEs of the integrity basis are given by

$$\begin{aligned} &1. \quad a, \\ &2. \quad bb', cc', dd', \\ &3. \quad bcd, \end{aligned} \tag{7.5}$$

from which the actual elements of the integrity basis follow.

Elements in E_{ij} only: (7)
Degree 1: $E_{11}, E_{22}, E_{33}, E_{12}$.
Degree 2: $E_{23}^2, E_{31}^2, E_{23}E_{31}$.

Elements in P_i only: (4)
Degree 2: $P_1^2, P_2^2, P_3^2, P_1P_2$.

Elements in M_i only: (4)
Degree 2: $M_1^2, M_2^2, M_3^2, M_1M_2$.

Elements in E_{ij} and P_i only: (4)
Degree 3: $P_1P_3E_{23}, P_2P_3E_{23},$
$P_1P_3E_{31}, P_2P_3E_{31}$.

Elements in E_{ij} and M_i only: (4)
Degree 3: $M_1M_3E_{23}, M_1M_3E_{31},$
$M_2M_3E_{23}, M_2M_3E_{31}$.

Elements in P_i and M_i only: (4)
Degree 2: $M_1P_3, M_2P_3, M_3P_1, M_3P_2$.

Elements in E_{ij}, P_i and M_i: (10)
Degree 3: $P_3M_3E_{23}, P_3M_3E_{31}, P_1M_1E_{23}, P_1M_1E_{31},$
$P_1M_2E_{23}, P_1M_2E_{31}, P_2M_1E_{23}, P_2M_1E_{31},$
$P_2M_2E_{23}, P_2M_2E_{31}$.

7.5. Magnetic Crystal Class $2/\underline{m} = \{I, D_3, \tau R_3, \tau C\}$

From Table 4.3(5) and Table B.3 (in Appendix B), we have

$$a, a', a'', a''' = E_{11}, E_{22}, E_{33}, E_{12},$$
$$b, b' = M_3, P_3,$$
$$c, c', c'', c''' = M_1, M_2, P_1, P_2,$$
$$d, d' = E_{23}, E_{31}.$$

The TMEs of the integrity basis are given by (7.5). The actual elements are

Elements in E_{ij} only: (7)
Degree 1: $E_{11}, E_{22}, E_{33}, E_{12}$.
Degree 2: $E_{23}^2, E_{31}^2, E_{23}E_{31}$.

Elements in P_i only: (4)
Degree 2: $P_1^2, P_2^2, P_3^2, P_1P_2$.

Elements in M_i only: (4)
Degree 2: $M_1^2, M_2^2, M_3^2, M_1M_2$.

Elements in E_{ij} and P_i only: (4)
Degree 3: $P_1P_3E_{32}, P_2P_3E_{32},$
$P_1P_3E_{31}, P_2P_3E_{31}.$

Elements in E_{ij} and M_i only: (4)
Degree 3: $M_1M_3E_{32}, M_2M_3E_{32},$
$M_1M_3E_{31}, M_2M_3E_{31}.$

Elements in P_i and M_i only: (5)
Degree 2: $M_1P_1, M_1P_2, M_2P_1, M_2P_2, M_3P_3.$

Elements in E_{ij}, P_i and M_i: (8)
Degree 3: $P_3M_1E_{32}, P_3M_2E_{32}, P_1M_3E_{32}, P_2M_3E_{32},$
$P_3M_1E_{31}, P_3M_2E_{31}, P_1M_3E_{31}, P_2M_3E_{31}.$

7.6. Magnetic Crystal Class $\underline{2}/\underline{m} = \{I, \tau D_3, \tau R_3, C\}$

From Table 4.3(6) and Table B.3 (in Appendix B), we have

$$a, a', a'', a''', a^{\mathrm{IV}}, a^{\mathrm{V}} = M_1, M_2, E_{11}, E_{22}, E_{33}, E_{12},$$
$$b = P_3,$$
$$c, c' = P_1, P_2,$$
$$d, d', d'' = M_3, E_{32}, E_{31}.$$

The TMEs of the integrity basis are given by (7.5).

Elements in E_{ij} only: (7)
Same as for the class $\underline{2}m$.

Elements in P_i only: (4)
Same as for the class $\underline{2}/m$.

Elements in M_i only: (3)
Degree 1: $M_1, M_2.$
Degree 2: $M_3^2.$

Elements in E_{ij} and P_i only: (4)
Same as for the class $\underline{2}/m$.

Elements in E_{ij} and M_i only: (2)
Degree 2: $M_3E_{32}, M_3E_{31}.$

Elements in P_i and M_i only: (2)
Degree 3: $P_1P_3M_3, P_2P_3M_3.$

Elements in E_{ij}, P_i and M_i: None.

7.7. Magnetic Crystal Class $\underline{2}m\underline{m} = \{I, \tau D_3, \tau R_1, R_2\}$

From Table 4.3(7) and Table B.3 (in Appendix B), we write

$$a, a', a'', a''', a^{\mathrm{IV}} = M_2, E_{11}, E_{22}, E_{33}, P_3,$$

$$b, b' = M_1, E_{12},$$

$$c, c', c'' = M_3, E_{23}, P_2,$$

$$d, d' = E_{31}, P_1.$$

The TMEs of the integrity basis are given by (7.5).

Elements in E_{ij} only: (7)
Degree 1: E_{11}, E_{22}, E_{33}.
Degree 2: $E_{12}^2, E_{23}^2, E_{31}^2$.
Degree 3: $E_{23}E_{31}E_{12}$.

Elements in P_i only: (3)
Degree 1: P_3.
Degree 2: P_1^2, P_2^2.

Elements in M_i only: (3)
Degree 1: M_2,
Degree 2: M_1^2, M_3^2.

Elements in E_{ij} and P_i only: (5)
Degree 2: $E_{23}P_2, E_{31}P_1$,
Degree 3: $E_{12}P_1P_2, E_{23}E_{12}P_1,$
$E_{31}E_{12}P_2$.

Elements in E_{ij} and M_i only: (5)
Degree 2: $E_{23}M_3, E_{12}M_1$.
Degree 3: $E_{31}M_1M_3, E_{31}E_{12}M_3, E_{23}E_{31}M_1$.

Elements in P_i and M_i only: (3)
Degree 2: P_2M_3,
Degree 3: $P_1M_1M_3, P_1P_2M_1$.

Elements in E_{ij}, P_i and M_i: (3)
Degree 3: $E_{31}P_2M_1, E_{23}P_1M_1, E_{12}P_1M_3$.

7.8. Magnetic Crystal Class $2\underline{m}\underline{m} = \{I, D_3, \tau R_1, \tau R_2\}$

From Table 4.3(8) and Table B.3 (in Appendix B), we write

$$a, a', a'', a''', a^{\mathrm{IV}} = M_3, E_{11}, E_{22}, E_{33}, P_3,$$

$$b = E_{12},$$

$$c, c', c'' = M_2, E_{23}, P_2,$$

$$d, d', d'' = M_1, E_{31}, P_1.$$

The TMEs of the integrity basis are given by (7.5).

Elements in E_{ij} only: (7)
Same as for the class $\underline{2}m\underline{m}$.

Elements in P_i only: (3)
Same as for the class $\underline{2}m\underline{m}$.

Elements in M_i only: (3)
Degree 1: M_3,
Degree 2: M_1^2, M_2^2.

Elements in E_{ij} and *P_i only*: (5)
Same as for the class $\underline{2}m\underline{m}$.

Elements in E_{ij} and *M_i only*: (5)
Degree 2: $E_{23}M_2, E_{31}M_1$.
Degree 3: $E_{12}M_1M_2, E_{31}E_{12}M_2, E_{23}E_{12}M_1$.

Elements in P_i and *M_i only*: (2)
Degree 2: P_1M_1, P_2M_2.

Elements in E_{ij}, P_i and *M_i*: (2)
Degree 3: $E_{12}P_1M_2, E_{12}P_2M_1$.

7.9. Magnetic Crystal Class $\underline{22}2 = \{I, \tau D_1, \tau D_2, D_3\}$

From Table 4.3(9) and Table B.3 (in Appendix B), we write

$$a, a', a'' a''' = M_3, E_{11}, E_{22}, E_{33},$$

$$b, b', b'' = M_2, E_{23}, P_1,$$

$$c, c', c'' = M_1, E_{31}, P_2,$$

$$d, d' = E_{12}, P_3.$$

The TMEs are given by (7.5). The actual elements are:

Elements in E_{ij} only: (7)
Degree 1: E_{11}, E_{22}, E_{33}.
Degree 2: $E_{23}^2, E_{31}^2, E_{12}^2$.
Degree 3: $E_{23}E_{31}E_{12}$.

Elements in P_i only: (4)
Degree 2: P_1^2, P_2^2, P_3^2.
Degree 3: $P_1P_2P_3$.

Elements in M_i only: (3)
Degree 1: M_3.
Degree 2: M_1^2, M_2^2.

Elements in E_{ij} and P_i only: (9)
Degree 2: $E_{23}P_1, E_{31}P_2, E_{12}P_3$.
Degree 3: $E_{23}E_{31}P_3, E_{31}E_{12}P_1, E_{23}E_{12}P_2,$
$P_1P_3E_{31}, P_1P_2E_{12}, P_2P_3E_{23}$.

Elements in E_{ij} and M_i only: (5)
Degree 2: $E_{23}M_2, E_{31}M_1,$
Degree 3: $M_1M_2E_{12}, E_{23}E_{12}M_1, E_{31}E_{12}M_2$.

Elements in P_i and M_i only: (5)
Degree 2: $P_1M_2, P_2M_1,$
Degree 3: $P_1P_3M_1, P_3M_1M_2, P_2P_3M_2$.

Elements in E_{ij}, P_i and M_i: (4)
Degree 3: $E_{12}P_1M_1, E_{23}P_3M_1, E_{31}P_3M_2, E_{12}P_2M_2$.

7.10. Magnetic Crystal Class $\underline{m}mm$ (see Table 3.3(4))

From Table 4.3(10) and Table B.4 (in Appendix B), we have

$$a, a', a'' = E_{11}, E_{22}, E_{33},$$
$$b = E_{23},$$
$$c = E_{31},$$
$$d = E_{21},$$
$$A = M_3,$$
$$B, B' = M_2, P_1,$$
$$C, C' = M_1, P_2,$$
$$D = P_3.$$

The TMEs of the integrity basis are given by (B.3). The actual elements are listed below.

Elements in E_{ij} only: (7)
Degree 1: E_{11}, E_{22}, E_{33}.
Degree 2: $E_{23}^2, E_{31}^2, E_{21}^2$.
Degree 3: $E_{23}E_{31}E_{21}$.

Elements in P_i only: (3)
Degree 2: P_1^2, P_2^2, P_3^2.

Elements in M_i only: (3)
Degree 2: M_1^2, M_2^2, M_3^2.

Elements in E_{ij} and P_i only: (6)
Degree 2: None.
Degree 3: $E_{23}P_2P_3$, $E_{31}P_1P_3$, $E_{21}P_1P_2$.
Degree 4: $E_{31}E_{12}P_2P_3$, $E_{23}E_{21}P_1P_3$, $E_{23}E_{31}P_1P_2$.

Elements in E_{ij} and M_i only: (6)
Degree 2: None.
Degree 3: $E_{23}M_2M_3$, $E_{31}M_1M_3$, $M_{21}M_1M_2$.
Degree 4: $E_{23}E_{31}M_1M_2$, $E_{23}E_{21}M_1M_3$, $E_{31}E_{12}M_2M_3$.

Elements in P_i and M_i only: (6)
Degree 2: P_1M_2, P_2M_1.
Degree 3: None.
Degree 4: $P_1P_2P_3M_3$, $M_1M_2M_3P_3$, $P_1P_3M_1M_3$, $P_2P_3M_2M_3$.

Elements in E_{ij}, P_i, and M_i: (14)
Degree 3: $E_{23}P_1M_3$, $E_{23}P_3M_1$, $E_{31}P_2M_3$, $E_{31}P_3M_2$, $E_{21}P_3M_3$, $E_{12}P_1M_1$, $E_{12}P_2M_2$.
Degree 4: $E_{23}E_{31}P_2M_2$, $E_{23}E_{31}P_1M_1$, $E_{23}E_{31}P_3M_3$, $E_{23}E_{21}P_3M_2$, $E_{23}E_{21}P_2M_3$, $E_{31}E_{12}P_3M_1$, $E_{31}E_{12}P_1M_3$.

7.11. Magnetic Crystal Class $\underline{m}\underline{m}\underline{m}$ (see Table 3.3(4))

From Table 4.3(11) and Table B.4, we have

$$
\begin{aligned}
a, a', a'' &= E_{11}, E_{22}, E_{33},\\
b &= E_{23},\\
c &= E_{31},\\
d &= E_{21},\\
B, B' &= M_1, P_1,\\
C, C' &= M_2, P_2,\\
D, D' &= M_3, P_3.
\end{aligned}
$$

The TMEs of the integrity basis are given by (B.3). The actual elements are listed below.

Elements in E_{ij} only: (7)
Same as for the class $\underline{m}mm$.

Elements in P_i only: (3)
Same as for the class $\underline{m}mm$.

Elements in M_i only: (3)
Degree 2: M_1^2, M_2^2, M_3^2.

Elements in E_{ij} and P_i only: (6)
Same as for the class $\underline{m}mm$.

Elements in E_{ij} and M_i only: (6)
Degree 2: None.
Degree 3: $E_{23}M_2M_3$, $E_{31}M_1M_3$, $E_{12}M_1M_2$.
Degree 4: $E_{23}E_{31}M_1M_2$, $E_{23}E_{21}M_1M_3$, $E_{31}E_{21}M_2M_3$.

Elements in P_i and M_i only: (3)
Degree 2: M_1P_1, P_2M_2, P_3M_3.

Elements in E_{ij}, P_i and M_i: (12)
Degree 3: $E_{23}P_3M_2$, $E_{23}P_2M_3$, $E_{31}P_1M_3$, $E_{31}P_3M_1$, $E_{21}P_1M_2$, $E_{21}P_2M_1$.
Degree 4: $E_{23}E_{31}P_1M_2$, $E_{23}E_{31}P_2M_1$, $E_{31}E_{21}P_2M_3$, $E_{23}E_{12}P_1M_3$, $E_{23}E_{12}P_3M_1$, $E_{31}E_{21}P_3M_2$.

7.12. Magnetic Crystal Class $m\underline{m}\underline{m}$ (see Table 3.3(4))

From Table 4.3(12) and Table B.4, we write

$$a, a', a'', a''' = M_3, E_{11}, E_{22}, E_{33},$$
$$b, b' = M_2, E_{23},$$
$$c, c' = M_1, E_{31},$$
$$d = E_{12},$$
$$B = P_1,$$
$$C = P_2,$$
$$D = P_3.$$

The TMEs of the integrity basis are given by (B.3),

Elements in E_{ij} only: (7)
Same as for the class $\underline{m}mm$.

Elements in P_i only: (3)
Same as for the class $\underline{m}mm$.

Elements in M_i only: (3)
Degree 1: M_3.
Degree 2: M_1^2, M_2^2.

Elements in E_{ij} and P_i only: (6)
Same as for the class $\underline{m}mm$.

Elements in E_{ij} and M_i only: (5)
Degree 2: $E_{23}M_2$, $E_{31}M_1$.
Degree 3: $E_{21}M_1M_2$, $E_{31}E_{12}M_2$, $E_{23}M_1E_{21}$.

Elements in P_i and M_i only: (3)
Degree 2: None.
Degree 3: $P_2P_3M_2, P_1P_3M_1$.
Degree 4: $P_1P_2M_1M_2$.

Elements in E_{ij}, P_i, and M_i: (4)
Degree 3: None.
Degree 4: $E_{31}P_1P_2M_2, E_{23}P_1P_2M_1,$
$E_{21}P_1P_3M_2, E_{12}P_2P_3M_1$.

7.13. Magnetic Crystal Class $\underline{\bar{4}} = \{I, D_3, \tau D_1 T_3, \tau D_2 T_3\}$

From Table 4.3(13) and Table B.5, we have

$$\phi, \phi' = E_{11} + E_{22}, E_{33},$$
$$\psi, \psi', \psi'', \psi''' = M_3, E_{11} - E_{22}, E_{12}, P_3,$$
$$a, b, c = M_1 - iM_2, E_{13} + iE_{23}, P_1 - iP_2.$$

The TMEs of the integrity basis are given by (B.4). The actual elements of the integrity basis are listed below.

Elements in E_{ij} only: (12)
Degree 1: $E_{11} + E_{22}, E_{33}$.
Degree 2: $E_{11}E_{22}, E_{12}^2, E_{12}(E_{11} - E_{22}), E_{13}^2 + E_{23}^2$.
Degree 3: $(E_{13}^2 - E_{23}^2)(E_{11} - E_{22}), E_{23}E_{13}(E_{11} - E_{22}),$
$(E_{13}^2 - E_{23}^2)E_{12}, E_{23}E_{13}E_{12}$.
Degree 4: $E_{13}^2E_{23}^2, E_{13}E_{23}(E_{13}^2 - E_{23}^2)$.

Elements in P_i only: (6)
Degree 1: None.
Degree 2: $P_1^2 + P_2^2, P_3^2$.
Degree 3: $(P_1^2 - P_2^2)P_3, P_1P_2P_3$.
Degree 4: $P_1^2P_2^2, P_1P_2(P_1^2 - P_2^2)$.

Elements in M_i only: (6)
Degree 1: None.
Degree 2: $M_1^2 + M_2^2, M_3^2$.
Degree 3: $(M_1^2 - M_2^2)M_3, M_1M_2M_3$.
Degree 4: $M_1^2M_2^2, M_1M_2(M_1^2 - M_2^2)$.

Elements in E_{ij} and P_i only: (22)
Degree 2: $(E_{11} - E_{22})P_3, E_{12}P_3,$
$(E_{31}P_1 - E_{32}P_2), (E_{23}P_1 + E_{13}P_2)$.
Degree 3: $P_3(E_{13}^2 - E_{23}^2), P_3E_{23}E_{13}, E_{12}(P_1E_{13} + P_2E_{23}),$
$(P_1E_{23} - P_2E_{13})E_{12}, (P_1E_{23} - P_2E_{13})P_3, P_3(P_1E_{13} + P_2E_{23}),$
$(E_{11} - E_{22})(P_1E_{23} - P_2E_{13}), (E_{11} - E_{22})(P_1E_{13} + P_2E_{23}),$
$(P_1^2 - P_2^2)(E_{11} - E_{22}), E_{12}(P_1^2 - P_2^2), P_1P_2E_{12},$
$P_1P_2(E_{11} - E_{22})$.

Degree 4: $P_1P_2(P_1E_{23} - P_2E_{13})$, $P_1P_2(P_1E_{13} + P_2E_{23})$,
$E_{23}E_{31}(P_1E_{23} - P_2E_{13})$, $E_{23}E_{31}(P_1E_{13} + P_2E_{23})$,
$P_1^2E_{23}^2 + P_2^2E_{31}^2$, $P_1P_2(E_{23}^2 - E_{31}^2)$.

Elements in E_{ij} and M_i only: (22)
Degree 2: $(E_{11} - E_{22})M_3$, $E_{12}M_3$,
$(M_1E_{13} - M_2E_{23})$, $(M_2E_{13} + M_1E_{23})$.
Degree 3: $(M_1^2 - M_2^2)(E_{11} - E_{22})$, $(M_1^2 - M_2^2)E_{12}$,
$M_1M_2(E_{11} - E_{22})$, $M_1M_2E_{12}$, $(M_1E_{13} + M_2E_{23})M_3$,
$(M_1E_{13} + M_2E_{23})(E_{11} - E_{22})$, $(M_1E_{13} + M_2E_{23})E_{12}$,
$(M_1E_{23} - M_2E_{13})M_3$, $(M_1E_{23} - M_2E_{13})(E_{11} - E_{22})$,
$(M_1E_{23} - M_2E_{13})E_{12}$, $(E_{13}^2 - E_{23}^2)M_3$, $E_{13}E_{23}M_3$.
Degree 4: $M_1M_2(M_1E_{23} - M_2E_{13})$, $M_1M_2(M_1E_{13} + M_2E_{23})$,
$E_{23}E_{31}(M_1E_{23} - M_2E_{13})$, $E_{23}E_{31}(M_1E_{13} + M_2E_{23})$,
$M_1^2E_{23}^2 + M_2^2E_{31}^2$, $M_1M_2(E_{23}^2 - E_{31}^2)$.

Elements in P_i and M_i only: (17)
Degree 2: $P_1M_1 + P_2M_2$, $P_1M_2 - P_2M_1$, P_3M_3.
Degree 3: $P_3M_1M_2$, $(M_1P_1 - M_2P_2)M_3$, $(M_1P_1 - M_2P_2)P_3$,
$(M_1P_2 + M_2P_1)M_3$, $(M_1P_2 + M_2P_1)P_3$, $(M_1^2 - M_2^2)P_3$,
$(P_1^2 - P_2^2)M_3$, $P_1P_2M_3$.
Degree 4: $M_1M_2(M_1P_2 + M_2P_1)$, $P_1P_2(M_1P_2 + M_2P_1)$,
$M_1M_2(M_1P_1 - M_2P_2)$, $P_1P_2(M_1P_1 - M_2P_2)$,
$M_1M_2(P_1 - P_2)$, $M_1M_2P_1P_2$.

Elements in E_{ij}, P_i, and M_i only: (11)
Degree 3: $P_3(M_1E_{13} + M_2E_{23})$, $P_3(M_1E_{23} - M_2E_{13})$,
$(M_1P_1 - M_2P_2)E_{12}$, $(M_1P_1 - M_2P_2)(E_{11} - E_{22})$,
$(M_1P_2 + M_2P_1)E_{12}$, $(M_1P_2 + M_2P_1)(E_{11} - E_{22})$,
$(P_1E_{13} + P_2E_{23})M_3$, $(P_1E_{23} - P_2E_{13})M_3$.
Degree 4: $E_{23}E_{31}(P_1M_2 + P_2M_1)$,
$P_1P_2(M_1E_{23} - M_2E_{13})$, $M_1M_2(P_1E_{23} - P_2E_{13})$.

7.14. Magnetic Crystal Class $\underline{4} = \{I, D_3, \tau R_1T_3, \tau R_2T_3\}$

From Table 4.3(14) and Table B.5, we have

$$\phi, \phi', \phi''' = E_{11} + E_{22}, E_{23}, P_3,$$

$$\psi, \psi', \psi'' = M_3, E_{11} - E_{22}, E_{12},$$

$$a, b, c = M_1 - iM_2, E_{13} + iE_{23}, P_1 + iP_2.$$

The TMEs of the integrity basis are given by (B.4). The actual elements are listed below.

Elements in E_{ij} only: (12)
Same as for the class $\underline{4}$.

Elements in P_i only: (4)
Degree 1: P_3.
Degree 2: $P_1^2 + P_2^2$.
Degree 3: None.
Degree 4: $P_1^2P_2^2, P_1P_2(P_1^2 - P_2^2)$.

Elements in M_i only: (6)
Same as for the class $\underline{4}$.

Elements in E_{ij} and P_i only: (16)
Degree 2: $(E_{31}P_1 + E_{32}P_2), (E_{23}P_1 - E_{13}P_2)$.
Degree 3: $E_{12}(P_1E_{13} - P_2E_{23}), E_{12}(P_1E_{23} + P_2E_{13})$,
$(E_{11} - E_{22})(P_1E_{23} + P_2E_{13}), (E_{11} - E_{22})(P_1E_{13} - P_2E_{23})$,
$(P_1^2 - P_2^2)(E_{11} - E_{22}), E_{12}(P_1^2 - P_2^2), P_1P_2E_{12}, P_1P_2(E_{11} - E_{22})$.
Degree 4: $P_1P_2(P_1E_{23} + P_2E_{13}), P_1P_2(P_1E_{13} - P_2E_{23})$,
$E_{23}E_{31}(P_1E_{23} + P_2E_{13}), E_{23}E_{31}(P_1E_{13} - P_2E_{23})$,
$P_1^2E_{23}^2 + P_2^2E_{31}^2, P_1P_2(E_{23}^2 - E_{31}^2)$.

Elements in E_{ij} and M_i only: (22)
Same as for the class $\underline{4}$.

Elements in P_i and M_i only: (12)
Degree 2: $P_1M_1 - P_2M_2, P_1M_2 + P_2M_1$.
Degree 3: $(M_1P_1 + M_2P_2)M_3, (M_1P_2 - M_2P_1)M_3$,
$(P_1^2 - P_2^2)M_3, P_1P_2M_3$.
Degree 4: $M_1M_2(M_1P_2 - M_2P_1), P_1P_2(M_1P_2 - M_2P_1)$,
$M_1M_2(M_1P_1 + M_2P_2), P_1P_2(M_1P_1 + M_2P_2)$,
$M_1M_2(P_1^2 - P_2^2), M_1M_2P_1P_2$.

Elements in E_{ij}, P_i and M_i only: (9)
Degree 3: $E_{12}(M_1P_1 + M_2P_2), (E_{11} - E_{22})(M_1P_1 + M_2P_2)$,
$E_{12}(M_1P_2 - M_2P_1), (E_{11} - E_{22})(M_1P_2 - M_2P_1)$,
$M_3(P_1E_{13} - P_2E_{23}), M_3(P_1E_{23} + P_2E_{13})$.
Degree 4: $E_{23}E_{31}(P_1M_2 - P_2M_1), P_1P_2(M_1E_{23} - M_2E_{13})$,
$M_1M_2(P_1E_{23} + P_2E_{13})$.

7.15. Magnetic Crystal Class $\underline{4}/\underline{m}$ (see Table 3.3(6))

From Table 4.3(15) and Table B.6, we have

$$\phi, \phi' = E_{11} + E_{22}, E_{33},$$

$$\psi, \psi' = E_{11} - E_{22}, E_{12},$$

$$a = E_{31} + iE_{23},$$

$$\xi = P_3,$$

$$\eta = M_3,$$

$$A, B = M_1 - iM_2, P_1 + iP_2.$$

The TMEs of the integrity basis are given by (B.5). The actual elements are listed below.

Elements in E_{ij} only: (12)
Same as for the class $\bar{4}$.

Elements in P_i only: (4)
Degree 2: P_3^2, $P_1^2 + P_2^2$.
Degree 3: None.
Degree 4: $P_1P_2(P_1^2 - P_2^2)$, $P_1^2P_2^2$.

Elements in M_i only: (4)
Degree 2: M_3^2, $M_1^2 + M^2$.
Degree 3: None.
Degree 4: $M_1M_2(M_1^2 - M_2^2)$, $M_1^2M_2^2$.

Elements in E_{ij} and P_i only: (18)
Degree 2: None.
Degree 3: $(E_{11} - E_{22})(P_1^2 - P_2^2)$, $P_3(P_1E_{31} + P_2E_{32})$, $(E_{11} - E_{22})P_1P_2$, $P_3(P_1E_{32} - P_2E_{31})$, $E_{12}(P_1^2 - P_2^2)$, $E_{12}P_1P_2$.
Degree 4: $P_1P_2E_{31}E_{32}$, $E_{31}E_{32}(P_1^2 - P_2^2)$, $P_1P_2(E_{32}^2 - E_{31}^2)$, $P_1^2E_{23}^2 + P_2^2E_{31}^2$, $P_3E_{12}(P_1E_{31} - P_2E_{32})$, $P_3E_{12}(P_1E_{23} + P_2E_{31})$, $P_3(P_1E_{22}E_{31} + P_2E_{11}E_{23})$, $P_3(P_1E_{22}E_{23} - P_2E_{11}E_{31})$.
Degree 5: $P_3E_{23}E_{31}(P_1E_{31} - P_2E_{32})$, $P_3E_{23}E_{31}(P_1E_{23} + P_2E_{31})$, $P_1P_2P_3(P_1E_{31} - P_2E_{23})$, $P_1P_2P_3(P_1E_{23} + P_2E_{31})$.

Elements in E_{ij} and M_i only: (18)
Degree 2: None.
Degree 3: $(E_{11} - E_{22})(M_1^2 - M_2^2)$, $E_{12}(M_1^2 - M_2^2)$, $E_{12}M_1M_2$, $(E_{11} - E_{22})M_1M_2$, $M_3(E_{31}M_1 + E_{32}M_2)$, $M_3(E_{23}M_1 - E_{31}M_2)$.
Degree 4: $M_1M_2E_{31}E_{32}$, $E_{31}E_{32}(M_1^2 - M_2^2)$, $M_1M_2(E_{32}^2 - E_{31}^2)$, $M_1^2E_{23}^2 + M_2^2E_{31}^2$, $M_3E_{12}(M_1E_{31} - M_2E_{32})$, $M_3E_{12}(M_1E_{23} + M_2E_{31})$, $M_3(M_1E_{22}E_{31} + M_2E_{11}E_{23})$, $M_3(M_1E_{22}E_{23} - M_2E_{11}E_{31})$.
Degree 5: $M_3E_{23}E_{31}(M_1E_{31} - M_2E_{32})$, $M_3E_{23}E_{31}(M_1E_{23} + M_2E_{31})$, $M_1M_2M_3(M_1E_{31} - M_2E_{23})$, $M_1M_2M_3(M_1E_{23} + M_2E_{31})$.

Elements in P_i and M_i only: (14)
Degree 2: $M_1P_1 - M_2P_2$, $P_1M_2 + P_2M_1$.
Degree 3: None.

Degree 4: $P_3M_3(M_1^2 - M_2^2)$, $P_3M_3M_1M_2$,
$P_3M_3(P_1^2 - P_2^2)$, $P_3M_3P_1P_2$,
$M_3P_3(M_1P_1 + M_2P_2)$, $P_3M_3(P_1M_2 - P_2M_1)$,
$P_1P_2M_1M_2$, $P_1^2M_1^2 + P_2^2M_2^2$,
$(M_1^2 - M_2^2)(M_1P_1 + M_2P_2) - 2M_1M_2(M_1P_2 - M_2P_1)$,
$(M_1^2 - M_2^2)(M_1P_2 - M_2P_1) - 2M_1M_2(M_1P_1 + M_2P_2)$,
$(P_1^2 - P_2^2)(M_1P_1 + M_2P_2) - 2P_1P_2(M_1P_2 - M_2P_1)$,
$(P_1^2 - P_2^2)(M_1P_2 - M_2P_1) + 2P_1P_2(M_1P_1 + M_2P_2)$.

Elements in E_{ij}, P_i, and M_i: (32)

Degree 3: $(E_{11} - E_{22})(M_1P_1 + M_2P_2)$, $E_{12}(M_1P_1 + M_2P_2)$,
$(E_{11} - E_{22})(M_1P_2 - M_2P_1)$, $E_{12}(M_1P_2 - M_2P_1)$,
$(E_{11} - E_{22})P_3M_3$, $E_{12}P_3M_3$,
$P_3(E_{31}M_1 - E_{32}M_2)$, $P_3(E_{31}M_2 + E_{32}M_1)$,
$M_3(E_{31}P_1 - E_{32}P_2)$, $M_3(E_{31}P_2 + E_{32}P_1)$.

Degree 4: $(E_{11} - E_{22})P_3(E_{31}M_1 - E_{32}M_2)$, $E_{12}P_3(E_{31}M_1 - E_{32}M_2)$,
$(E_{11} - E_{22})P_3(E_{31}M_2 + E_{32}M_1)$, $E_{12}P_3(E_{31}M_2 + E_{32}M_1)$,
$(E_{11} - E_{22})M_3(E_{31}P_1 + E_{32}P_2)$, $E_{12}M_3(E_{31}P_1 + E_{32}P_2)$,
$(E_{11} - E_{22})M_3(E_{32}P_1 - E_{31}P_2)$, $E_{12}M_3(E_{32}P_1 - E_{31}P_2)$,
$P_3M_3(E_{31}^2 - E_{32}^2)$, $P_3M_3E_{31}E_{32}$,
$(E_{31}^2 - E_{32}^2)(M_1P_1 + M_2P_2)$, $E_{31}E_{32}(M_1P_2 - M_2P_1)$,
$(E_{31}^2 - E_{32}^2)(M_1P_2 - M_2P_1)$, $E_{31}E_{32}(M_1P_1 + M_2P_2)$.

Degree 5: $P_3(E_{31} + iE_{23})(M_1 - iM_2)^3$,
$P_3(E_{31} + iE_{32})(P_1 + iP_2)(M_1 - iM_2)^2$,
$P_3(E_{31} + iE_{32})(M_1 - iM_2)(P_1 + iP_2)^2$,
$P_3(M_1 - iM_2)(E_{31} + iE_{32})^3$,
$M_3(E_{31} - iE_{32})(P_1 + iP_2)(M_1 - iM_2)^2$,
$M_3(E_{31} - iE_{32})(M_1 - iM_2)(P_1 + iP_2)^2$,
$M_3(E_{31} - iE_{32})(P_1 + iP_2)^3$,
$M_3(P_1 - iP_2)(E_{31} + iE_{32})^3$.

Real and imaginary parts of these elements will give the elements of the integrity basis.

7.16. Magnetic Crystal Class $4/\underline{m}$ (see Table 3.3(6))

From Table 4.3(16) and Table B.6, we have:

$$\phi, \phi' = E_{11} + E_{22}, E_{33},$$

$$\psi, \psi' = E_{11} - E_{22}, E_{12},$$

$$a = E_{31} + iE_{23},$$

$$\xi, \xi' = P_3, M_3,$$

$$A, B = P_1 + iP_2, M_1 + iM_2.$$

The TMEs of the integrity basis are given by (B.5). The actual elements of the integrity basis are obtained accordingly. They are

Elements in E_{ij} only: (12)
Same as for the class $\underline{4}$.

Elements in P_i only: (4)
Same as in the class $\underline{4}/\underline{m}$.

Elements in M_i only: (4)
Same as in the class $\underline{4}/\underline{m}$.

Elements in E_{ij} and P_i only: (18)
Same as in the class $\underline{4}/\underline{m}$.

Elements in E_{ij} and M_i only: (18)

Degree 2: None.

Degree 3: $(E_{11} - E_{22})(M_1^2 - M_2^2)$, $M_3(M_1E_{31} + M_2E_{32})$,
$(E_{11} - E_{22})M_1M_2$, $M_3(M_1E_{32} - M_2E_{31})$,
$E_{12}(M_1^2 - M_2^2)$, $E_{12}M_1M_2$.

Degree 4: $M_1M_2E_{31}E_{32}$, $E_{31}E_{32}(M_1^2 - M_2^2)$, $M_1M_2(E_{32}^2 - E_{31}^2)$,
$M_1^2E_{23}^2 + M_2^2E_{31}^2$, $M_3E_{12}(M_1E_{31} - M_2E_{32})$,
$M_3E_{12}(M_1E_{23} + M_2E_{31})$, $M_3(M_1E_{22}E_{31} + M_2E_{11}E_{23})$,
$M_3(M_1E_{22}E_{23} - M_2E_{11}E_{31})$.

Degree 5: $M_3E_{23}E_{31}(M_1E_{31} - M_2E_{32})$,
$M_3E_{23}E_{31}(M_1E_{23} + M_2E_{31})$,
$M_1M_2M_3(M_1E_{23} + M_2E_{31})$, $M_1M_2M_3(M_1E_{31} - M_2E_{23})$.

Elements in P_i and M_i only: (9)

Degree 2: P_3M_3, $P_1M_1 + P_2M_2$, $P_1M_2 - P_2M_1$.

Degree 3: None.

Degree 4: $P_1P_2M_1M_2$, $P_1P_2(M_1^2 - M_2^2)$, $P_1P_2(P_1M_2 + P_2M_1)$,
$P_1P_2(P_1M_1 - P_2M_2)$, $M_1M_2(P_1M_2 + P_2M_1)$,
$M_1M_2(P_1M_1 - P_2M_2)$.

Elements in E_{ij}, P_i and M_i: (26)

Degree 3: $(E_{11} - E_{22})(M_1P_1 - M_2P_2)$, $E_{12}(M_1P_1 - M_2P_2)$,
$(E_{11} - E_{22})(P_1M_2 + P_2M_1)$, $E_{12}(P_1M_2 + P_2M_1)$,
$M_3(E_{31}P_1 + E_{23}P_2)$, $M_3(P_1E_{23} - P_2E_{31})$,
$P_3(E_{31}M_1 + E_{23}M_2)$, $P_3(M_1E_{23} - M_2E_{31})$.

Degree 4: $(E_{11} - E_{22})P_3(E_{31}M_1 - E_{32}M_2)$,
$(E_{11} - E_{22})P_3(E_{31}M_2 + E_{32}M_1)$,
$E_{12}P_3(E_{31}M_1 - E_{32}M_2)$,
$E_{12}P_3(E_{31}M_2 + E_{32}M_1)$,
$(E_{11} - E_{22})M_3(E_{31}P_1 - E_{32}P_2)$,
$(E_{11} - E_{22})M_3(E_{31}P_2 + E_{32}P_1)$,
$E_{12}M_3(E_{31}P_1 - E_{32}P_2)$,

$E_{12}M_3(E_{31}P_2 + E_{32}P_1)$,
$(E_{31}^2 + E_{32}^2)(M_1P_1 - M_2P_2)$, $E_{31}E_{32}(M_1P_1 - M_2P_2)$,
$(E_{31}^2 + E_{32}^2)(M_1P_2 + M_2P_1)$, $E_{31}E_{32}(M_1P_2 + M_2P_1)$.

Degree 5: $M_3(E_{31} + iE_{32})(M_1 + iM_2)^2(P_1 + iP_2)$,
$M_3(E_{31} + iE_{32})(M_1 + iM_2)(P_1 + iP_2)^2$,
$P_3(E_{31} + iE_{32})(M_1 + iM_2)^3$,
$M_3(E_{31} + iE_{32})(P_1 + iP_2)^3$,
$M_3(P_1 + iP_2)(E_{31} + iE_{32})^3$,
$P_3(M_1 + iM_2)(E_{31} + iE_{32})^3$.

Both real and imaginary parts give the elements of the integrity basis of total degree five.

7.17. Magnetic Crystal Class $\underline{4}/m$ (see Table 3.3(6))

From Table 4.3(17) and Table B.6, we have

$$\phi, \phi' = E_{11} + E_{22}, E_{33},$$
$$\psi, \psi', \psi'' = M_3, E_{11} - E_{22}, E_{12},$$
$$a, b = M_1 - iM_2, E_{31} + iE_{23},$$
$$\xi = P_3,$$
$$A = P_1 + iP_2.$$

The TMEs of the integrity basis are given by (B.5). The actual elements are listed below.

Elements in E_{ij} only: (12)
Same as for the class $\underline{4}$.

Elements in P_i only: (4)
Same as for the class $\underline{4}/\underline{m}$.

Elements in M_i only: (6)
Degree 2: M_3^2, $M_1^2 + M_2^2$.
Degree 3: $M_3(M_1^2 - M_2^2)$, $M_3M_1M_2$.
Degree 4: $M_1^2M_2^2$, $M_1M_2(M_1^2 - M_2^2)$.

Elements in E_{ij} and P_i only: (18)
Same as for the class $\underline{4}/\underline{m}$.

Elements in E_{ij} and M_i only: (22)
Degree 2: $M_3(E_{11} - E_{22})$, M_3E_{12}, $(E_{31}M_1 - E_{23}M_2)$,
$(E_{23}M_1 + E_{31}M_2)$.

Degree 3: $(E_{11} - E_{22})(M_1^2 - M_2^2)$, $(E_{11} - E_{22})M_1M_2$,
$E_{12}(M_1^2 - M_2^2)$, $E_{12}M_1M_2$, $M_3(E_{31}^2 - E_{32}^2)$, $M_3E_{31}E_{32}$,
$M_3(M_1E_{31} + M_2E_{32})$, $M_3(M_1E_{23} - M_2E_{31})$,
$(E_{11} - E_{22})(M_1E_{31} + M_2E_{32})$, $(E_{11} - E_{22})(M_1E_{23} - M_2E_{31})$,
$E_{12}(M_1E_{31} + M_2E_{23})$, $E_{12}(M_1E_{23} - M_2E_{31})$.

Degree 4: $M_1E_{31}^3 - M_2E_{23}^3$,
$M_1^2E_{31}^2 + M_2^2E_{23}^2$, $M_1^3E_{31} - M_2^3E_{23}$,
$K(M, E, E, E)$, $K(M, M, E, E)$, $K(M, M, M, E)$,

where

$$K(A, B, C, D) = A_{31}B_{31}C_{31}D_{23} - B_{31}C_{31}D_{31}A_{23} + C_{31}D_{31}A_{31}B_{23}$$
$$+ D_{31}A_{31}B_{31}C_{23} - A_{23}B_{23}C_{23}D_{31} - B_{23}C_{23}D_{23}A_{31}$$
$$- C_{23}D_{23}A_{23}B_{31} - D_{23}A_{23}B_{23}C_{31},$$

and note that $(\cdot)_{31} = M_1$, $(\cdot)_{32} = -M_2$.

Elements in P_i and M_i only: (14)

Degree 2: None.

Degree 3: $P_3(P_1M_1 - P_2M_2)$, $P_3(P_1M_2 + P_2M_1)$,
$M_3(P_1^2 - P_2^2)$, $M_3P_1P_2$.

Degree 4: $P_1P_2M_1M_2$, $M_1M_2(P_1^2 - P_2^2)$, $P_1P_2(M_1^2 - M_2^2)$,
$P_1^2M_2^2 + P_2^2M_1^2$, $P_3M_3(P_1M_1 + P_2M_2)$,
$P_3M_3(P_2M_1 - P_1M_2)$.

Degree 5: $P_3M_1M_2(P_1M_1 + P_2M_2)$,
$P_3M_1M_2(P_2M_1 - P_1M_2)$,
$P_1P_2P_3(P_1M_1 + P_2M_2)$,
$P_1P_2P_3(P_2M_1 - P_1M_2)$.

Elements in E_{ij}, P_i and M_i: (16)

Degree 3: None.

Degree 4: $(P_1^2 - P_2^2)(M_1E_{31} + M_2E_{23})$,
$(P_1 - P_2)(M_1E_{23} - M_2E_{31})$,
$P_1P_2(M_1E_{31} + M_2E_{23})$,
$P_1P_2(M_1E_{23} - M_2E_{31})$,
$P_3(E_{11} - E_{22})(M_1P_1 + M_2P_2)$, $P_3(E_{11} - E_{22})(M_1P_2 - M_2P_1)$,
$P_3E_{12}(M_1P_1 + M_2P_2)$, $P_3E_{12}(M_1P_2 - M_2P_1)$,
$M_3P_3(E_{31}P_1 - E_{23}P_2)$, $M_3P_3(E_{31}P_2 + E_{32}P_1)$,
$(E_{11} - E_{22})P_3(E_{31}P_1 - E_{23}P_2)$, $(E_{11} - E_{22})P_3(E_{31}P_2 + E_{32}P_1)$,
$E_{12}P_3(E_{31}P_1 - E_{23}P_2)$, $E_{12}P_3(E_{31}P_2 + E_{32}P_1)$,

Degree 5: $P_3(P_1 + iP_2)(M_1 - iM_2)^2(E_{31} + iE_{23})$,
$P_3(P_1 + iP_2)(M_1 - iM_2)(E_{31} + iE_{23})^2$.

Both real and imaginary parts of these quantities yield the elements of the integrity basis of total degree five.

7.18. Magnetic Crystal Class $4\underline{m}\underline{m}$ (see Table 3.3(7))

From Table 4.3(18) and Table B.7, we have

$$\phi, \phi', \phi'', \phi''' = M_3, E_{11} + E_{22}, E_{33}, P_3,$$
$$v = E_{12},$$
$$\tau = E_{11} - E_{22},$$
$$\mathbf{a}, \mathbf{b}, \mathbf{c} = \begin{pmatrix} M_1 \\ M_2 \end{pmatrix}, \begin{pmatrix} E_{31} \\ E_{23} \end{pmatrix}, \begin{pmatrix} P_1 \\ P_2 \end{pmatrix}.$$

The TME of the integrity basis are given by (B.6). The actual elements of the integrity basis are listed below.

Elements in E_{ij} only: (8)
Degree 1: $E_{11} + E_{22}, E_{33}$.
Degree 2: $E_{23}^2 + E_{31}^2, E_{12}^2, E_{11}E_{22}$.
Degree 3: $E_{23}E_{21}E_{31}, E_{11}E_{23}^2 + E_{22}E_{31}^2$.
Degree 4: $E_{23}^2E_{31}^2$.

Elements in P_i only: (3)
Degree 1: P_3.
Degree 2: $P_1^2 + P_2^2$
Degree 3: None.
Degree 4: $P_1^2P_2^2$.

Elements in M_i only: (3)
Degree 1: M_3.
Degree 2: $M_1^2 + M_2^2$.
Degree 3: None.
Degree 4: $M_1^2M_2^2$.

Elements in E_{ij} and P_i only: (9)
Degree 2: $P_1E_{31} + P_2E_{23}$.
Degree 3: $P_1P_2E_{12}, P_1^2E_{22} + P_2^2E_{11}, E_{12}(P_1E_{23} + P_2E_{31})$,
$P_1E_{22}E_{31} + P_2E_{11}E_{32}$.
Degree 4: $P_1^2E_{23}^2 + P_2^2E_{31}^2, E_{12}(P_1E_{22}E_{23} + P_2E_{11}E_{31})$,
$E_{23}E_{31}(P_1E_{23} + P_2E_{31}), P_1P_2(P_1E_{23} + P_2E_{31})$.

Elements in E_{ij} and M_i only: (9)
Degree 2: $M_1E_{31} + M_2E_{23}$.
Degree 3: $M_1M_2E_{12}, M_1^2E_{22} + M_2^2E_{11}, E_{12}(M_1E_{23} + M_2E_{31})$,
$M_1E_{22}E_{31} + M_2E_{11}E_{23}$.
Degree 4: $M_1^2E_{23}^2 + M_2^2E_{31}^2, E_{12}(M_1E_{22}E_{23} + M_2E_{11}E_{31})$,
$E_{23}E_{31}(M_1E_{23} + M_2E_{31}), M_1M_2(M_1E_{23} + M_2E_{31})$.

Elements in P_i and M_i only: (4)
Degree 2: $P_1M_1 + P_2M_2$.
Degree 3: None.
Degree 4: $P_1P_2M_1M_2$, $P_1P_2(P_1M_2 + P_2M_1)$,
$M_1M_2(P_1M_2 + P_2M_1)$.

Elements in E_{ij}, P_i and M_i: (6)
Degree 3: $E_{12}(M_1P_2 + M_2P_1)$, $(E_{11} - E_{22})(M_1P_1 - M_2P_2)$.
Degree 4: $M_1^2E_{31}P_1 + M_2^2E_{23}P_2$, $E_{31}^2M_1P_1 + E_{23}^2M_2P_2$,
$P_1^2M_1E_{31} + P_2^2M_2E_{23}$,
$E_{12}(E_{11} - E_{22})(M_1P_2 - M_2P_1)$.

7.19. Magnetic Crystal Class $\underline{4}m\underline{m}$ (see Table 3.3(7))

From Table 4.3(19) and Table B.7, we write

$$\phi, \phi', \phi'' = E_{11} + E_{22}, E_{33}, P_3,$$

$$\nu, \nu' = M_3, E_{12},$$

$$\tau = E_{11} - E_{22},$$

$$\mathbf{a}, \mathbf{b}, \mathbf{c} = \begin{pmatrix} M_2 \\ M_1 \end{pmatrix}, \begin{pmatrix} E_{31} \\ E_{32} \end{pmatrix}, \begin{pmatrix} P_1 \\ P_2 \end{pmatrix}.$$

The TMEs of the integrity basis are given by (B.6). The actual elements are listed below.

Elements in E_{ij} only: (8)
Same as in the class $4\underline{m}\underline{m}$.

Elements in P_i only: (3)
Same as in the class $4\underline{m}\underline{m}$.

Elements in M_i only: (4)
Degree 2: $M_1^2 + M_2^2$, M_3^2.
Degree 3: $M_1M_2M_3$.
Degree 4: $M_1^2M_2^2$.

Elements in E_{ij} and P_i only: (9)
Same as in the class $4\underline{m}\underline{m}$.

Elements in E_{ij} and M_i only: (12)
Degree 2: M_3E_{12}, $M_2E_{31} + M_1E_{32}$.
Degree 3: $E_{12}M_1M_2$, $E_{12}(M_2E_{32} + M_1E_{31})$,
$M_3(M_2E_{23} + M_1E_{31})$, $M_3E_{31}E_{32}$,
$(E_{11} - E_{22})(M_1^2 - M_2^2)$, $(E_{11} - E_{22})(M_2E_{31} - M_1E_{32})$.
Degree 4: $M_3(E_{11} - E_{22})(M_2E_{32} - M_1E_{31})$,
$E_{12}(E_{11} - E_{22})(M_2E_{32} - M_1E_{31})$,

$M_2^3E_{31} + M_1^3E_{23}$, $M_2E_{31}^3 + M_1E_{23}^3$,
$M_2^2E_{31}^2 + M_1^2E_{32}^2$.

Elements in P_i and M_i only: (6)
Degree 2: $P_1M_2 + P_2M_1$.
Degree 3: $M_3(M_2P_2 + M_1P_1)$, $M_3P_1P_2$.
Degree 4: $M_2^3P_1 + M_1^3P_2$, $P_1^3M_2 + P_2^3M_1$, $M_2^2P_1^2 + M_1^2P_2^2$.

Elements in E_{ij}, P_i and M_i: (9)
Degree 3: $E_{12}(M_2P_2 + M_1P_1)$, $M_3(E_{31}P_2 + E_{23}P_1)$,
$(E_{11} - E_{22})(M_2P_1 - M_1P_2)$.
Degree 4: $M_2^2E_{31}P_1 + M_1^2E_{23}P_2$,
$E_{31}^2M_2P_1 + E_{23}^2M_1P_2$,
$P_1^2M_2E_{31} + P_2^2M_1E_{23}$,
$M_3(E_{11} - E_{22})(M_2P_2 - M_1P_1)$,
$E_{12}(E_{11} - E_{22})(M_2P_2 - M_1P_1)$,
$M_3(E_{11} - E_{22})(E_{31}P_2 - E_{23}P_1)$.

7.20. Magnetic Crystal Class 4$\underline{2}\underline{2}$ (see Table 3.3(7))

From Table 4.3(20) and Table B.7, we write

$$\begin{aligned} \phi, \phi', \phi'' &= M_3, E_{11} + E_{22}, E_{33}, \\ \psi &= P_3, \\ \nu &= E_{12}, \\ \tau &= E_{11} - E_{22}, \\ \mathbf{a}, \mathbf{b}, \mathbf{c} &= \begin{pmatrix} M_2 \\ -M_1 \end{pmatrix}, \begin{pmatrix} E_{23} \\ -E_{31} \end{pmatrix}, \begin{pmatrix} P_1 \\ P_2 \end{pmatrix}. \end{aligned} \tag{7.6}$$

The TMEs of the integrity basis are given by (B.6). The actual elements are listed below.

Elements in E_{ij} only: (8)
Same as in the class 4$\underline{m}\underline{m}$.

Elements in P_i only: (4)
Degree 2: $P_1^2 + P_2^2$, P_3^2.
Degree 3: None.
Degree 4: $P_1^2P_2^2$.
Degree 5: $P_1P_2P_3(P_1^2 - P_2^2)$.

Elements in M_i only: (3)
Degree 1: M_3.
Degree 2: $M_1^2 + M_2^2$.
Degree 3: $M_1^2M_2^2$.

Elements in E_{ij} and P_i only: (21)

Degree 2: $P_1 E_{23} - P_2 E_{31}$.

Degree 3: $P_1 P_2 E_{12}, P_1^2 E_{22} + P_2^2 E_{11}, E_{12}(P_1 E_{13} - P_2 E_{23})$,
$P_3 E_{12}(E_{11} - E_{22}), P_1 E_{22} E_{23} - P_2 E_{11} E_{13}, P_3(P_1 E_{31} + P_2 E_{23})$.

Degree 4: $P_1^2 E_{23}^2 + P_2^2 E_{31}^2$,
$E_{12}(P_1 E_{22} E_{31} - P_2 E_{11} E_{23}), P_3 E_{12}(E_{23}^2 - E_{31}^2)$,
$P_3 E_{23} E_{31}(E_{11} - E_{22}), P_1 E_{23}^3 - P_2 E_{31}^3$,
$P_1 P_2 P_3(E_{11} - E_{22}), P_3 E_{12}(P_1 E_{23} + P_2 E_{31})$,
$P_3 E_{12}(P_1^2 - P_2^2), P_1 P_2(P_1 E_{31} - P_2 E_{23})$,
$P_3(P_1 E_{22} E_{31} + P_2 E_{11} E_{23})$.

Degree 5: $P_3 E_{23} E_{31}(E_{23}^2 - E_{31}^2)$,
$P_1 P_2 P_3(E_{23}^2 - E_{31}^2)$,
$P_3 E_{23} E_{31}(P_1 E_{23} + P_2 E_{31})$,
$P_1 P_2 P_3(P_1 E_{23} + P_2 E_{31})$.

Elements in E_{ij} and M_i only: (9)

Degree 2: $M_2 E_{23} + M_1 E_{31}$.

Degree 3: $E_{12} M_1 M_2, (E_{11} - E_{22})(M_1^2 - M_2^2)$,
$(E_{11} - E_{22})(M_2 E_{23} - M_1 E_{31}), E_{12}(M_2 E_{31} + M_1 E_{23})$.

Degree 4: $M_2^3 E_{23} + M_1^3 E_{31}, E_{23}^3 M_2 + E_{31}^3 M_1$,
$E_{23}^2 M_2^2 + E_{31}^2 M_1^2, E_{12}(E_{11} - E_{22})(M_2 E_{31} - M_1 E_{32})$.

Elements in P_i and M_i only: (9)

Degree 2: $M_2 P_1 - M_1 P_2$.

Degree 3: $P_3(M_2 P_2 + M_1 P_1)$.

Degree 4: $M_2^3 P_1 - M_1^3 P_2, P_1^3 M_2 - P_2^3 M_1$,
$M_2^2 P_1^2 + M_1^2 P_2^2$.

Degree 5: $P_3 K(\mathbf{a}, \mathbf{a}, \mathbf{a}, \mathbf{a}), P_3 K(\mathbf{a}, \mathbf{a}, \mathbf{a}, \mathbf{c})$,
$P_3 K(\mathbf{a}, \mathbf{a}, \mathbf{c}, \mathbf{c}), P_3 K(\mathbf{c}, \mathbf{c}, \mathbf{c}, \mathbf{a})$.

Elements in E_{ij}, P_i and M_i: (19)

Degree 3: $P_3(M_2 E_{31} - M_1 E_{23}), E_{12}(M_2 P_2 - M_1 P_1)$,
$(E_{11} - E_{22})(M_2 P_1 + M_1 P_2)$.

Degree 4: $M_2^2 E_{23} P_1 - M_1^2 E_{31} P_1$,
$E_{23}^2 M_2 P_1 - E_{31}^2 M_1 P_2$,
$P_1^2 M_2 E_{31} + P_2^2 M_1 E_{31}$,
$P_3 E_{12}(M_1^2 - M_2^2), P_3 E_{12}(M_2 E_{23} - M_1 E_{31})$,
$P_3 E_{12}(M_2 P_1 + M_1 P_2), P_3(E_{11} - E_{22}) M_1 M_2$,
$P_3(E_{11} - E_{22})(M_2 E_{31} + M_1 E_{23})$,
$P_3(E_{11} - E_{22})(M_2 P_2 - M_1 P_1)$,
$E_{12}(E_{11} - E_{22})(M_2 P_2 + M_1 P_1)$.

Degree 5: $P_3 K(\mathbf{a}, \mathbf{a}, \mathbf{a}, \mathbf{b}), P_3 K(\mathbf{a}, \mathbf{a}, \mathbf{b}, \mathbf{b})$,
$P_3 K(\mathbf{b}, \mathbf{b}, \mathbf{b}, \mathbf{a}), P_3 K(\mathbf{a}, \mathbf{a}, \mathbf{b}, \mathbf{c})$,
$P_3 K(\mathbf{b}, \mathbf{b}, \mathbf{a}, \mathbf{c}), P_3 K(\mathbf{c}, \mathbf{c}, \mathbf{a}, \mathbf{b})$.

where $K(\mathbf{a}, \mathbf{b}, \mathbf{c}, \mathbf{d})$ is defined by

$$
\begin{aligned}
K(\mathbf{a}, \mathbf{b}, \mathbf{c}, \mathbf{d}) = {} & a_1 b_1 c_1 d_2 + a_1 b_1 d_1 c_2 + a_1 c_1 d_1 b_2 + b_1 c_1 d_1 a_2 \\
& - a_2 b_2 c_2 d_1 - a_2 b_2 d_2 c_1 - a_2 c_2 d_2 b_1 - b_2 c_2 d_2 a_1,
\end{aligned} \tag{7.7}
$$

and $\mathbf{a}$, $\mathbf{b}$ and $\mathbf{c}$ are defined by (7.6).

7.21. Magnetic Crystal Class $\underline{4}2\underline{2}$ (see Table 3.3(7))

From Table 4.3(21) and Table B.7, we have

$$
\begin{aligned}
\phi, \phi' &= E_{11} + E_{22}, E_{33}, \\
\psi &= P_3, \\
\nu, \nu' &= M_3, E_{12}, \\
\tau &= E_{11} - E_{22}, \\
\mathbf{a}, \mathbf{b}, \mathbf{c} &= \begin{pmatrix} M_1 \\ -M_2 \end{pmatrix}, \begin{pmatrix} E_{23} \\ -E_{31} \end{pmatrix}, \begin{pmatrix} P_1 \\ P_2 \end{pmatrix}.
\end{aligned} \tag{7.8}
$$

The TMEs of the integrity basis are given by (B.6). The actual elements are listed below.

Elements in E_{ij} only: (8)
Same as in the class $4\underline{m}\underline{m}$.

Elements in P_i only: (4)
Same as in the class $4\underline{2}\underline{2}$.

Elements in M_i only: (4)
Degree 2: M_3^2, $M_1^2 + M_2^2$.
Degree 3: $M_1 M_2 M_3$.
Degree 4: $M_1^2 M_2^2$.

Elements in E_{ij} and P_i only: (21)
Same elements as in the class $4\underline{2}\underline{2}$.

Elements in E_{ij} and M_i only: (12)
Degree 2: $M_1 E_{23} + M_2 E_{31}$.
Degree 3: $E_{12} M_1 M_2$, $E_{12}(M_1 E_{31} + M_2 E_{23})$,
$M_3 E_{23} E_{31}$, $M_3(M_1 E_{31} + M_2 E_{23})$,
$(E_{11} - E_{22})(M_1^2 - M_2^2)$, $(E_{11} - E_{22})(M_1 E_{23} - M_2 E_{32})$.
Degree 4: $M_1^3 E_{23} + M_2^3 E_{31}$, $E_{31}^3 M_1 + E_{32}^3 M_2$,
$M_1^2 E_{23}^2 + M_2^2 E_{31}^2$
$M_3(E_{11} - E_{22})(M_1 E_{31} - M_2 E_{23})$,
$E_{12}(E_{11} - E_{22})(M_1 E_{31} - M_2 E_{23})$.

Elements in P_i and M_i only: (14)
Degree 2: $M_1P_1 - M_2P_2$.
Degree 3: $P_3(M_1P_2 + M_2P_1)$, $M_3(M_1P_2 - M_2P_1)$,
$M_3P_1P_2$.
Degree 4: $M_1^3P_1 - M_2^3P_2$, $P_1^3M_1 - P_2^3M_2$,
$M_1^2P_1^2 + M_2^2P_2^2$, $P_3M_3(M_1^2 - M_2^2)$,
$P_3M_3(M_1P_1 + M_2P_2)$, $P_3M_3(P_1^2 - P_2^2)$.
Degree 5: $P_3K(\mathbf{a}, \mathbf{a}, \mathbf{a}, \mathbf{a})$, $P_3K(\mathbf{a}, \mathbf{a}, \mathbf{a}, \mathbf{c})$,
$P_3K(\mathbf{c}, \mathbf{c}, \mathbf{c}, \mathbf{a})$, $P_3K(\mathbf{a}, \mathbf{a}, \mathbf{c}, \mathbf{c})$.

Elements in E_{ij}, P_i and M_i: (26)
Degree 3: $P_3(M_1E_{31} - M_2E_{23})$, $E_{12}(M_1P_2 - M_2P_1)$,
$M_3(E_{23}P_2 - E_{31}P_1)$, $(E_{11} - E_{22})(M_1P_1 + M_2P_2)$,
$P_3M_3(E_{11} - E_{22})$.
Degree 4: $M_1^2E_{23}P_1 - M_2^2E_{31}P_2$,
$E_{23}^2M_1P_1 - E_{31}^2M_2P_2$,
$P_1^2M_1E_{23} + P_2^2M_2E_{31}$,
$P_3M_3(M_1E_{23} - M_2E_{31})$,
$P_3M_3(E_{23}^2 - E_{31}^2)$, $P_3M_3(P_1E_{23} + P_2E_{31})$,
$P_3E_{12}(M_1^2 - M_2^2)$, $P_3E_{12}(M_1E_{23} - M_2E_{31})$,
$P_3E_{12}(M_1P_1 + M_2P_2)$, $P_3(E_{11} - E_{22})M_1M_2$,
$P_3(E_{11} - E_{22})(M_1E_{31} + M_2E_{23})$,
$P_3(E_{11} - E_{22})(M_1P_2 + M_2P_1)$,
$M_3(E_{11} - E_{22})(M_1P_2 + M_2P_1)$,
$M_3(E_{11} - E_{22})(E_{23}P_2 + E_{31}P_1)$,
$E_{12}(E_{11} - E_{22})(M_1P_2 + M_2P_1)$.
Degree 5: $P_3K(\mathbf{a}, \mathbf{a}, \mathbf{b}, \mathbf{b})$, $P_3K(\mathbf{a}, \mathbf{a}, \mathbf{b}, \mathbf{c})$,
$P_3K(\mathbf{b}, \mathbf{b}, \mathbf{a}, \mathbf{c})$, $P_3K(\mathbf{c}, \mathbf{c}, \mathbf{a}, \mathbf{b})$,
$P_3K(\mathbf{a}, \mathbf{a}, \mathbf{a}, \mathbf{b})$, $P_3K(\mathbf{b}, \mathbf{b}, \mathbf{b}, \mathbf{a})$,

where $K(\mathbf{a}, \mathbf{b}, \mathbf{c}, \mathbf{d})$ is defined by (7.7) and $\mathbf{a}, \mathbf{b}, \mathbf{c}$ are given by (7.8).

7.22. Magnetic Crystal Class $\bar{4}\underline{2}\underline{m}$ (see Table 3.3(7))

From Table 4.3(22) and Table B.7, we have

$$\phi, \phi' = E_{11} + E_{22}, E_{33},$$

$$\nu, \nu', \nu'' = M_3, E_{12}, P_3,$$

$$\tau = E_{11} - E_{22},$$

$$\mathbf{a}, \mathbf{b}, \mathbf{c} = \begin{pmatrix} M_1 \\ M_2 \end{pmatrix}, \begin{pmatrix} E_{23} \\ E_{31} \end{pmatrix}, \begin{pmatrix} P_1 \\ P_2 \end{pmatrix}.$$

The TMEs of the integrity basis are given by (B.6). The actual elements are

Elements in E_{ij} only: (8)
Same elements as in the class $4\underline{m}\underline{m}$.

Elements in P_i only: (4)
Degree 2: $P_1^2 + P_2^2, P_3^2$.
Degree 3: $P_1 P_2 P_3$.
Degree 4: $P_1^2 P_2^2$.

Elements in M_i only: (4)
Degree 2: $M_1^2 + M_2^2, M_3^2$.
Degree 3: $M_1 M_2 M_3$.
Degree 4: $M_1^2 M_2^2$.

Elements in E_{ij} and P_i only: (13)
Degree 2: $P_3 E_{12}, P_1 E_{23} + P_2 E_{31}$.
Degree 3: $P_1 P_2 E_{12}, P_1^2 E_{22} + P_2^2 E_{11}, E_{12}(P_1 E_{31} + P_2 E_{23})$,
$P_3 E_{23} E_{31}, P_1 E_{22} E_{23} + P_2 E_{11} E_{31}, P_3(P_1 E_{31} + P_2 E_{23})$.
Degree 4: $P_1^2 E_{23}^2 + P_2^2 E_{31}^2, P_1 E_{23}^3 + P_2 E_{31}^3$,
$E_{12}(P_1 E_{22} E_{31} + P_2 E_{11} E_{23})$,
$P_3(P_1 E_{22} E_{31} + P_2 E_{11} E_{23}), P_1 P_2(P_1 E_{31} + P_2 E_{23})$.

Elements in E_{ij} and M_i only: (13)
Degree 2: $M_3 E_{12}, M_1 E_{23} + M_2 E_{31}$.
Degree 3: $M_1 M_2 E_{12}, M_1^2 E_{22} + M_2^2 E_{11}, E_{12}(M_1 E_{31} + M_2 E_{23})$
$M_3 E_{23} E_{31}, M_1 E_{22} E_{23} + M_2 E_{11} E_{31}, M_3(M_1 E_{31} + M_2 E_{23})$.
Degree 4: $M_1^2 E_{23}^2 + M_2^2 E_{31}^2, M_1 E_{23}^3 + M_2 E_{31}^3$,
$E_{12}(M_1 E_{22} E_{31} + M_2 E_{11} E_{23})$,
$M_3(M_1 E_{22} E_{31} + M_2 E_{11} E_{23}), M_1 M_2(M_1 E_{31} + M_2 E_{23})$.

Elements in P_i and M_i only: (9)
Degree 2: $P_1 M_1 + P_2 M_2, M_3 P_3$.
Degree 3: $M_3(M_1 P_2 + M_2 P_1), M_3 P_1 P_2, P_3 M_1 M_2$,
$P_3(M_1 P_2 + M_2 P_1)$.
Degree 4: $M_1^3 P_1 + M_2^3 P_2, P_1^3 M_1 + P_2^3 M_2, M_1^2 P_1^2 + M_2^2 P_2^2$.

Elements in E_{ij}, P_i and M_i: (12)
Degree 3: $M_3(E_{23} P_2 + E_{31} P_1), (E_{11} - E_{22})(M_1 P_1 - M_2 P_2)$,
$E_{12}(M_1 P_2 + M_2 P_1), P_3(M_1 E_{31} + M_2 E_{23})$.
Degree 4: $M_1^2 E_{23} P_1 + M_2^2 E_{31} P_2$,
$E_{23}^2 M_1 P_1 + E_{31}^2 M_2 P_2$,
$P_1^2 M_1 E_{23} + P_2^2 M_2 E_{31}$,
$E_{12}(E_{11} - E_{22})(M_1 P_2 - M_2 P_1)$,
$P_3(E_{11} - E_{22})(M_1 E_{31} - M_2 E_{23})$,
$P_3(E_{11} - E_{22})(M_1 P_2 - M_2 P_1)$,
$M_3(E_{11} - E_{22})(M_1 P_2 - M_2 P_1)$,
$M_3(E_{11} - E_{22})(E_{23} P_2 - E_{31} P_1)$.

7.23. Magnetic Crystal Class $\underline{\bar{4}}\underline{2}m$ (see Table 3.3(7))

From Table 4.3(22) and Table B.7, we write

$$\phi, \phi' = E_{11} + E_{22}, E_{33},$$

$$v, v' = E_{12}, P_3,$$

$$\tau, \tau' = M_3, E_{11} - E_{22},$$

$$\mathbf{a}, \mathbf{b}, \mathbf{c} = \begin{pmatrix} M_2 \\ -M_1 \end{pmatrix}, \begin{pmatrix} E_{23} \\ E_{31} \end{pmatrix}, \begin{pmatrix} P_1 \\ P_2 \end{pmatrix}.$$

The TMEs of the integrity basis are given by (B.6). The elements of the integrity basis are

Elements in E_{ij} only: (18)
Same elements as in the class $\bar{4}2\underline{m}$.

Elements in P_i only: (4)
Same elements as in the class $\bar{4}2\underline{m}$.

Elements in M_i only: (4)
Degree 2: $M_3^2, M_1^2 + M_2^2$.
Degree 3: $M_3(M_1^2 - M_2^2)$.
Degree 4: $M_1^2 M_2^2$.

Elements in E_{ij} and P_i only: (13)
Same elements as in the class $\bar{4}2\underline{m}$.

Elements in E_{ij} and M_i only: (13)
Degree 2: $M_2 E_{23} - M_1 E_{31}, M_3(E_{11} - E_{22})$.
Degree 3: $E_{12} M_1 M_2, E_{12}(M_2 E_{31} - M_1 E_{23})$,
$M_3(M_2 E_{23} + M_1 E_{31}), M_3(E_{23}^2 - E_{31}^2)$,
$(E_{11} - E_{22})(M_1^2 - M_1^2), (E_{11} - E_{22})(M_2 E_{23} + M_1 E_{31})$.
Degree 4: $M_2^3 E_{23} - M_1^3 E_{31}, E_{23}^3 M_2 - E_{31}^3 M_1$,
$M_2^2 E_{23}^2 + M_1^2 E_{31}^2, E_{12} M_3(M_2 E_{31} + M_1 E_{23})$,
$E_{12}(E_{11} - E_{22})(M_2 E_{31} + M_1 E_{23})$.

Elements in P_i and M_i only: (9)
Degree 2: $M_2 P_1 - M_1 P_2$.
Degree 3: $M_3(M_2 P_1 + M_1 P_2), M_3(P_1^2 - P_2^2), P_3 M_1 M_2$,
$P_3(M_2 P_2 - M_1 P_1)$.
Degree 4: $M_2^3 P_1 - M_1^3 P_2, P_1^3 M_2 - P_2^3 M_1$
$M_2^2 P_1^2 + M_1^2 P_2^2, P_3 M_3(M_2 P_2 + M_1 P_1)$.

Elements in E_{ij}, P_i and M_i: (14)
Degree 3: $E_{12}(M_2 P_2 - M_1 P_1), M_3(E_{23} P_1 - E_{31} P_2)$,
$P_3(M_2 E_{31} - M_1 E_{23}), (E_{11} - E_{22})(M_2 P_1 + M_1 P_2)$.
Degree 4: $M_2^2 E_{23} P_1 + M_1^2 E_{31} P_2$,
$E_{23}^2 M_2 P_1 - E_{31}^2 M_1 P_2$,

$P_1^2 M_2 E_{23} - P_2^2 M_1 E_{31},$
$E_{12} M_3 (M_1 P_1 + M_2 P_2),$
$E_{12} M_3 (E_{23} P_2 - E_{31} P_1),$
$P_3 (E_{11} - E_{22})(M_2 E_{31} + M_1 E_{23}),$
$P_3 (E_{11} - E_{22})(M_2 P_2 + M_1 P_1),$
$P_3 M_3 (M_2 E_{31} + M_1 E_{23}),$
$P_3 M_3 (E_{23} P_2 - E_{31} P_1),$
$E_{12} (E_{11} - E_{22})(M_2 P_2 + M_1 P_1).$

7.24. Magnetic Crystal Class $\bar{4}\underline{2}\underline{m}$ (see Table 3.3(7))

From Table 4.3(24) and Table B.7, we write

$$\phi, \phi, \phi' = M_3, E_{11} + E_{22}, E_{33},$$

$$v, v' = E_{12}, P_3,$$

$$\tau = E_{11} - E_{22},$$

$$\mathbf{a}, \mathbf{b}, \mathbf{c} = \begin{pmatrix} M_2 \\ M_1 \end{pmatrix}, \begin{pmatrix} E_{23} \\ E_{31} \end{pmatrix}, \begin{pmatrix} P_1 \\ P_2 \end{pmatrix}.$$

The TMEs of the integrity basis are given by (B.6). The actual elements are listed below.

Elements in E_{ij} only: (8)
Same elements as in the class $\bar{4}2\underline{m}$.

Elements in P_i only: (4)
Same elements as in the class $\bar{4}2\underline{m}$.

Elements in M_i only: (3)
Degree 1: M_3.
Degree 2: $M_1^2 + M_2^2$.
Degree 3: None.
Degree 4: $M_1^2 M_2^2$.

Elements in E_{ij} and P_i only: (13)
Same elements as in the class $\bar{4}2\underline{m}$.

Elements in E_{ij} and M_i only: (9)
Degree 2: $M_2 E_{23} + M_1 E_{31}$.
Degree 3: $M_1 M_2 E_{12}, M_2^2 E_{22} + M_1^2 E_{11}, E_{12}(M_2 E_{31} + M_1 E_{23}),$
$M_2 E_{22} E_{23} + M_1 E_{11} E_{31}$.
Degree 4: $M_2^2 E_{23}^2 + M_1^2 E_{31}^2, M_2 E_{23}^3 + M_1 E_{31}^3,$
$E_{12}(M_2 E_{22} E_{31} + M_1 E_{11} E_{23}), M_1 M_2 (M_2 E_{31} + M_1 E_{23})$.

Elements in P_i and M_i only: (6)
Degree 2: $P_1 M_2 + P_2 M_1$.
Degree 3: $P_3 M_1 M_2, P_3(M_2 P_2 + M_1 P_1)$.
Degree 4: $M_2^3 P_1 + M_1^3 P_2, P_1^3 M_2 + P_2^3 M_1, P_1^2 M_2^2 + P_2^2 M_1^2$.

Elements in E_{ij}, P_i and M_i: (9)
Degree 3: $(E_{11} - E_{22})(M_2P_1 - M_1P_2)$, $E_{12}(M_2P_2 + M_1P_1)$, $P_3(M_2E_{31} + M_1E_{23})$.
Degree 4: $M_2^2E_{23}P_1 + M_1^2E_{31}P_2$, $E_{23}^2M_2P_1 + E_{31}^2M_1P_2$, $P_1^2M_2E_{23} + P_2^2M_1E_{31}$, $E_{12}(E_{11} - E_{22})(M_2P_2 - M_1P_1)$, $P_3(E_{11} - E_{22})(M_2E_{31} - M_1E_{23})$, $P_3(E_{11} - E_{22})(M_2P_2 - M_1P_1)$.

7.25. Magnetic Crystal Class $\underline{4}/m\underline{m}\underline{m}$ (see Table 3.3(8))

From Table 4.3(35) and Table B.8, we see that

$$\phi, \phi' = E_{11} + E_{22}, E_{33},$$
$$\nu = E_{12},$$
$$\tau = E_{11} - E_{22},$$
$$a = \begin{pmatrix} E_{23} \\ -E_{31} \end{pmatrix},$$
$$\eta = P_3,$$
$$\theta = M_3,$$
$$\mathbf{A}, \mathbf{B} = \begin{pmatrix} M_1 \\ -M_2 \end{pmatrix}, \begin{pmatrix} P_1 \\ P_2 \end{pmatrix}.$$

TMEs of the integrity basis are given by (B.7). The actual elements are

Elements in E_{ij} only: (Same as in the class $4\underline{m}\underline{m}$) (8)
Degree 1: $E_{11} + E_{22}$, E_{33}.
Degree 2: E_{12}^2, $E_{11}E_{22}$, $E_{32}^2 + E_{31}^2$.
Degree 3: $E_{23}E_{31}E_{12}$, $E_{11}E_{23}^2 + E_{22}E_{31}^2$.
Degree 4: $E_{23}^2E_{31}^2$.

Elements in P_i only: (3)
Degree 2: P_3^2, $P_1^2 + P_2^2$.
Degree 3: None.
Degree 4: $P_1^2P_2^2$.

Elements in M_i only: (3)
Degree 2: M_3^2, $M_1^2 + M_2^2$.
Degree 3: None.
Degree 4: $M_1^2M_2^2$.

Elements in E_{ij} and P_i only: (10)
Degree 2: None.
Degree 3: $P_1P_2E_{12}$, $P_1^2E_{22} + P_2^2E_{11}$, $P_3(P_1E_{31} + P_2E_{23})$.

Degree 4: $P_1^2E_{23}^2 + P_2^2E_{31}^2, P_3E_{12}(P_1E_{23} + P_2E_{31}),$
$P_3(P_1E_{22}E_{31} + P_2E_{11}E_{23}), P_1P_2E_{23}E_{31}.$
Degree 5: $P_3E_{23}E_{31}(P_1E_{23} + P_2E_{31}),$
$P_3E_{12}(P_1E_{22}E_{23} + P_2E_{11}E_{31}),$
$P_1P_2P_3(P_1E_{23} + P_2E_{31}).$

Elements in E_{ij} and M_i only: (10)
Degree 2: None.
Degree 3: $E_{12}M_1M_2, M_1^2E_{22} + M_2^2E_{11}, M_3(M_1E_{31} + M_2E_{23}).$
Degree 4: $E_{23}E_{31}M_1M_2, (E_{23}^2 - E_{31}^2)(M_1^2 - M_2^2),$
$E_{12}M_3(E_{23}M_1 + M_2E_{31}),$
$M_3(E_{11} - E_{22})(M_1E_{31} - M_2E_{23}).$
Degree 5: $E_{12}(E_{11} - E_{22})M_3(E_{23}M_1 - E_{31}M_2),$
$M_3(E_{23}^3M_2 + E_{31}^3M_1),$
$M_3(M_1^3E_{31} + M_2^3E_{23}).$

Elements in P_i and M_i only: (7)
Degree 2: $M_1P_1 - M_2P_2.$
Degree 3: None.
Degree 4: $M_1^3P_1 - M_2^3P_2, P_1^3M_1 - P_2^3M_2, P_1P_2M_1M_2,$
$M_3P_3(M_1^2 - M_2^2), M_3P_3(M_1P_1 + M_2P_2), M_3P_3(P_1^2 - P_2^2).$

Elements in E_{ij}, P_i, and M_i: (27)
Degree 3: $E_{12}(M_1P_2 - M_2P_1), (E_{11} - E_{22})(M_1P_1 + M_2P_2),$
$P_3(E_{23}M_2 - E_{31}M_1), M_3(E_{23}P_2 - E_{31}P_1),$
$P_3M_3(E_{11} - E_{22}).$
Degree 4: $E_{23}E_{31}(M_1P_2 - M_2P_1), M_3P_3(E_{23}^2 - E_{31}^2),$
$(E_{23}^2 - E_{31}^2)(M_1P_1 + M_2P_2),$
$E_{12}(E_{11} - E_{22})(M_1P_2 + M_2P_1),$
$E_{12}M_3(E_{23}P_1 - E_{31}P_2),$
$E_{12}P_3(E_{23}M_1 - E_{31}M_2),$
$P_3(E_{11} - E_{22})(E_{23}M_2 + E_{31}M_1),$
$M_3(E_{11} - E_{22})(E_{23}P_2 + E_{31}P_1).$
Degree 5: $E_{12}(E_{23}^2 - E_{31}^2)(M_1P_2 + M_2P_1),$
$E_{23}E_{31}(E_{11} - E_{22})(M_1P_2 + M_2P_1),$
$E_{12}P_3M_3(M_1P_2 + M_2P_1),$
$E_{12}P_3(E_{11} - E_{22})(E_{23}M_1 + E_{31}M_2),$
$E_{12}M_3(E_{11} - E_{22})(E_{23}P_1 + E_{31}P_2),$
$P_3(E_{23}^3M_2 - E_{31}^3M_1),$
$M_3(E_{23}^3P_2 - E_{31}^3P_1),$
$P_3(M_1^3E_{31} - M_2^3E_{23}), M_3(P_1^3E_{31} - P_2^3E_{23}),$
$P_3(M_1^2P_1E_{31} + M_2^2P_2E_{23}),$
$P_3(P_1^2M_1E_{31} - P_2^2M_2E_{23}),$
$M_3(M_1^2P_1E_{31} - M_2^2P_2E_{23}),$
$M_3(P_1^2M_1E_{31} - P_2^2M_2E_{23}).$
Degree 6: $E_{23}E_{31}(E_{23}^2 - E_{31}^2)(M_1P_2 + M_2P_1).$

7.26. Magnetic Crystal Class $4/\underline{m}mm$ (see Table 3.3(8))

From Table 4.3(26) and Table B.8, we write

$$
\begin{aligned}
\phi, \phi' &= E_{11} + E_{22}, E_{33}, \\
v &= E_{12}, \\
\tau &= E_{11} - E_{22}, \\
\mathbf{a} &= \begin{pmatrix} E_{23} \\ -E_{31} \end{pmatrix}, \\
\xi &= M_3, \\
\eta &= P_3, \\
\mathbf{A}, \mathbf{B} &= \begin{pmatrix} M_2 \\ -M_1 \end{pmatrix}, \begin{pmatrix} P_1 \\ P_2 \end{pmatrix}.
\end{aligned}
$$

The TMEs of the integrity basis are again given by (B.7). The actual elements are

Elements in E_{ij} only: (8)
Same elements as in the class $\underline{4}/\underline{m}m\underline{m}$.

Elements in P_i only: (3)
Same elements as in the class $\underline{4}/\underline{m}m\underline{m}$.

Elements in M_i only: (3)
Degree 2: $M_3^2, M_1^2 + M_2^2$.
Degree 3: None.
Degree 4: $M_1^2 M_2^2$.

Elements in E_{ij} and P_i only: (10)
Same elements as in the class $\underline{4}/\underline{m}m\underline{m}$.

Elements in E_{ij} and M_i only: (9)
Degree 2: None.
Degree 3: $E_{12}M_1M_2, (E_{11} - E_{22})(M_1^2 - M_2^2)$,
$M_3(E_{23}M_2 + E_{31}M_1)$.
Degree 4: $E_{23}E_{31}M_1M_2, (E_{23}^2 - E_{31}^2)(M_1^2 - M_2^2)$,
$M_3(E_{11} - E_{22})(E_{23}M_2 - E_{31}M_1), M_3E_{12}(E_{23}M_1 + E_{31}M_2)$.
Degree 5: $M_3(E_{23}^3 M_2 + E_{31}^3 M_1), M_3(M_2^3 E_{23} + M_1^3 E_{31})$.

Elements in P_i and M_i: (10)
Degree 2: $M_2P_1 - M_1P_2$.
Degree 3: None.
Degree 4: $M_2^3P_1 - M_1^3P_2, P_1^3M_2 - P_2^3M_1, P_1P_2M_1M_2$,
$P_3M_3(M_2P_2 + M_1P_1)$.
Degree 6: $P_3M_3M_1M_2(M_1^2 - M_2^2), P_3M_3P_1P_2(P_1^2 - P_2^2)$,
$M_3P_3[M_2^2(M_2P_2 - 3M_1P_1) + M_1^2(M_1P_1 - 3M_2P_2)]$,

$$M_3P_3[-M_1M_2(P_1^2 - P_2^2) + P_1P_2(M_2^2 - M_1^2)],$$
$$M_3P_3[P_1^2(-M_1P_1 + 3M_2P_2) - P_2^2(M_2P_2 - 3M_1P_1)].$$

Elements in E_{ij}, P_i, *and* M_i: (38)

Degree 3: $E_{12}(M_2P_2 - M_1P_1)$, $(E_{11} - E_{22})(M_2P_1 + M_1P_2)$, $M_3(E_{23}P_1 - E_{31}P_2)$, $P_3(E_{23}M_1 - E_{31}M_2)$.

Degree 4: $E_{23}E_{31}(M_2P_2 - M_1P_1)$, $(E_{23}^2 - E_{31}^2)(M_2P_1 + M_1P_2)$, $E_{12}(E_{11} - E_{22})(M_2P_2 + M_1P_1)$, $E_{12}P_3(E_{23}M_2 - E_{31}M_1)$, $M_3(E_{11} - E_{22})(E_{23}P_1 + E_{31}P_2)$, $E_{12}M_3(E_{23}P_2 - E_{31}P_1)$, $(E_{11} - E_{22})P_3(E_{23}M_1 + E_{31}M_2)$, $E_{12}(E_{11} - E_{22})P_3M_3$.

Degree 5: $E_{23}E_{31}M_3P_3(E_{11} - E_{22})$, $E_{12}(E_{23}^2 - E_{31}^2)(M_2P_2 + M_1P_1)$, $E_{23}E_{31}(E_{11} - E_{22})(M_1P_1 + M_2P_2)$, $M_3(E_{23}^3P_1 - E_{31}^3P_2)$, $P_3(E_{23}^3M_1 - E_{31}^3M_2)$, $P_3(M_2^3E_{31} - M_1^3E_{23})$, $M_3(P_1^3E_{23} - P_2^3E_{31})$, $P_3(M_2^2P_1E_{31} + M_1^2P_2E_{23})$, $P_3(P_1^2M_2E_{31} - P_2^2M_1E_{23})$, $M_3(M_2^2P_1E_{23} - M_1^2P_2E_{31})$, $M_3(P_1^2M_2E_{23} + P_2^2M_1E_{31})$, $E_{12}M_3P_3(M_1^2 - M_2^2)$, $E_{12}M_3P_3(M_2P_1 + M_1P_2)$, $E_{12}M_3P_3(P_1^2 - P_2^2)$, $M_3P_3(E_{11} - E_{22})M_1M_2$, $(E_{11} - E_{22})M_3P_3P_1P_2$, $M_3P_3(E_{11} - E_{22})(M_2P_2 - M_1P_1)$, $E_{12}P_3(E_{11} - E_{22})(E_{23}M_2 + E_{31}M_1)$, $E_{12}P_3(E_{11} - E_{22})(E_{23}P_1 - E_{31}P_2)$, $M_3E_{12}(E_{11} - E_{22})(E_{23}M_1 - E_{31}M_2)$, $M_3E_{12}(E_{11} - E_{22})(E_{23}P_2 + E_{31}P_1)$.

Degree 6: $P_3M_3M_1M_2(E_{23}^2 - E_{31}^2)$, $P_3M_3P_1P_2(E_{23}^2 - E_{31}^2)$, $P_3M_3(E_{23}^2 - E_{31}^2)(M_2P_2 - M_1P_1)$, $P_3M_3E_{23}E_{31}(E_{23}^2 - E_{31}^2)$, $E_{23}E_{31}(E_{23}^2 - E_{31}^2)(M_1P_1 + M_2P_2)$.

7.27. Magnetic Crystal Class $\underline{4}/m\underline{m}\underline{m}$ (see Table 3.3(8))

From Table 4.3(27) and Table B.8, we get

$$\phi, \phi', \phi'' = M_3, E_{11} + E_{22}, E_{33},$$
$$\nu = E_{12},$$
$$\tau = E_{11} - E_{22},$$
$$\mathbf{a}, \mathbf{b} = \begin{pmatrix} M_2 \\ -M_1 \end{pmatrix}, \begin{pmatrix} E_{23} \\ -E_{31} \end{pmatrix},$$

$$\eta = P_3$$

$$\mathbf{A} = \begin{pmatrix} P_1 \\ P_2 \end{pmatrix}.$$

The TMEs of the integrity basis are given by (B.7). The actual elements are

Elements in E_{ij} only: (8)
Same as in the class $\underline{4}/\underline{m}m\underline{m}$.

Elements in P_i only: (3)
Same elements as in the class $\underline{4}/\underline{m}m\underline{m}$.

Elements in M_i only: (3)
Degree 1: M_3.
Degree 2: $M_1^2 + M_2^2$.
Degree 3: None.
Degree 4: $M_1^2 M_2^2$.

Elements in E_{ij} and P_i only: (10)
Same elements as in the class $\underline{4}/\underline{m}m\underline{m}$.

Elements in E_{ij} and M_i only: (9)
Degree 2: $M_2 E_{23} + M_1 E_{31}$.
Degree 3: $E_{12} M_1 M_2$, $E_{12}(M_2 E_{31} + M_1 E_{23})$,
$(E_{11} - E_{22})(M_1^2 - M_2^2)$, $(E_{11} - E_{22})(M_2 E_{23} - M_1 E_{31})$.
Degree 4: $M_2^3 E_{23} + M_1^3 E_{31}$, $E_{23}^3 M_2 + E_{31}^3 M_1$,
$M_1 M_2 E_{23} E_{31}$, $E_{12}(E_{11} - E_{22})(M_2 E_{31} - M_1 E_{23})$.

Elements in P_i and M_i only: (5)
Degree 2: None.
Degree 3: $P_3(P_2 M_2 + M_1 P_1)$.
Degree 4: $M_1 M_2 P_1 P_2$, $P_1^2 M_2^2 + P_2^2 M_1^2$.
Degree 5: $P_3(P_1^3 M_1 + P_2^3 M_2)$,
$P_3(M_2^3 P_2 + M_1^3 P_1)$.

Elements in E_{ij}, P_i and M_i: (10)
Degree 3: None.
Degree 4: $P_1 P_2(M_2 E_{31} + M_1 E_{23})$,
$(P_1^2 - P_2^2)(M_2 E_{23} - M_1 E_{31})$,
$E_{12} P_3(M_2 P_1 + M_1 P_2)$,
$P_3(E_{11} - E_{22})(M_2 P_2 - M_1 P_1)$.
Degree 5: $E_{12}(P_1^2 - P_2^2)(M_2 E_{31} - M_1 E_{23})$,
$P_1 P_2(E_{11} - E_{22})(M_2 E_{31} - M_1 E_{23})$,
$E_{12}(E_{11} - E_{22}) P_3(M_2 P_1 - M_1 P_2)$,
$P_3(M_2^2 E_{23} P_2 + M_1^2 E_{31} P_1)$,
$P_3(E_{23}^2 M_2 P_2 + E_{31}^2 M_1 P_1)$.
Degree 6: $(M_2 E_{31} - M_1 E_{23}) P_1 P_2 (P_1^2 - P_2^2)$.

7.28. Magnetic Crystal Class $4/\underline{m}\underline{m}\underline{m}$ (see Table 3.3(8))

From Table 4.3(28) and Table B.8, we get

$$\phi, \phi' = E_{11} + E_{22}, E_{33},$$
$$\nu = E_{12},$$
$$\tau = E_{11} - E_{22},$$
$$\mathbf{a} = \begin{pmatrix} E_{23} \\ -E_{31} \end{pmatrix},$$
$$\eta, \eta' = M_3, P_3,$$
$$\mathbf{A}, \mathbf{B} = \begin{pmatrix} M_1 \\ M_2 \end{pmatrix}, \begin{pmatrix} P_1 \\ P_2 \end{pmatrix}.$$

The TMEs of the integrity basis are given by (B.7). The actual elements are:

Elements in E_{ij} only: (8)
Same elements as in the class $\underline{4}/\underline{m}m\underline{m}$.

Elements in P_i only: (3)
Same as in the class $\underline{4}/\underline{m}m\underline{m}$.

Elements in M_i only:
Degree 2: $M_3^2, M_1^2 + M_2^2$.
Degree 3: None.
Degree 4: $M_1^2 M_2^2$.

Elements in E_{ij} and P_i only: (10)
Same as in the class $\underline{4}/\underline{m}m\underline{m}$.

Elements in E_{ij} and M_i only: (10)
Degree 2: None.
Degree 3: $M_1 M_2 E_{12}, M_1^2 E_{22} + M_2^2 E_{11}, M_3(M_1 E_{31} + M_2 E_{23})$.
Degree 4: $M_1^2 E_{23}^2 + M_2^2 E_{31}^2, M_3 E_{12}(M_1 E_{23} + M_2 E_{31})$,
$M_3(M_1 E_{22} E_{31} + M_2 E_{11} E_{23}), M_1 M_2 E_{23} E_{31}$.
Degree 5: $M_3 E_{23} E_{31}(M_1 E_{23} + M_2 E_{31})$,
$M_3 E_{12}(M_1 E_{22} E_{23} + M_2 E_{11} E_{31})$,
$M_1 M_2 M_3(M_1 E_{23} + M_2 E_{31})$.

Elements in P_i and M_i only: (5)
Degree 2: $P_3 M_3, P_1 M_1 + P_2 M_2$.
Degree 3: None.
Degree 4: $P_1 P_2 M_1 M_2, P_1 P_2(P_1 M_2 + P_2 M_1)$,
$M_1 M_2(P_1 M_2 + P_2 M_1)$.

Elements in E_{ij}, P_i, and M_i: (24)
Degree 3: $E_{12}(M_1 P_2 + M_2 P_1), (E_{11} - E_{22})(M_1 P_1 - M_2 P_2)$,
$M_3(E_{23} P_2 + E_{31} P_1), P_3(E_{23} M_2 + E_{31} M_1)$.

Degree 4: $E_{23}E_{31}(M_1P_2 + M_2P_1)$,
$(E_{23}^2 - E_{31}^2)(M_1P_1 - M_2P_2)$,
$E_{12}(E_{11} - E_{22})(M_1P_2 - M_2P_1)$,
$E_{12}M_3(E_{23}P_1 + E_{31}P_2)$,
$E_{12}P_3(E_{23}M_1 + E_{31}M_2)$,
$M_3(E_{11} - E_{22})(E_{23}P_2 - E_{31}P_1)$,
$P_3(E_{11} - E_{22})(E_{23}M_2 - E_{31}M_1)$.

Degree 5: $E_{12}(E_{23}^2 - E_{31}^2)(M_1P_2 - M_2P_1)$,
$E_{23}E_{31}(E_{11} - E_{22})(M_1P_2 - M_2P_1)$,
$M_3(E_{23}^3P_2 + E_{31}^3P_1)$,
$P_3(E_{23}^3M_2 + E_{31}^3M_1)$,
$E_{12}M_3(E_{11} - E_{22})(E_{23}P_1 - E_{31}P_2)$,
$E_{12}P_3(E_{11} - E_{22})(E_{23}M_1 - E_{31}M_2)$,
$M_3(M_1^2P_1E_{31} + M_2^2P_2E_{23})$,
$M_3(P_1^2M_1E_{31} + P_2^2M_2E_{23})$,
$M_3(P_1^3E_{31} + P_2^3E_{23})$,
$P_3(M_1^3E_{31} + M_2^3E_{23})$,
$P_3(M_1^2P_1E_{31} + M_2^2P_2E_{23})$,
$P_3(P_1^2M_1E_{31} + P_2^2M_2E_{23})$.

Degree 6: $(M_1P_2 - M_2P_1)E_{23}E_{31}(E_{23}^2 - E_{31}^2)$.

7.29. Magnetic Crystal Class $\underline{4}/mm\underline{m}$ (see Table 3.3(8))

From Table 4.3(29) and Table B.8, we have

$$\phi, \phi' = E_{11} + E_{22}, E_{33},$$

$$\nu, \nu' = M_3, E_{12},$$

$$\tau = E_{11} - E_{12},$$

$$\mathbf{a}, \mathbf{b} = \begin{pmatrix} M_1 \\ -M_2 \end{pmatrix}, \begin{pmatrix} E_{23} \\ -E_{31} \end{pmatrix},$$

$$\eta = P_3,$$

$$\mathbf{A} = \begin{pmatrix} P_1 \\ P_2 \end{pmatrix}.$$

The TMEs of the integrity basis are given by (B.7). The actual elements are

Elements in E_{ij} only: (8)
Same elements as in the class $\underline{4}/mm\underline{m}$.

Elements in P_i only: (3)
Same elements as in the class $\underline{4}/mm\underline{m}$.

Elements in M_i only: (4)
Degree 2: $M_3^2, M_1^2 + M_2^2$.
Degree 3: $M_1M_2M_3$.
Degree 4: $M_1^2M_2^2$.

Elements in E_{ij} and P_i only: (10)
Same elements as in the class $\underline{4}/\underline{m}m\underline{m}$.

Elements in E_{ij} and M_i only: (13)
Degree 2: $M_1E_{23} + M_2E_{31}, M_3E_{12}$.
Degree 3: $M_3(M_1E_{31} + M_2E_{23}), E_{12}M_1M_2, M_3E_{23}E_{31}$,
$E_{12}(M_1E_{31} + M_2E_{23}), (E_{11} - E_{22})(M_1^2 - M_2^2)$,
$(E_{11} - E_{22})(M_1E_{23} - M_2E_{31})$.
Degree 4: $M_1^3E_{23} + M_2^3E_{31}, E_{23}^3M_1 + E_{31}^3M_2, M_1M_2E_{23}E_{31}$,
$M_3(E_{11} - E_{22})(M_1E_{31} - M_2E_{23})$,
$E_{12}(E_{11} - E_{22})(M_1E_{31} - M_2E_{23})$.

Elements in P_i and M_i: (7)
Degree 2: None.
Degree 3: $P_1P_2M_3, P_3(P_1M_2 + P_2M_1)$.
Degree 4: $M_1M_2P_1P_2, P_3M_3(M_1P_1 + M_2P_2), P_1^2M_1^2 + P_2^2M_2^2$.
Degree 5: $P_3(M_1^3P_2 + M_2^3P_1)$,
$P_3(P_1^3M_2 + P_2^3M_1)$.

Elements in E_{ij}, P_i, and M_i: (15)
Degree 3: None.
Degree 4: $P_1P_2(M_1E_{31} + M_2E_{23}), (P_1^2 - P_2^2)(M_1E_{23} - M_2E_{31})$,
$M_3P_3(E_{23}P_1 + E_{31}P_2)$,
$E_{12}P_3(E_{23}P_1 + E_{31}P_2)$,
$P_3(E_{11} - E_{22})(M_1P_2 - M_2P_1), E_{12}P_3(P_1M_1 + P_2M_2)$.
Degree 5: $M_3(P_1^2 - P_2^2)(M_1E_{31} - M_2E_{23})$,
$E_{12}(P_1^2 - P_2^2)(M_1E_{31} - M_2E_{23})$,
$P_1P_2(E_{11} - E_{22})(M_1E_{31} - M_2E_{23})$,
$P_3(M_1^2E_{23}P_2 + M_2^2E_{31}P_1)$,
$P_3(E_{23}^2M_1P_2 + E_{31}^2M_2P_1)$,
$M_3P_3(E_{11} - E_{22})(M_1P_1 - M_2P_2)$,
$M_3P_3(E_{11} - E_{22})(E_{23}P_1 - E_{31}P_2)$,
$E_{12}P_3(E_{11} - E_{22})(M_1P_1 - M_2P_2)$.
Degree 6: $P_1P_2(P_1^2 - P_2^2)(M_1E_{31} - M_2E_{23})$.

7.30. Magnetic Crystal Class $3\underline{m} = \{I_1, S_1, S_2, \tau R_1, \tau R_1 S_1, \tau R_1 S_2\}$

From Table 4.3(30) and Table B.10, we write

$$\phi, \phi', \phi'', \phi''' = M_3, E_{11} + E_{22}, E_{33}, P_3,$$

$$\mathbf{a}, \mathbf{b}, \mathbf{c}, \mathbf{d} = \begin{pmatrix} M_1 \\ M_2 \end{pmatrix}, \begin{pmatrix} E_{13} \\ E_{23} \end{pmatrix}, \begin{pmatrix} 2E_{12} \\ E_{11} - E_{22} \end{pmatrix}, \begin{pmatrix} P_1 \\ P_2 \end{pmatrix}.$$

The TMEs of the integrity basis are given by

$$
\begin{aligned}
&1.\quad \phi.\\
&2.\quad a_1b_1 + a_2b_2. \qquad (7.9)\\
&3.\quad K_2(\mathbf{a}, \mathbf{b}, \mathbf{c}) \equiv a_2b_2c_2 - a_1b_1c_2 - b_1c_1a_2 - c_1a_1b_2.
\end{aligned}
$$

The actual elements of the integrity basis are listed below.

Elements in E_{ij} only: (9)

Degree 1: $E_{11} + E_{22} \equiv R_1,\ E_{33} \equiv R_0.$ (7.9a)

Degree 2: $E_{13}^2 + E_{23}^2 \equiv R_3,$
$E_{11}E_{22} - E_{12}^2 \equiv R_2,$
$(E_{22} - E_{11})E_{23} - 2E_{13}E_{12} \equiv R_5.$ (7.9b)

Degree 3: $E_{23}(E_{23}^2 - 3E_{31}^2) \equiv R_9,$
$E_{11}E_{23}^2 + E_{22}E_{13}^2 - E_{13}E_{23}E_{12} \equiv R_{10}$
$E_{11}(E_{11}^2 + 6E_{11}E_{22} - 12E_{12}^2 + 9E_{22}^2) \equiv R_6,$
$E_{23}(E_{11}^2 + 2E_{11}E_{22} - 3E_{22}^2 + 4E_{12}) + 8E_{11}E_{12}E_{13} \equiv R_{12}.$ (7.9c)

Elements in P_i only: (3)

Degree 1: $P_3.$

Degree 2: $P_1^2 + P_2^2 \equiv S_1.$ (7.10)

Degree 3: $P_2(P_2^2 - 3P_1^2) \equiv S_3.$ (7.11)

Elements in M_i only: (3)

Degree 1: $M_3.$

Degree 2: $M_1^2 + M_2^2 \equiv Q_1.$ (7.12)

Degree 3: $M_2(M_2^2 - 3M_1^2) \equiv Q_3.$ (7.13)

Elements in E_{ij} and P_i only: (7)

Degree 2: $E_{31}P_1 + E_{23}P_2 \equiv N_3,$ (7.14)
$2E_{12}P_1 + P_2(E_{11} - E_{22}) \equiv N_2.$ (7.15)

Degree 3: $(P_1^2 - P_2^2)E_{23} + 2P_1P_2E_{31} \equiv N_{14},$ (7.16)
$P_2(E_{23}^2 - E_{31}^2) - 2P_1E_{23}E_{31} \equiv N_8,$ (7.17)
$E_{11}P_2^2 + E_{22}P_1^2 - 2P_1P_2E_{12} \equiv N_{11},$ (7.18)
$P_2(E_{11}^2 - 3E_{22}^2 + 2E_{11}E_{22} + 4E_{12}^2) + 8P_1E_{11}E_{12} \equiv N_6.$ (7.19)
$(P_1E_{23} + P_2E_{31})E_{12} - P_1E_{22}E_{31} - P_2E_{11}E_{23} \equiv N_9.$ (7.20)

Elements in E_{ij} and M_i only: (7)

Degree 2: $E_{31}M_1 + E_{23}M_2 \equiv T_3,$ (7.21)
$2E_{12}M_1 + M_2(E_{11} - E_{22}) \equiv T_2.$ (7.22)

Degree 3: $(M_1^2 - M_2^2)E_{23} + 2M_1M_2E_{31} \equiv T_{14},$ (7.23)
$M_2(E_{23}^2 - E_{31}^2) - 2M_1E_{23}E_{31} \equiv T_8,$ (7.24)
$E_{11}M_2^2 + E_{22}M_1^2 - 2M_1M_2E_{12} \equiv T_{11}$ (7.25)
$M_2(E_{11}^2 - 3E_{22}^2 + 2E_{11}E_{22} + 4E_{12}^2) + 8M_1E_{11}E_{12} \equiv T_6,$ (7.26)
$(M_1E_{23} + M_2E_{31})E_{12} - M_1E_{22}E_{31} - M_2E_{11}E_{23} \equiv T_9.$ (7.27)

Elements in P_i and M_i only: (3)

Degree 2: $M_1 P_1 + M_2 P_2$.

Degree 3: $M_2(P_2^2 - P_1^2) - 2P_1 P_2 M_1$,
$P_2(M_2^2 - M_1^2) - 2M_1 M_2 P_1$.

Elements in E_{ij}, P_i, and M_i: (2)

Degree 3: $M_2 E_{23} P_2 - M_1 E_{13} P_2 - E_{13} P_1 M_2 - M_1 P_1 E_{23}$,
$M_2 P_2(E_{11} - E_{22}) - 2M_1 E_{12} P_2 - 2E_{12} P_1 M_2 - M_1 P_1(E_{11} - E_{22})$.

7.31. Magnetic Crystal Class $3\underline{2} = \{I, S_1, S_2, \tau D_1, \tau D_1 S_1, \tau D_1 S_2\}$

From Table 4.3(31) and Table B.10, we have

$$\phi, \phi', \phi'' = M_3, E_{11} + E_{22}, E_{33},$$

$$\psi = P_3,$$

$$\mathbf{a}, \mathbf{b}, \mathbf{c}, \mathbf{d} = \begin{pmatrix} M_1 \\ M_2 \end{pmatrix}, \begin{pmatrix} E_{13} \\ E_{23} \end{pmatrix}, \begin{pmatrix} 2E_{12} \\ E_{11} - E_{22} \end{pmatrix}, \begin{pmatrix} P_2 \\ -P_1 \end{pmatrix}.$$

The TMEs of the integrity basis are given by B.10. The actual elements are

Elements in E_{ij} only: (Same as in the class $3\underline{m}$) (9)

Degree 1: R_0, R_1.

Degree 2: R_2, R_3, R_5.

Degree 3: R_6, R_9, R_{10}, R_{12}.

Elements in P_i only: (4)

Degree 1: None.

Degree 2: P_3^2, S_1.

Degree 3: $P_1(P_1^2 - 3P_2^2) \equiv S_2$. (7.28)

Degree 4: $P_3 S_3$.

Elements in M_i only: (Same as in the class $3\underline{m}$) (3)

Degree 1: M_3.

Degree 2: Q_1.

Degree 3: Q_3.

Elements in E_{ij} and P_i only: (19)

Degree 2: $P_1(E_{11} - E_{22}) - 2P_2 E_{12} \equiv N_1$, (7.29)
$P_1 E_{23} - P_2 E_{31} \equiv N_4$. (7.30)

Degree 3: N_{14}, N_{11},
$P_1(E_{31}^2 - E_{23}^2) - 2P_2 E_{23} E_{31} \equiv N_7$, (7.31)
$P_1(E_{11}^2 - 3E_{22}^2 + 2E_{11}E_{22} + 4E_{12}^2) - 8P_2 E_{11} E_{12} \equiv N_5$, (7.32)
$(P_1 E_{31} - P_2 E_{23})E_{12} + P_1 E_{22} E_{23} - P_2 E_{11} E_{31} \equiv N_{10}$,
$P_3 R_4, P_3 N_2, P_3 N_3$, (7.33)

where

$$R_4 \equiv (E_{11} - E_{22})E_{31} - 2E_{12}E_{23}. \tag{7.34}$$

Degree 4: $P_3R_7, P_3R_8, P_3R_{11}, P_3R_{13},$
$P_3N_6, P_3N_8, P_3N_9, P_3N_{12}, P_3N_{13},$

where, R_7, R_8, R_{11}, R_{13}, and N_{12}, N_{13} are defined by

$$\begin{aligned} R_7 &\equiv 3E_{12}(E_{11} - E_{22})^2 - 4E_{12}^3, \\ R_8 &\equiv E_{31}(E_{31}^2 - E_{23}^2), \\ R_{11} &\equiv E_{31}(E_{11}^2 + 2E_{11}E_{22} - 3E_{22}^2 + 4E_{12}^2) - 8E_{11}E_{12}E_{23}, \\ R_{13} &\equiv (E_{11} - E_{22})E_{23}E_{31} + E_{12}(E_{23}^2 - E_{31}^2), \end{aligned} \tag{7.35}$$

and

$$\begin{aligned} N_{12} &\equiv P_1P_2(E_{11} - E_{22}) + E_{12}(P_2^2 - P_1^2), \\ N_{13} &\equiv (P_1^2 - P_2^2)E_{31} - 2P_1P_2E_{23}. \end{aligned} \tag{7.36}$$

Elements in E_{ij} and M_i only: (Same as in the class $3\underline{m}$) (7)

Degree 2: $T_2, T_3.$

Degree 3: $T_6, T_8, T_9, T_{11}, T_{14}.$

Elements in P_i and M_i only: (7)

Degree 2: $M_1P_2 - M_2P_1.$

Degree 3: $M_2(P_1^2 - P_2^2) + 2P_1P_2M_1,$
$P_1(M_1^2 - M_2^2) - 2M_1M_2P_2,$
$P_3(P_1M_1 + P_2M_2).$

Degree 4: $P_3Q_2,$
$P_3(M_1P_2^2 - M_1P_1^2 + 2P_1P_2M_2),$
$P_3(P_2M_1^2 - P_2M_2^2 + 2M_1M_2P_1),$

where

$$Q_2 \equiv M_1(M_1^2 - 3M_2^2). \tag{7.37}$$

Elements in E_{ij}, P_i, and *M_i*: (11)

Degree 3: $P_3(M_1E_{23} - M_2E_{13}),$
$P_3(M_1E_{11} - M_1E_{22} - 2M_2E_{12}),$
$K_2\left[\begin{pmatrix} M_1 \\ M_2 \end{pmatrix}, \begin{pmatrix} E_{13} \\ E_{23} \end{pmatrix}, \begin{pmatrix} P_2 \\ -P_1 \end{pmatrix}\right],$
$K_2\left[\begin{pmatrix} M_1 \\ M_2 \end{pmatrix}, \begin{pmatrix} 2E_{12} \\ E_{11} - E_{22} \end{pmatrix}, \begin{pmatrix} P_2 \\ -P_1 \end{pmatrix}\right],$

where $K_2(\mathbf{x}, \mathbf{y}, \mathbf{z})$ has been defined by (7.9).

Degree 4: $P_3J_1\left[\begin{pmatrix} M_1 \\ M_2 \end{pmatrix}, \begin{pmatrix} E_{13} \\ E_{23} \end{pmatrix}\right],$
$P_3J_1\left[\begin{pmatrix} E_{13} \\ E_{23} \end{pmatrix}, \begin{pmatrix} M_1 \\ M_2 \end{pmatrix}\right],$
$P_3J_1\left[\begin{pmatrix} M_1 \\ M_2 \end{pmatrix}, \begin{pmatrix} 2E_{12} \\ E_{11} - E_{22} \end{pmatrix}\right],$
$P_3J_1\left[\begin{pmatrix} 2E_{12} \\ E_{11} - E_{22} \end{pmatrix}, \begin{pmatrix} M_1 \\ M_2 \end{pmatrix}\right],$

and

$$P_3K_1\left[\begin{pmatrix} M_1 \\ M_2 \end{pmatrix}, \begin{pmatrix} E_{13} \\ E_{23} \end{pmatrix}, \begin{pmatrix} 2E_{12} \\ E_{11} - E_{22} \end{pmatrix}\right],$$
$$PK_1\left[\begin{pmatrix} M_1 \\ M_2 \end{pmatrix}, \begin{pmatrix} E_{13} \\ E_{23} \end{pmatrix}, \begin{pmatrix} P_2 \\ -P_1 \end{pmatrix}\right],$$
$$PK_1\left[\begin{pmatrix} M_1 \\ M_2 \end{pmatrix}, \begin{pmatrix} 2E_{12} \\ E_{11} - E_{22} \end{pmatrix}, \begin{pmatrix} P_2 \\ -P_1 \end{pmatrix}\right],$$

where $J_1(\mathbf{x}, \mathbf{y})$ and $K_1(\mathbf{x}, \mathbf{y}, \mathbf{z})$ are defined by

$$J_1\left[\begin{pmatrix} x_1 \\ x_2 \end{pmatrix}, \begin{pmatrix} y_1 \\ y_2 \end{pmatrix}\right] = x_1y_1^2 - x_1y_2^2 - 2y_1y_2x_2, \tag{7.38}$$

$$K_1\left[\begin{pmatrix} x_1 \\ x_2 \end{pmatrix}, \begin{pmatrix} y_1 \\ y_2 \end{pmatrix}, \begin{pmatrix} z_1 \\ z_2 \end{pmatrix}\right] = x_1y_1z_1 - x_1y_2z_2 - y_1x_2z_2 - z_1x_2y_2. \tag{7.39}$$

7.32. Magnetic Crystal Class $\underline{\bar{3}} = \{I, S_1, S_2, \tau C, \tau CS_1, \tau CS_2\}$

From Table 4.3(32) and Table B.11, we have

$$\phi, \phi' = E_{11} + E_{22}, E_{33},$$
$$a, b = E_{13} - iE_{23}, E_{11} - E_{22} + 2iE_{12},$$
$$\xi, \xi' = M_3, P_3,$$
$$A, B = M_1 - iM_2, P_1 - iP_2.$$

The TMEs of the integrity basis are listed in (B.11). The actual elements are listed below

Elements in E_{ij} only: (14)
Degree 1: R_0, R_1.
Degree 2: R_2, R_3, R_4, R_5.
Degree 3: $R_6, R_7, R_8, R_9, R_{10}, R_{11}, R_{12}, R_{13}$.

Elements in P_i only: (6)
Degree 1: None.
Degree 2: P_3^2, S_1.
Degree 3: None.
Degree 4: P_3S_2, P_3S_3.
Degree 5: None.
Degree 6: S_2S_2, S_2S_3.

Elements in M_i only: (6)
Degree 1: None.
Degree 2: M_3^2, Q_1.
Degree 3: None.
Degree 4: M_3Q_2, M_3Q_3.
Degree 5: None.
Degree 6: Q_2Q_2, Q_2Q_3.

Elements in E_{ij} and P_i only: (24)
Degree 2: None.
Degree 3: $P_3N_1, P_3N_2, P_3N_3, P_3N_4, N_{11}, N_{12}, N_{13}, N_{14}$.
Degree 4: $P_3N_5, P_3N_6, P_3N_7, P_3N_8, P_3N_9, P_3N_{10}, N_1N_1, N_1N_2, N_1N_3, N_2N_3, N_3N_3, N_3N_4$.
Degree 5: $S_2N_1, S_2N_2, S_2N_3, S_2N_4$.

Elements in E_{ij} and M_i only: (24)
Degree 2: None.
Degree 3: $M_3T_1, M_3T_2, M_3T_3, M_3T_4, T_{11}, T_{12}, T_{13}, T_{14}$.
Degree 4: $M_3T_5, M_3T_6, M_3T_7, M_3T_8, M_3T_9, M_3T_{10}, T_1T_1, T_1T_2, T_1T_3, T_2T_3, T_3T_3, T_3T_4$.
Degree 5: $Q_2T_1, Q_2T_2, Q_2T_3, Q_2T_4$,

where

$$\begin{aligned} T_1 &\equiv M_1(E_{11} - E_{22}) - 2M_2E_{12}, \\ T_4 &\equiv M_1E_{23} - M_2E_{31}, \\ T_5 &\equiv M_1(E_{11}^2 - 3E_{22}^2 + 2E_{11}E_{22} + 4E_{12}^2) - 8M_2E_{11}E_{12}, \\ T_7 &\equiv M_1(E_{31}^2 - E_{23}^2) - 2M_2E_{23}E_{31}, \\ T_{10} &\equiv (M_1E_{31} - M_2E_{23})E_{12} + M_1E_{22}E_{23} - M_2E_{11}E_{31}, \\ T_{12} &\equiv M_1M_2(E_{11} - E_{22}) + E_{12}(M_2^2 - M_1^2), \\ T_{13} &\equiv (M_1^2 - M_2^2)E_{31} - 2P_1P_2E_{23}. \end{aligned} \tag{7.40}$$

Elements in P_i and M_i only: (25)
Degree 2: $P_3M_3, P_1M_1 + P_2M_2, P_1M_2 - P_2M_1$.
Degree 3: None.
Degree 4: $P_3Q_2, P_3Q_3, M_3S_2, M_3S_3$,

$$P_3J_1\left[\begin{pmatrix}M_1\\M_2\end{pmatrix},\begin{pmatrix}P_1\\P_2\end{pmatrix}\right], \qquad P_3J_2\left[\begin{pmatrix}M_1\\M_2\end{pmatrix},\begin{pmatrix}P_1\\P_2\end{pmatrix}\right],$$

$$P_3J_1\left[\begin{pmatrix}P_1\\P_2\end{pmatrix},\begin{pmatrix}M_1\\M_2\end{pmatrix}\right], \qquad P_3J_2\left[\begin{pmatrix}P_1\\P_2\end{pmatrix},\begin{pmatrix}M_1\\M_2\end{pmatrix}\right],$$

and

$$M_3J_1\left[\begin{pmatrix}M_1\\M_2\end{pmatrix},\begin{pmatrix}P_1\\P_2\end{pmatrix}\right], \qquad M_3J_2\left[\begin{pmatrix}M_1\\M_2\end{pmatrix},\begin{pmatrix}P_1\\P_2\end{pmatrix}\right],$$

$$M_3J_1\left[\begin{pmatrix}P_1\\P_2\end{pmatrix},\begin{pmatrix}M_1\\M_2\end{pmatrix}\right], \qquad M_3J_2\left[\begin{pmatrix}P_1\\P_2\end{pmatrix},\begin{pmatrix}M_1\\M_2\end{pmatrix}\right],$$

where $J_2(\mathbf{x}, \mathbf{y})$ is defined by

$$J_2\left[\begin{pmatrix}x_1\\x_2\end{pmatrix},\begin{pmatrix}y_1\\y_2\end{pmatrix}\right] \equiv x_2y_2^2 - x_2y_1^2 - 2y_1y_2x_1. \tag{7.41}$$

Degree 5: $Q_2 J_1(\mathbf{P}, \mathbf{M})$, $Q_2 J_2(\mathbf{P}, \mathbf{M})$, $Q_2 J_1(\mathbf{M}, \mathbf{P})$,
$Q_2 J_2(\mathbf{M}, \mathbf{P})$, $Q_2 S_2$, $Q_3 S_2$,
$S_2 J_1(\mathbf{P}, \mathbf{M})$, $S_2 J_2(\mathbf{P}, \mathbf{M})$,
$S_2 J_1(\mathbf{M}, \mathbf{P})$, $S_2 J_2(\mathbf{M}, \mathbf{P})$,

where $J_1(\mathbf{x}, \mathbf{y})$ and $J_2(\mathbf{x}, \mathbf{y})$ are defined by (7.38) and (7.41).

Elements in E_{ij}, P_i, and M_i: (42)

Degree 3: $(P_1 M_1 - P_2 M_2)E_{13} - (P_1 M_2 + P_2 M_1)E_{23}$,
$(P_1 M_1 - P_2 M_2)E_{23} + (P_1 M_2 + P_2 M_1)E_{13}$,
$(P_1 M_1 - P_2 M_2)(E_{11} - E_{22}) + 2(P_1 M_2 + P_2 M_1)E_{12}$,
$2(P_1 M_1 - P_2 M_2)E_{12} - (P_1 M_2 + P_2 M_1)(E_{11} - E_{22})$,
$M_3(E_{13} P_1 + E_{23} P_2)$, $M_3(E_{13} P_2 - E_{23} P_1)$,
$P_3(E_{13} M_1 + E_{23} M_2)$, $P_3(E_{13} M_2 - E_{23} M_1)$,
$M_3(E_{11} - E_{22})P_1 - 2E_{12} P_2 M_3$,
$M_3(E_{11} - E_{22})P_2 + 2E_{12} P_1 M_3$,
$P_3(E_{11} - E_{22})M_1 - 2E_{12} M_2 P_3$,
$P_3(E_{11} - E_{22})M_2 + 2E_{12} M_1 P_3$.

Degree 4: $M_3 N_7$, $M_3 N_8$, $P_3 T_7$, $P_3 T_8$,
$M_3 N_9$, $M_3 N_{10}$, $P_3 T_9$, $P_3 T_{10}$,
$M_3 N_5$, $M_3 N_6$, $P_3 T_5$, $P_3 T_6$,

and

$T_1 N_1$, $T_1 N_2$, $T_1 N_3$, $T_2 N_3$, $T_3 N_3$, $T_3 N_4$.

Degree 5: $N_3 Q_2 - N_4 Q_3$, $N_3 Q_3 + N_4 Q_2$,
$T_3 S_2 - T_4 S_3$, $T_3 S_3 + T_4 S_2$,
$N_1 Q_2 + N_2 Q_3$, $N_1 Q_3 - N_2 Q_2$,
$T_1 S_2 + T_2 S_3$, $T_1 S_3 - T_2 S_2$,
$N_1 J_1(\mathbf{P}, \mathbf{M}) + N_2 J_2(\mathbf{P}, \mathbf{M})$,
$N_1 J_2(\mathbf{P}, \mathbf{M}) - N_2 J_1(\mathbf{P}, \mathbf{M})$,
$T_3 J_1(\mathbf{M}, \mathbf{P}) - T_4 J_2(\mathbf{M}, \mathbf{P})$,
$T_3 J_2(\mathbf{M}, \mathbf{P}) + T_4 J_1(\mathbf{M}, \mathbf{P})$.

7.33. Magnetic Crystal Class $\underline{\bar{6}} = \{I, S_1, S_2, \tau R_3, \tau R_3 S_1, \tau R_3 S_2\}$

From Table 4.3(33) and Table B.11, we have

$$\phi, \phi' = E_{11} + E_{22}, E_{33},$$

$$a, b, c = M_1 - iM_2, E_{11} - E_{22} + 2iE_{12}, P_1 - iP_2,$$

$$\xi, \xi' = M_3, P_3,$$

$$A = E_{13} - iE_{23}.$$

The TMEs of the integrity basis are given by (B.11). The actual elements are listed below.

Elements in E_{ij} only: (14)

Degree 1: R_0, R_1.

Degree 2: R_2, R_3.

Degree 3: R_6, R_7, R_{10}, R_{13}.
Degree 4: R_4R_4, R_4R_5.
Degree 5: R_4R_8, R_5R_8.
Degree 6: R_8R_8, R_8R_9.

Elements in P_i only: (4)
Degree 1: None.
Degree 2: P_3^2, S_1.
Degree 3: S_2, S_3.

Elements in M_i only: (4)
Degree 1: None.
Degree 2: M_3, Q_1.
Degree 3: Q_2, Q_3.

Elements in E_{ij} and P_i only: (26)
Degree 2: N_1, N_2.
Degree 3: $N_5, N_6, N_7, N_8, N_{11}, N_{12}, P_3N_3, P_3N_4, P_3R_4, P_3R_5$.
Degree 4: $N_3N_3, N_3N_4, R_5N_3, R_5N_4,$
$P_3N_9, P_3N_{10}, P_3N_{13}, P_3N_{14},$
$P_3R_8, P_3R_9, P_3R_{11}, P_3R_{12}$.
Degree 5: R_8N_3, R_8N_4.

Elements in E_{ij} and M_i only: (26)
Degree 2: T_1, T_2.
Degree 3: $T_5, T_6, T_7, T_8, T_{11}, T_{12},$
$M_3T_3, M_3T_4, M_3R_4, M_3R_5$.
Degree 4: $T_3T_3, T_3T_4, R_5T_3, R_5T_4,$
$M_3T_9, M_3T_{10}, M_3T_{13}, M_3T_{14},$
$M_3R_8, M_3R_9, M_3R_{11}, M_3R_{12}$.
Degree 5: R_8N_3, R_8N_4.

Elements in E_{ij} and M_i only: (26)
Degree 2: T_1, T_2.
Degree 3: $T_5, T_6, T_7, T_8, T_{11}, T_{12},$
$M_3T_3, M_3T_4, M_3R_4, M_3R_5$.
Degree 4: $T_3T_3, T_3T_4, R_5T_3, R_5T_4,$
$M_3T_9, M_3T_{10}, M_3T_{13}, M_3T_{14},$
$M_3R_8, M_3R_9, M_3R_{11}, M_3R_{12}$.
Degree 5: R_8T_3, R_8T_4.

Elements in P_i and M_i only: (7)
Degree 1: None.
Degree 2: $P_3M_3, P_1M_1 + P_2M_2, P_1M_2 - P_2M_1$.
Degree 3: $J_1\left[\begin{pmatrix}P_1\\P_2\end{pmatrix}, \begin{pmatrix}M_1\\M_2\end{pmatrix}\right], J_1\left[\begin{pmatrix}M_1\\M_2\end{pmatrix}, \begin{pmatrix}P_1\\P_2\end{pmatrix}\right],$
$J_2\left[\begin{pmatrix}P_1\\P_2\end{pmatrix}, \begin{pmatrix}M_1\\M_2\end{pmatrix}\right], J_2\left[\begin{pmatrix}M_1\\M_2\end{pmatrix}, \begin{pmatrix}P_1\\P_2\end{pmatrix}\right]$.

Elements in E_{ij}, P_i and M_i: (20)

Degree 3:
$$K_1\left[\begin{pmatrix} M_1 \\ M_2 \end{pmatrix}, \begin{pmatrix} E_{11} - E_{22} \\ -2E_{12} \end{pmatrix}, \begin{pmatrix} P_1 \\ P_2 \end{pmatrix}\right]$$
$$K_2\left[\begin{pmatrix} M_1 \\ M_2 \end{pmatrix}, \begin{pmatrix} E_{11} - E_{22} \\ -2E_{12} \end{pmatrix}, \begin{pmatrix} P_1 \\ P_2 \end{pmatrix}\right]$$

and

$$M_3 N_3,\ M_3 N_4,\ P_3 T_3,\ P_3 T_4.$$

Degree 4:
$$N_3 T_3 - N_3 T_4,\ N_3 T_4 + N_4 T_3,$$
$$M_3 K_1\left[\begin{pmatrix} E_{13} \\ E_{23} \end{pmatrix}, \begin{pmatrix} M_1 \\ M_2 \end{pmatrix}, \begin{pmatrix} P_1 \\ P_2 \end{pmatrix}\right],$$
$$M_3 K_2\left[\begin{pmatrix} E_{13} \\ E_{23} \end{pmatrix}, \begin{pmatrix} M_1 \\ M_2 \end{pmatrix}, \begin{pmatrix} P_1 \\ P_2 \end{pmatrix}\right],$$
$$P_3 K_1\left[\begin{pmatrix} E_{13} \\ E_{23} \end{pmatrix}, \begin{pmatrix} M_1 \\ M_2 \end{pmatrix}, \begin{pmatrix} P_1 \\ P_2 \end{pmatrix}\right],$$
$$P_3 K_2\left[\begin{pmatrix} E_{13} \\ E_{23} \end{pmatrix}, \begin{pmatrix} M_1 \\ M_2 \end{pmatrix}, \begin{pmatrix} P_1 \\ P_2 \end{pmatrix}\right],$$
$$M_3 K_1\left[\begin{pmatrix} E_{13} \\ E_{23} \end{pmatrix}, \begin{pmatrix} E_{11} - E_{22} \\ -2E_{12} \end{pmatrix}, \begin{pmatrix} P_1 \\ P_2 \end{pmatrix}\right],$$
$$M_3 K_2\left[\begin{pmatrix} E_{13} \\ E_{23} \end{pmatrix}, \begin{pmatrix} E_{11} - E_{22} \\ -2E_{12} \end{pmatrix}, \begin{pmatrix} P_1 \\ P_2 \end{pmatrix}\right],$$
$$P_3 K_1\left[\begin{pmatrix} E_{13} \\ E_{23} \end{pmatrix}, \begin{pmatrix} M_1 \\ M_2 \end{pmatrix}, \begin{pmatrix} E_{11} - E_{22} \\ -2E_{12} \end{pmatrix}\right],$$
$$P_3 K_2\left[\begin{pmatrix} E_{13} \\ E_{23} \end{pmatrix}, \begin{pmatrix} M_1 \\ M_2 \end{pmatrix}, \begin{pmatrix} E_{11} - E_{22} \\ -2E_{12} \end{pmatrix}\right],$$
$$M_3 J_1\left[\begin{pmatrix} E_{13} \\ E_{23} \end{pmatrix}, \begin{pmatrix} P_1 \\ P_2 \end{pmatrix}\right],\ M_3 J_2\left[\begin{pmatrix} E_{13} \\ E_{23} \end{pmatrix}, \begin{pmatrix} P_1 \\ P_2 \end{pmatrix}\right],$$
$$P_3 J_1\left[\begin{pmatrix} E_{13} \\ E_{23} \end{pmatrix}, \begin{pmatrix} M_1 \\ M_2 \end{pmatrix}\right],\ P_3 J_2\left[\begin{pmatrix} E_{13} \\ E_{23} \end{pmatrix}, \begin{pmatrix} M_1 \\ M_2 \end{pmatrix}\right],$$

where $J_1(\mathbf{x}, \mathbf{y})$, $J_2(\mathbf{x}, \mathbf{y})$ and $K_1(\mathbf{x}, \mathbf{y}, \mathbf{z})$, $K_2(\mathbf{x}, \mathbf{y}, \mathbf{z})$ are defined earlier by (7.38), (7.41), (7.39), and (7.9) respectively.

7.34. Magnetic Crystal Class $\underline{6} = \{I, S_1, S_2, \tau D_3, \tau D_3 S_1, \tau D_3 S_2\}$

From Table 4.3(34) and Table B.11, we write

$$\phi, \phi', \phi'' = E_{11} + E_{22}, E_{33}, P_3,$$
$$a, b = M_1 - iM_2, E_{11} - E_{22} + 2iE_{12},$$
$$\xi = M_3,$$
$$A, B = E_{13} - iE_{23}, P_1 - iP_2.$$

The TMEs of the integrity basis are given by (B.11). The actual elements are listed below.

Elements in E_{ij} only: (14). Same elements as in the class $\underline{\bar{6}}$.

Elements in P_i only: (4)
Degree 1: P_3.
Degree 2: S_1.
Degree 3: None.
Degree 4: None.
Degree 5: None.
Degree 6: S_2S_2, S_2S_3.

Elements in M_i only: (4)
Degree 1: None.
Degree 2: M_3^2, Q_1.
Degree 3: Q_2, Q_3.

Elements in E_{ij} and P_i only: (28)
Degree 2: N_3, N_4.
Degree 3: N_9, N_{10}, N_{11}, N_{12}.
Degree 4: R_5N_1, R_5N_2, N_1N_1, N_1N_2.
Degree 5: R_9N_2, R_9N_1, N_1N_7, N_2N_7, S_3R_4, S_3R_5, S_2N_1, S_2N_2.
Degree 6: R_8N_7, R_8N_8, N_7N_7, N_7N_8, N_7N_{13}, N_7N_{14}, S_2N_7, S_2N_8, S_3N_{13}, S_2N_{14}.

Elements in E_{ij} and M_i only: (26)
Same elements as in the class $\underline{\bar{6}}$.

Elements in P_i and M_i only: (11)
Degree 2: None.
Degree 3: $J_1(\mathbf{M}, \mathbf{P})$, $J_2(\mathbf{M}, \mathbf{P})$, $M_3(P_1M_1 + P_2M_2)$, $M_3(P_1M_2 - P_2M_1)$.
Degree 4: $(M_1P_2^2 + M_2P_1^2)$, $(M_1P_1 + M_2P_2)(M_1P_2 - M_2P_1)$, $M_3J_1(\mathbf{P}, \mathbf{M})$, $M_3J_2(\mathbf{P}, \mathbf{M})$, M_3S_2, M_3S_3.
Degree 5: $(M_1P_1 + M_2P_2)S_2 - (P_1M_2 - P_2M_1)S_3$, $(M_1P_1 + M_2P_2)S_3 + (P_1M_2 - P_2M_1)S_2$.

Elements in E_{ij}, P_i, and M_i: (24)
Degree 3: $K_1\left[\begin{pmatrix} M_1 \\ M_2 \end{pmatrix}, \begin{pmatrix} E_{13} \\ E_{23} \end{pmatrix}, \begin{pmatrix} P_1 \\ P_2 \end{pmatrix}\right]$, $K_2\left[\begin{pmatrix} M_1 \\ M_2 \end{pmatrix}, \begin{pmatrix} E_{13} \\ E_{23} \end{pmatrix}, \begin{pmatrix} P_1 \\ P_2 \end{pmatrix}\right]$, M_3N_1, M_3N_2.
Degree 4: $M_3J_1\left[\begin{pmatrix} P_1 \\ P_2 \end{pmatrix}, \begin{pmatrix} E_{13} \\ E_{23} \end{pmatrix}\right]$, $M_3J_2\left[\begin{pmatrix} P_1 \\ P_2 \end{pmatrix}, \begin{pmatrix} E_{13} \\ E_{23} \end{pmatrix}\right]$, $M_3J_1\left[\begin{pmatrix} E_{13} \\ E_{23} \end{pmatrix}, \begin{pmatrix} P_1 \\ P_2 \end{pmatrix}\right]$,

$$M_3J_2\left[\begin{pmatrix}E_{13}\\E_{23}\end{pmatrix},\begin{pmatrix}P_1\\P_2\end{pmatrix}\right],$$
$$M_3J_1\left[\begin{pmatrix}P_1\\P_2\end{pmatrix},\begin{pmatrix}E_{11}-E_{22}\\-2E_{12}\end{pmatrix}\right],$$
$$M_3J_2\left[\begin{pmatrix}P_1\\P_2\end{pmatrix},\begin{pmatrix}E_{11}-E_{22}\\-2E_{12}\end{pmatrix}\right],$$
$$M_3K_1\left[\begin{pmatrix}M_1\\M_2\end{pmatrix},\begin{pmatrix}E_{11}-E_{22}\\-2E_{12}\end{pmatrix},\begin{pmatrix}P_1\\P_2\end{pmatrix}\right],$$
$$M_3K_2\left[\begin{pmatrix}M_1\\M_2\end{pmatrix},\begin{pmatrix}E_{11}-E_{22}\\-2E_{12}\end{pmatrix},\begin{pmatrix}P_1\\P_2\end{pmatrix}\right],$$
$T_3N_1 - T_4N_2,\ T_3N_2 + N_1T_4,$
$(M_1P_1 + M_2P_2)N_1 - (M_1P_2 - M_2P_1)N_2,$
$(M_1P_1 + M_2P_2)N_2 + (M_1P_2 - M_2P_1)N_1,$
$(M_1P_1 + M_2P_2)T_3 - (M_1P_2 - M_2P_1)T_4,$
$(M_1P_1 + M_2P_2)T_4 - (M_1P_2 - M_2P_1)T_3.$

Degree 5: $(M_1P_1 + M_2P_2)R_8 - (P_1M_2 - P_2M_1)R_9,$
$(M_1P_1 + M_2P_2)R_9 + (P_1M_2 - P_2M_1)R_8,$
$T_3S_2 - T_4S_4,\ T_3S_3 + T_4S_2,$
$$T_3J_1\left[\begin{pmatrix}E_{13}\\E_{23}\end{pmatrix},\begin{pmatrix}P_1\\P_2\end{pmatrix}\right]-T_4J_2\left[\begin{pmatrix}E_{13}\\E_{23}\end{pmatrix},\begin{pmatrix}P_1\\P_2\end{pmatrix}\right],$$
$$T_3J_2\left[\begin{pmatrix}E_{13}\\E_{23}\end{pmatrix},\begin{pmatrix}P_1\\P_2\end{pmatrix}\right]+T_4J_1\left[\begin{pmatrix}E_{13}\\E_{23}\end{pmatrix},\begin{pmatrix}P_1\\P_2\end{pmatrix}\right].$$

7.35. Magnetic Crystal Class $\bar{6}\underline{m}\underline{2}$ (see Table 3.3(13))

From Table 4.3(35) and Table B.12, we have

$$\phi, \phi'\ \phi'' = M_3,\ E_{11} + E_{22},\ E_{33},$$
$$\xi = P,$$
$$\mathbf{A}, \mathbf{B} = \begin{pmatrix}M_2\\-M_1\end{pmatrix},\begin{pmatrix}E_{23}\\-E_{31}\end{pmatrix},$$
$$\mathbf{a}, \mathbf{b} = \begin{pmatrix}2E_{12}\\E_{11}-E_{22}\end{pmatrix},\begin{pmatrix}P_1\\P_2\end{pmatrix}.$$

The TMEs of the integrity basis are given by (B.12). The actual elements are shown below.

Elements in E_{ij} only: (9)
Degree 1: $R_0, R_1.$
Degree 2: $R_2, R_3.$
Degree 3: $R_6, R_{10}.$
Degree 4: $R_4R_4.$

Degree 5: $R_4 R_8$.
Degree 6: $R_8 R_8$.

Elements in P_i only: (3)
Degree 1: None.
Degree 2: P_3^2, S_1.
Degree 3: S_3.

Elements in M_i only: (3)
Degree 1: M_3.
Degree 2: Q_1.
Degree 3: None.
Degree 4: None.
Degree 5: None.
Degree 6: $Q_2 Q_2$.

Elements in E_{ij} and P_i only: (13)
Degree 2: N_2.
Degree 3: $P_3 R_5$, N_6, N_8, $P_3 N_3$, N_{11}.
Degree 4: $P_3 R_9$, $P_3 R_{12}$, $R_5 N_3$, $N_3 N_3$, $P_3 N_9$, $P_3 N_{14}$.
Degree 5: $R_8 N_4$.

Elements in E_{ij} and M_i only: (15)
Degree 2: T_3.
Degree 3: T_{11}, T_9.
Degree 4: $T_1^2 - T_2^2$, $T_1 R_4 + T_2 R_5$.
Degree 5: $\mathrm{Re}[2E_{12} + i(E_{11} - E_{22})]^3$. $\mathrm{Im}[(M_2 - iM_1)(E_{23} + iE_{31})]$,
$\mathrm{Im}\{[2E_{12} + i(E_{11} - E_{22})](M_2 + iM_1)^4\}$,
$\mathrm{Im}\{[2E_{12} + i(E_{11} - E_{22})](E_{23} + iE_{31})(M_2 + iM_1)^3\}$,
$\mathrm{Im}\{[2E_{12} + i(E_{11} - E_{22})](E_{23} + iE_{31})^2(M_2 + iM_1)^2\}$,
$\mathrm{Im}\{[2E_{12} + i(E_{11} - E_{22})](E_{23} + iE_{31})^3(M_2 + iM_1)\}$.
Degree 6: $\mathrm{Re}\{(M_2 - iM_1)^5(E_{23} - iE_{31})\}$,
$\mathrm{Re}\{(M_2 - iM_1)^4(E_{23} - iE_{31})^2\}$,
$\mathrm{Re}\{(M_2 - iM_1)^3(E_{23} - iE_{31})^3\}$,
$\mathrm{Re}\{(M_2 - iM_1)^2(E_{23} - iE_{31})^4\}$,
$\mathrm{Re}\{(M_2 - iM_1)(E_{23} - iE_{31})^5\}$.

In the above equations $\mathrm{Re}(I_1 + iI_2)$ indicates that I_1 is the element of the integrity basis, and $\mathrm{Im}(I_1 + iI_2)$ indicates that I_2 is the element of the integrity basis.

Elements in P_i and M_i only: (6)
Degree 2: None.
Degree 3: $P_3(M_1 P_1 + M_2 P_2)$, $P_2(M_2^2 - M_1^2) - 2P_1 M_1 M_2$.
Degree 4: $P_3[(P_1^2 - P_2^2)M_2 + 2P_1 P_2 M_1]$,
$(P_1 M_1 + P_2 M_2)^2$, $P_3 Q_3$.
Degree 5: $(P_1 M_2 - P_2 M_1)Q_2$.

Elements in E_{ij}, P_i, and M_i: (16)

Degree 3: $P_3 T_2$,
$\mathrm{Im}\{(P_1 + iP_2)(M_2 - iM_1)(E_{23} - iE_{31})\}$.

Degree 4: $N_1 T_4$,
$\mathrm{Re}\{[2E_{12} + i(E_{11} - E_{22})](P_1 + iP_2)(M_2 + iM_1)^2\}$,
$\mathrm{Re}\{[2E_{12} + i(E_{11} - E_{22})](P_1 + iP_2)(M_2 + iM_1)(E_{23} + iE_{31})\}$,
$\mathrm{Re}\{(P_1 + iP_2)^2(M_2 + iM_1)(E_{23} + iE_{31})\}$,
$P_3 \cdot \mathrm{Re}\{[2E_{12} + i(E_{11} - E_{22})]^2(M_2 - iM_1)\}$,
$P_3 \cdot \mathrm{Re}\{[2E_{12} + i(E_{11} - E_{22})](P_1 + iP_2)(M_2 - iM_1)\}$,
$P_3 \cdot \mathrm{Re}\{(M_2 - iM_1)^2(E_{23} - iE_{31})\}$,
$P_3 \cdot \mathrm{Re}\{(M_2 - iM_1)(E_{23} - iE_{31})^2\}$.

Degree 5: $S_2 T_4$,
$T_4 \cdot \mathrm{Re}\{[2E_{12} + i(E_{11} - E_{22})]^2(P_1 + iP_2)\}$,
$T_4 \cdot \mathrm{Re}\{[2E_{12} + i(E_{11} - E_{22})](P_1 + iP_2)^2\}$,
$\mathrm{Im}\{(P_1 + iP_2)(M_2 + iM_1)^3(E_{23} + iE_{31})\}$,
$\mathrm{Im}\{(P_1 + iP_2)(M_2 + iM_1)^2(E_{23} + iE_{31})^2\}$,
$\mathrm{Im}\{(P_1 + iP_2)(M_2 + iM_1)(E_{23} + iE_{31})^3\}$,

7.36. Magnetic Crystal Class $\underline{\bar{6}}m\underline{2}$ (see Table 3.3(13))

From Table 4.3(36) and Table B.12, we have

$$\phi, \phi' = E_{11} + E_{22}, E_{33},$$
$$\xi = P_3,$$
$$\eta = M_3,$$
$$\mathbf{A} = \begin{pmatrix} E_{23} \\ -E_{31} \end{pmatrix},$$
$$\mathbf{a}, \mathbf{b}, \mathbf{c} = \begin{pmatrix} M_2 \\ -M_1 \end{pmatrix}, \begin{pmatrix} 2E_{12} \\ E_{11} - E_{22} \end{pmatrix}, \begin{pmatrix} P_1 \\ P_2 \end{pmatrix}.$$

The TMEs of the integrity basis are given by (B.12). The actual elements are listed below.

Elements in E_{ij} only: (9)
Same elements as in the class $\bar{6}\underline{m}\underline{2}$.

Elements in P_i only: (3)
Same as in the class $\bar{6}\underline{m}\underline{2}$.

Elements in M_i only: (3)
Degree 1: None.
Degree 2: M_3^2, $M_1^2 + M_2^2 = Q_1$.
Degree 3: Q_2.

Elements in E_{ij} and P_i only: (13)
Same as in the class $\underline{6}m\underline{2}$.

Elements in E_{ij} and M_i only: (13)
Degree 2: T_1.
Degree 3: M_3T_3, M_3R_4,
$\text{Im}\{(M_2 - iM_1)^2[2E_{12} + i(E_{11} - E_{22})]\}$.
$\text{Im}\{[2E_{12} + i(E_{11} - E_{22})]^2(M_2 - iM_1)\}$,
$\text{Im}\{(M_2 - iM_1)(E_{23} - iE_{31})^2\}$.
Degree 4: M_3R_8,
$\text{Re}\{(M_2 - iM_1)^2(E_{23} + iE_{31})^2\}$,
$\text{Re}\{(M_2 - iM_1)[2E_{12} + i(E_{11} - E_{22})](E_{23} + iE_{31})^2\}$,
$M_3\,\text{Im}\{(M_2 - iM_1)^2(E_{23} - iE_{31})\}$,
$M_3\,\text{Im}\{(M_2 - iM_1)[2E_{12} + i(E_{11} - E_{22})](E_{23} - iE_{31})\}$,
$M_3\,\text{Im}\{[2E_{12} + i(E_{11} - E_{22})]^2(E_{23} - iE_{31})\}$.
Degree 5: $\text{Im}\{(M_2 - iM_1)(E_{23} + iE_{31})^4\}$.

Elements in P_i and M_i only: (8)
Degree 2: $P_1M_2 - P_2M_1$.
Degree 3: $\text{Im}\{(M_2 - iM_1)^2(P_1 + iP_2)\}$,
$\text{Im}\{(M_2 - iM_1)(P_1 + iP_2)^2\}$.
Degree 4: $P_3M_3(P_1M_1 + P_2M_2)$.
Degree 5: $M_3P_3Q_3$, $P_3M_3S_2$,
$P_3M_3\cdot\text{Re}\{(M_2 - iM_1)^2(P_1 + iP_2)\}$,
$P_3M_3\cdot\text{Re}\{(M_2 - iM_1)(P_1 + iP_2)^2\}$.

Elements in E_{ij}, P_i, and M_i: (21)
Degree 3: P_3T_4, M_3N_4,
$\text{Im}\{(M_2 - iM_1)[2E_{12} + i(E_{11} - E_{22})](P_1 + iP_2)\}$.
Degree 4: $\text{Re}\{(M_2 - iM_1)(P_1 + iP_2)(E_{23} + iE_{31})^2\}$,
$P_3\cdot\text{Re}\{(M_2 - iM_1)^2(E_{23} - iE_{31})\}$,
$P_3\cdot\text{Re}\{[2E_{12} - i(E_{11} - E_{22})](E_{23} - iE_{31})(M_2 - iM_1)\}$,
$P_3\cdot\text{Re}\{(M_2 - iM_1)(P_1 + iP_2)(E_{23} - iE_{31})\}$,
$M_3\cdot\text{Im}\{(P_1 + iP_2)^2(E_{23} - iE_{31})\}$,
$M_3\cdot\text{Im}\{(P_1 + iP_2)[2E_{12} - i(E_{11} - E_{22})](E_{23} - iE_{31})\}$,
$M_3\cdot\text{Im}\{(P_1 + iP_2)(M_2 - iM_1)(E_{23} - iE_{31})\}$,
$P_3M_3T_2$, $P_3M_3N_1$.
Degree 5: $P_3M_3\cdot\text{Re}\{[2E_{12} + i(E_{11} - E_{22})]^3\}$,
$P_3M_3\cdot\text{Re}\{(M_2 - iM_1)^2[2E_{12} + i(E_{11} - E_{22})]\}$,
$P_3M_3\cdot\text{Re}\{(M_2 - iM_1)[2E_{12} + i(E_{11} - E_{22})]^2\}$,
$P_3M_3\cdot\text{Re}\{(P_1 + iP_2)[2E_{12} + i(E_{11} - E_{22})]^2\}$,
$P_3M_3\cdot\text{Re}\{(P_1 + iP_2)^2[2E_{12} + i(E_{11} - E_{22})]\}$,
$P_3M_3\cdot\text{Re}\{(M_2 - iM_1)[2E_{12} + i(E_{11} - E_{22})](P_1 + iP_2)\}$,
$P_3M_3\cdot\text{Re}\{(M_2 - iM_1)(E_{23} - iE_{31})^2\}$,
$P_3M_3\cdot\text{Re}\{[2E_{12} + i(E_{11} - E_{22})](E_{23} - iE_{31})^2\}$,
$P_3M_3\cdot\text{Re}\{(P_1 + iP_2)(E_{23} - iE_{31})^2\}$.

7.37. Magnetic Crystal Class $\underline{\bar{6}}\underline{m}2$ (see Table 3.3(13))

From Table 4.3(37) and Table B.12, we have

$$\phi, \phi' = E_{11} + E_{22}, E_{33},$$

$$\xi, \xi' = M_3, P_3,$$

$$\mathbf{A} = \begin{pmatrix} E_{23} \\ -E_{31} \end{pmatrix},$$

$$\mathbf{a}, \mathbf{b}, \mathbf{c} = \begin{pmatrix} M_1 \\ M_2 \end{pmatrix}, \begin{pmatrix} 2E_{12} \\ E_{11} - E_{22} \end{pmatrix}, \begin{pmatrix} P_1 \\ P_2 \end{pmatrix}.$$

The TMEs of the integrity basis are given by (B.12). The actual elements are listed below.

Elements in E_{ij} only: (9)
Same elements as in the class $\bar{6}\underline{m}\underline{2}$.

Elements in P_i only: (3)
Same elements as in the class $\bar{6}\underline{m}\underline{2}$.

Elements in M_i only: (3)
Degree 1: None.
Degree 2: M_3^2, Q_1.
Degree 3: Q_3.

Elements in E_{ij} and P_i only: (13)
Same as in the class $\bar{6}\underline{m}\underline{2}$.

Elements in E_{ij} and M_i only: (13)
Degree 2: T_2.
Degree 3: $M_3R_5, T_6, T_8, M_3T_3, T_4$.
Degree 4: $M_3R_9, M_3R_{12}, R_5T_3, T_3T_3, M_3T_9, M_3T_{14}$.
Degree 5: R_8T_4.

Elements in P_i and M_i only: (4)
Degree 2: $P_3M_3, P_1M_1 + P_2M_2$.
Degree 3: $P_2M_2^2 - P_2M_1^2 - 2M_1M_2P_1$,
$M_2P_2^2 - M_2P_1^2 - 2P_1P_2M_1$.

Elements in E_{ij}, P_i and M_i: (10)
Degree 3: $\text{Im}\{(M_1 + iM_2)[2E_{12} + i(E_{11} - E_{22})](P_1 + iP_2)\}$,
$\text{Re}\{(M_1 + iM_2)(P_1 + iP_2)(E_{23} - iE_{31})\}$,
M_3N_3, P_3T_3.
Degree 4: $M_3 \cdot \text{Re}\{(M_1 + iM_2)(P_1 + iP_2)(E_{23} - iE_{31})\}$,
$M_3 \cdot \text{Re}\{[2E_{12} + i(E_{11} - E_{22})](P_1 + iP_2)(E_{23} - iE_{31})\}$,
$M_3 \cdot \text{Re}\{(P_1 + iP_2)^2(E_{23} - iE_{31})\}$,
$P_3 \cdot \text{Re}\{(M_1 + iM_2)^2(E_{23} - iE_{31})\}$,

$$P_3 \cdot \mathrm{Re}\{(M_1 + iM_2)[2E_{12} + i(E_{11} - E_{22})](E_{23} - iE_{31})\},$$
$$P_3 \cdot \mathrm{Re}\{(M_1 + iM_2)(P_1 + iP_2)(E_{23} - iE_{31})\}.$$

7.38. Magnetic Crystal Class $\bar{3}\underline{m}$ (see Table 3.3(13))

From Table 4.3(38) and Table B.12, we have

$$\phi, \phi', \phi'' = M_3, E_{11} + E_{22}, E_{33},$$
$$\eta = P_3,$$
$$\mathbf{A} = \begin{pmatrix} P_1 \\ P_2 \end{pmatrix},$$
$$\mathbf{a}, \mathbf{b}, \mathbf{c} = \begin{pmatrix} M_1 \\ M_2 \end{pmatrix}, \begin{pmatrix} E_{13} \\ E_{23} \end{pmatrix}, \begin{pmatrix} 2E_{12} \\ E_{11} - E_{22} \end{pmatrix}.$$

The TMEs of the integrity basis are given by (B.12). The actual elements are

Elements in E_{ij} only: (9)
Degree 1: R_0, R_1.
Degree 2: R_2, R_3, R_5.
Degree 3: R_6, R_9, R_{10}, R_{12}.

Elements in P_i only: (4)
Degree 1: None.
Degree 2: P_3P_3, S_1.
Degree 3: None.
Degree 4: P_3, S_3.
Degree 5: None.
Degree 6: S_2S_2.

Elements in M_i only: (3)
Degree 1: M_3.
Degree 2: Q_1.
Degree 3: Q_3.

Elements in E_{ij} and P_i only: (12)
Degree 2: None.
Degree 3: $N_{11}, N_{14}, P_3N_2, P_3N_3$.
Degree 4: $P_3N_6, P_3N_8, P_3N_9, N_1N_1, N_2N_3, N_3N_3$.
Degree 5: S_2N_1, S_2N_4.

Elements in E_{ij} and M_i only: (7)
Degree 2: T_3, T_2.
Degree 3: $\mathrm{Im}\{(M_1 + iM_2)^2(E_{13} + iE_{23})\}$,
$\mathrm{Im}\{(M_1 + iM_2)^2[2E_{12} + i(E_{11} - E_{22})]\}$,
$\mathrm{Im}\{(E_{13} + iE_{23})^2(M_1 + iM_2)\}$,
$\mathrm{Im}\{[2E_{12} + i(E_{11} - E_{22})]^2(M_1 + iM_2)\}$,
$\mathrm{Im}\{(M_1 + iM_2)(E_{13} + iE_{23})[2E_{12} + i(E_{11} - E_{22})]\}$,

Elements in P_i and M_i only: (5)
Degree 2: None.
Degree 3: $P_3(P_1M_1 + P_2M_2)$,
$\operatorname{Im}\{(M_1 + iM_2)(P_1 + iP_2)^2\}$.
Degree 4: $\operatorname{Re}\{(M_1 + iM_2)^2(P_1 - iP_2)^2\}$,
$P_3 \cdot \operatorname{Im}\{(M_1 + iM_2)^2(P_1 + iP_2)\}$.
Degree 5: $\operatorname{Im}\{(M_1 + iM_2)(P_1 - iP_2)^4\}$.

Elements in E_{ij}, P_i, and M_i: (4)
Degree 3: None.
Degree 4: $\operatorname{Re}\{(M_1 + iM_2)(E_{13} + iE_{23})(P_1 - iP_2)^2\}$,
$\operatorname{Re}\{(M_1 + iM_2)[2E_{12} + i(E_{11} - E_{22})](P_1 - iP_2)^2\}$,
$P_3 \cdot \operatorname{Im}\{(M_1 + iM_2)(E_{13} + iE_{23})(P_1 + iP_2)\}$,
$P_3 \cdot \operatorname{Im}\{(M_1 + iM_2)[2E_{12} + i(E_{11} - E_{22})](P_1 + iP_2)\}$.

7.39. Magnetic Crystal Class $\underline{\bar{3}}\underline{m}$ (see Table 3.3(13))

From Table 4.3(40) and Table B.12, we have

$$\phi, \phi' = E_{11} + E_{22}, E_{33},$$

$$\eta, \eta' = M_3, P_3,$$

$$\mathbf{A}, \mathbf{B} = \begin{pmatrix} M_1 \\ M_2 \end{pmatrix}, \begin{pmatrix} P_1 \\ P_2 \end{pmatrix},$$

$$\mathbf{a}, \mathbf{b} = \begin{pmatrix} E_{31} \\ E_{23} \end{pmatrix}, \begin{pmatrix} 2E_{12} \\ E_{11} - E_{22} \end{pmatrix}.$$

The TMEs of the integrity basis are given by (B.12). The actual elements are

Elements in E_{ij} only: (9)
Same elements as in the class $\bar{3}\underline{m}$.

Elements in P_i only: (4)
Same as in the class $\bar{3}\underline{m}$.

Elements in M_i only: (4)
Degree 1: None.
Degree 2: M_3^2, Q_1.
Degree 3: None.
Degree 4: M_3Q_3.
Degree 5: None.
Degree 6: Q_2Q_2.

Elements in E_{ij} and P_i only: (12)
Same as in the class $\bar{3}\underline{m}$.

Elements in E_{ij} and M_i only: (12)
Degree 2: None.
Degree 3: T_{11}, T_{14}, M_3T_2, M_3T_3.

Degree 4: $M_3T_6, M_3T_8, M_3T_9, T_1T_1, T_2T_3, T_3T_3$.
Degree 5: Q_2T_1, Q_2T_4.

Elements in P_i and M_i only: (13)
Degree 2: $P_1M_1 + P_2M_2, P_3M_3$.
Degree 3: None.
Degree 4: $M_3S_3, P_3Q_3,$
$P_3J_2(\mathbf{M}, \mathbf{P}), P_3J_2(\mathbf{P}, \mathbf{M}),$
$M_3J_2(\mathbf{M}, \mathbf{P}), M_3J_2(\mathbf{P}, \mathbf{M})$.
Degree 5: None.
Degree 6: $S_2J_1(\mathbf{M}, \mathbf{P}), S_2J_1(\mathbf{P}, \mathbf{M}),$
$Q_2J_1(\mathbf{P}, \mathbf{M}), Q_2J_1(\mathbf{M}, \mathbf{P}), S_2Q_2,$
where $J_1(\mathbf{x}, \mathbf{y})$ and $J_2(\mathbf{x}, \mathbf{y})$ are defined by (7.38) and (7.41).

Elements in E_{ij}, P_i, and M_i only: (26)
Degree 3: $E_{23}(P_1M_1 - P_2M_2) + E_{31}(P_1M_2 + P_2M_1),$
$2E_{12}(P_1M_2 + P_2M_1) + (E_{11} - E_{22})(P_1M_1 - P_2M_2),$
$M_3N_3, M_3N_2, P_3T_3, P_3T_2$.
Degree 4: $(P_1M_2 - P_2M_1)R_4,$
$\mathrm{Re}\{(E_{31} + iE_{23})^2(M_1 - iM_2)(P_1 - iP_2)\},$
$\mathrm{Re}\{(E_{31} + iE_{23})[2E_{12} + i(E_{11} - E_{22})](M_1 - iM_2)(P_1 - iP_2)\},$
$\mathrm{Re}\{[2E_{12} + i(E_{11} - E_{22})]^2(M_1 - iM_2)(P_1 - iP_2)\},$
$M_3 \cdot \mathrm{Im}\{(E_{31} + iE_{23})^2(P_1 + iP_2)\},$
$M_3 \cdot \mathrm{Im}\{(E_{31} + iE_{23})[2E_{12} + i(E_{11} - E_{22})](P_1 + iP_2)\},$
$M_3 \cdot \mathrm{Im}\{(2E_{12} + i(E_{11} - E_{22})]^2(P_1 + iP_2)\},$
$P_3 \cdot \mathrm{Im}\{(E_{31} + iE_{23})^2(M_1 + iM_2)\},$
$P_3 \cdot \mathrm{Im}\{(E_{31} + iE_{23})[2E_{12} + i(E_{11} - E_{22})](M_1 + iM_2)\},$
$P_3 \cdot \mathrm{Im}\{[2E_{12} + i(E_{11} - E_{22})]^2(M_1 + iM_2)\}$.
Degree 5: $(P_1M_2 - P_2M_1)R_8, (P_1M_2 - P_2M_1)R_7,$
$(P_1M_2 - P_2M_1) \cdot \mathrm{Re}\{(E_{31} + iE_{23})^2[2E_{12} + i(E_{11} - E_{22})]\},$
$(P_1M_2 - P_2M_1) \cdot \mathrm{Re}\{(E_{31} + iE_{23})[2E_{12} + i(E_{11} - E_{22})]^2\},$
$\mathrm{Im}\{(E_{31} + iE_{23})(P_1 + iP_2)(M_1 + iM_2)^3\},$
$\mathrm{Im}\{(E_{31} + iE_{23})(P_1 + iP_2)^2(M_1 + iM_2)^2\},$
$\mathrm{Im}\{(E_{31} + iE_{23})(P_1 + iP_2)^3(M_1 + iM_2)\},$
$\mathrm{Im}\{[2E_{12} + i(E_{11} - E_{22})](P_1 + iP_2)(M_i + iM_2)^3\},$
$\mathrm{Im}\{[2E_{12} + i(E_{11} - E_{22})](P_1 + iP_2)^2(M_1 + iM_2)^2\},$
$\mathrm{Im}\{[2E_{12} + i(E_{11} - E_{22})](P_1 + iP_2)^3(M_1 + iM_2)\}$.

7.40. Magnetic Crystal Class $\underline{\bar{3}}m$ (see Table 3.3(13))

From Table 4.3(39) and Table B.12, we write

$$\phi, \phi' = E_{11} + E_{22}, E_{33},$$

$$\xi = M_3,$$

$$\eta = P_3,$$

$$\mathbf{A}, \mathbf{B} = \begin{pmatrix} M_2 \\ -M_1 \end{pmatrix}, \begin{pmatrix} P_1 \\ P_2 \end{pmatrix},$$

$$\mathbf{a}, \mathbf{b} = \begin{pmatrix} E_{31} \\ E_{23} \end{pmatrix}, \begin{pmatrix} 2E_{12} \\ E_{11} - E_{22} \end{pmatrix}.$$

The TMEs of the integrity basis are given by (B.12). The actual elements are

Elements in E_{ij} only: (9)
Same as in the class $\bar{3}\underline{m}$.

Elements in P_i only: (4)
Same as in the class $\bar{3}\underline{m}$.

Elements in M_i only: (4)
Degree 1: None.
Degree 2: M_3^2, Q_1.
Degree 3: None.
Degree 4: $M_3 Q_3$.
Degree 5: None.
Degree 6: $Q_2 Q_2$.

Elements in E_{ij} and P_i only: (12)
Same as in the class $\bar{3}\underline{m}$.

Elements in E_{ij} and M_i only: (12)
Degree 2: None.
Degree 3: $M_3 T_3, M_3 T_2,$
$\mathrm{Im}\{(E_{31} + iE_{23})(M_2 - iM_1)^2\},$
$\mathrm{Im}\{[2E_{12} + i(E_{11} - E_{22})](M_2 - iM_1)^2\}.$
Degree 4: $\mathrm{Re}\{(E_{31} + iE_{23})^2(M_2 + iM_1)^2\},$
$\mathrm{Re}\{(E_{31} + iE_{23})[2E_{12} + i(E_{11} - E_{22})](M_2 + iM_1)^2\},$
$\mathrm{Re}\{[2E_{12} + i(E_{11} - E_{22})]^2(M_2 + iM_1)^2\},$
$M_3 \cdot \mathrm{Re}\{[E_{13} + iE_{23})^2(M_2 - iM_1)\},$
$M_3 \cdot \mathrm{Re}\{(E_{13} + iE_{23})]2E_{12} + i(E_{11} - E_{22})](M_2 - iM_1)\}.$
$M_3 \cdot \mathrm{Re}\{[2E_{13} + i(E_{11} - E_{22})]^2(M_2 - iM_1)\}.$
Degree 5: $\mathrm{Im}\{(E_{13} + iE_{23})(M_2 + iM_1)^4\},$
$\mathrm{Im}\{[2E_{12} + i(E_{11} - E_{22}](M_2 + iM_1)^4\}.$

Elements in P_i and M_i only: (13)
Degree 2: $P_1 M_2 - P_2 M_1$.
Degree 3: None.
Degree 4: $P_3 M_3(P_1 M_1 + P_2 M_2),$
$M_3 \cdot \mathrm{Re}\{(M_2 - iM_1)^2(P_1 + iP_2)\},$
$M_3 \cdot \mathrm{Re}\{(M_2 - iM_1)(P_1 + iP_2)^2\},$
$M_3 \cdot \mathrm{Re}\{(P_1 + iP_2)^3\},$
$P_3 \cdot \mathrm{Im}\{(M_2 - iM_1)^3\},$
$P_3 \cdot \mathrm{Im}\{(M_2 - iM_1)^2(P_1 + iP_2)\},$
$P_3 \cdot \mathrm{Im}\{(M_2 - iM_1)(P_1 + iP_2)^2\}.$
Degree 5: None.

Degree 6: $\operatorname{Re}\{(M_2 - iM_1)^5(P_1 + iP_2)\}$,
$\operatorname{Re}\{(M_2 - iM_1)^4(P_1 + iP_2)^2\}$,
$\operatorname{Re}\{(M_2 - iM_1)^3(P_1 + iP_2)^3\}$,
$\operatorname{Re}\{(M_2 - iM_1)^2(P_1 + iP_2)^4\}$,
$\operatorname{Re}\{(M_2 - iM_1)(P_1 + iP_2)^5\}$.

Elements in E_{ij}, P_i, *and* M_i: (37)

Degree 3: $M_3N_4, M_3N_1, P_3T_4, P_3T_1$,
$\operatorname{Im}\{(E_{13} + iE_{23})(M_2 - iM_1)(P_1 + iP_2)\}$,
$\operatorname{Im}\{[2E_{12} + i(E_{11} - E_{22}](M_2 - iM_1)(P_1 + iP_2)\}$.

Degree 4: $(P_1M_1 + P_2M_2)R_4, P_3M_3R_4$,
$\operatorname{Re}\{(E_{31} + iE_{23})^2(M_2 + iM_1)(P_1 - iP_2)\}$,
$\operatorname{Re}\{(E_{31} + iE_{23})[2E_{12} + i(E_{11} - E_{22})](M_2 + iM_1)(P_1 - iP_2)\}$,
$\operatorname{Re}\{[2E_{12} + i(E_{11} - E_{22})]^2(M_2 + iM_1)(P_1 - iP_2)\}$,
$M_3 \cdot \operatorname{Re}\{(E_{31} + iE_{23})^2(P_1 + iP_2)\}$,
$M_3 \cdot \operatorname{Re}\{(E_{31} + iE_{23})[2E_{12} + i(E_{11} - E_{22})](P_1 + iP_2)\}$,
$M_3 \cdot \operatorname{Re}\{[2E_{12} + i(E_{11} - E_{22})]^2(P_1 + iP_2)\}$,
$P_3 \cdot \operatorname{Im}\{(E_{31} + iE_{32})^2(M_2 - iM_1)\}$,
$P_3 \cdot \operatorname{Im}\{(E_{31} + iE_{32})[2E_{12} + i(E_{11} - E_{22})](M_2 - iM_1)\}$,
$P_3 \cdot \operatorname{Im}\{[2E_{12} + i(E_{11} - E_{22})](M_2 - iM_1)\}$.

Degree 5: $(P_1M_1 + P_2M_2) \cdot \operatorname{Re}\{(E_{13} + iE_{23})^3\}$,
$(P_1M_1 + P_2M_2) \cdot \operatorname{Re}\{(E_{13} + iE_{23})^2[2E_{12} + i(E_{11} - E_{22})]\}$,
$(P_1M_1 + P_2M_2) \cdot \operatorname{Re}\{(E_{13} + iE_{23})[2E_{12} + i(E_{11} - E_{22})]^2\}$,
$(P_1M_1 + P_2M_2) \cdot \operatorname{Re}\{[2E_{13} + i(E_{11} - E_{22})]^3\}$,
$P_3M_3 \cdot \operatorname{Re}\{(E_{13} + iE_{23})^3\}$,
$P_3M_3 \cdot \operatorname{Re}\{(E_{13} + iE_{23})^2[2E_{12} + i(E_{11} - E_{22})]\}$,
$P_3M_3 \cdot \operatorname{Re}\{(E_{13} + iE_{23})[2E_{12} + i(E_{11} - E_{22})^2]\}$,
$P_3M_3 \cdot \operatorname{Re}\{[2E_{12} + i(E_{11} - E_{22})]^3\}$,
$\operatorname{Im}\{(E_{13} + iE_{23})(M_2 + iM_1)^3(P_1 - iP_2)\}$,
$\operatorname{Im}\{(E_{13} + iE_{23})(M_2 + iM_1)^2(P_1 - iP_2)^2\}$,
$\operatorname{Im}\{(E_{13} + iE_{23})(M_2 + iM_1)(P_1 - iP_2)^3\}$,
$\operatorname{Im}\{[2E_{12} + i(E_{11} - E_{22})](M_2 + iM_1)^3(P_1 - iP_2)\}$,
$\operatorname{Im}\{[2E_{12} + i(E_{11} - E_{22})](M_2 + iM_1)^2(P_1 - iP_2)^2\}$,
$\operatorname{Im}\{[2E_{12} + i(E_{11} - E_{22})](M_2 + iM_1)(P_1 - iP_2)^3\}$,
$P_3M_3 \cdot \operatorname{Re}\{(E_{13} + iE_{23})(M_2 - iM_1)^2\}$,
$P_3M_3 \cdot \operatorname{Re}\{(E_{13} + iE_{23})(M_2 - iM_1)(P_1 + iP_2)\}$,
$P_3M_3 \cdot \operatorname{Re}\{(E_{13} + iE_{23})(P_1 + iP_2)^2\}$,
$P_3M_3 \cdot \operatorname{Re}\{[2E_{12} + i(E_{11} - E_{22})](M_2 - iM_1)^2\}$,
$P_3M_3 \cdot \operatorname{Re}\{[2E_{12} + i(E_{11} - E_{22})](M_2 - iM_1)(P_1 + iP_2)\}$,
$P_3M_3 \cdot \operatorname{Re}\{[2E_{12} + i(E_{11} - E_{22})](P_1 + iP_2)^2\}$.

7.41. Magnetic Crystal Class $6\underline{2}\underline{2}$ (see Table 3.3(13))

From Table 4.3(41) and Table B.12, we write

$$\phi, \phi', \phi'' = M_3, E_{11} + E_{22}, E_{33},$$

$$\psi = P_3,$$

$$\mathbf{A}, \mathbf{B}, \mathbf{C} = \begin{pmatrix} M_2 \\ -M_1 \end{pmatrix}, \begin{pmatrix} E_{23} \\ -E_{31} \end{pmatrix}, \begin{pmatrix} P_1 \\ P_2 \end{pmatrix},$$

$$\mathbf{a} = \begin{pmatrix} 2E_{12} \\ E_{11} - E_{22} \end{pmatrix}.$$

The TMEs of the integrity basis are listed in (B.12). The actual elements are listed below.

Elements in E_{ij} only: (9)
Degree 1: R_0, R_1.
Degree 2: R_2, R_3.
Degree 3: R_6, R_{10}.
Degree 4: $R_4 R_4$.
Degree 5: $R_4 R_8$.
Degree 6: $R_8 R_8$.

Elements in P_i only: (4)
Degree 1: None.
Degree 2: $P_3 P_3, S_1$.
Degree 3, 4, 5: None.
Degree 6: S_2^2.
Degree 7: $P_3 S_2 S_3$.

Elements in M_i only: (3)
Degree 1: M_3.
Degree 2: $M_1^2 + M_2^2$.
Degree 3: None.
Degree 4: None.
Degree 5: None.
Degree 6: $Q_3 Q_3$.

Elements in E_{ij} and P_i only: (34)
Degree 2: N_4.
Degree 3: $N_{10}, N_{11}, P_3 N_3$.
Degree 4: $R_5 N_1, N_1 N_1, P_3 R_{13}, P_3 R_7, P_3 N_9, P_3 N_{12}$.
Degree 5: $R_7 N_3, R_9 N_1, S_3 R_4, S_2 N_1, P_3 R_4 R_5, P_3 R_5 N_2, P_3 N_1 N_2, N_1 N_7$.
Degree 6: $R_8 N_8, N_7 N_7, N_7 N_{14}, S_2 N_7, S_2 N_{14}, P_3 R_5 R_8, P_3 R_9 N_2, P_3 N_2 N_7, P_3 S_3 R_5, P_3 S_2 N_2$.
Degree 7: $P_3 R_8 R_9, P_3 R_8 N_7, P_3 N_7 N_8, P_3 N_7 N_{13}, P_3 S_2 N_8, P_3 S_2 N_{13}$.

Elements in E_{ij} and M_i only: (15)
Degree 2: T_3.
Degree 3: $\mathrm{Im}\{[2E_{12} + i(E_{11} - E_{22})](M_2 - iM_1)^2\}$,
$\mathrm{Im}\{[2E_{12} + i(E_{11} - E_{22})](M_2 - iM_1)(E_{23} - iE_{31})\}$.
Degree 4: $\mathrm{Re}\{[2E_{12} + i(E_{11} - E_{22})]^2(M_2 + iM_1)^2\}$,
$\mathrm{Re}\{[2E_{12} + i(E_{11} - E_{22})]^2(M_2 + iM_1)(E_{23} + iE_{31})\}$.
Degree 5: $T_4 \cdot \mathrm{Re}\{[2E_{12} + i(E_{11} - E_{22})^3\}$,
$\mathrm{Im}\{[2E_{12} - i(E_{11} - E_{22})](M_2 - iM_1)^4\}$,
$\mathrm{Im}\{[2E_{12} - i(E_{11} - E_{22})](M_2 - iM_1)^3(E_{23} - iE_{31})\}$,

$\text{Im}\{[2E_{12} - i(E_{11} - E_{22})](M_2 - iM_1)(E_{23} - iE_{31})^3\}$,
$\text{Im}\{[2E_{12} - i(E_{11} - E_{22})](M_2 - iM_1)^2(E_{23} - iE_{31})^2\}$.

Degree 6: $\text{Re}\{(M_2 - iM_1)^5(E_{23} - iE_{31})\}$,
$\text{Re}\{(M_2 - iM_1)^4(E_{23} - iE_{31})^2\}$,
$\text{Re}\{(M_2 - iM_1)^3(E_{23} - iE_{31})^3\}$,
$\text{Re}\{(M_2 - iM_1)^2(E_{23} - iE_{31})^4\}$,
$\text{Re}\{(M_2 - iM_1)(E_{23} - iE_{31})^5\}$.

Elements in P_i and M_i only: (13)

Degree 2: $P_1M_2 - P_2M_1$.
Degree 3: $P_3(P_1M_1 + P_2M_2)$.
Degree 4: None.
Degree 5: None.
Degree 6: $\text{Re}\{(M_2 - iM_1)^5(P_1 + iP_2)\}$,
$\text{Re}\{(M_2 - iM_1)^4(P_1 + iP_2)^2\}$,
$\text{Re}\{(M_2 - iM_1)^3(P_1 + iP_2)^3\}$,
$\text{Re}\{(M_2 - iM_1)^2(P_1 + iP_2)^4\}$,
$\text{Re}\{(M_2 - iM_1)(P_1 + iP_2)^5\}$.
Degree 7: $P_3 \cdot \text{Im}\{(M_2 - iM_1)^6\}$,
$P_3 \cdot \text{Im}\{(M_2 - iM_1)^5(P_1 + iP_2)\}$,
$P_3 \cdot \text{Im}\{(M_2 - iM_1)^4(P_1 + iP_2)^2\}$,
$P_3 \cdot \text{Im}\{(M_2 - iM_1)^3(P_1 + iP_2)^3\}$,
$P_3 \cdot \text{Im}\{(M_2 - iM_1)^2(P_1 + iP_2)^4\}$,
$P_3 \cdot \text{Im}\{(M_2 - iM_1)(P_1 + iP_2)^5\}$.

Elements in E_{ij}, P_i, and M_i: (51)

Degree 3: P_3T_4,
$\text{Im}\{[2E_{12} + i(E_{11} - E_{22})](M_2 - iM_1)(P_1 + iP_2)\}$.
Degree 4: $\text{Re}\{[2E_{12} + i(E_{11} - E_{22})]^2(M_2 + iM_1)(P_1 - iP_2)\}$,
$P_3 \cdot \text{Im}\{[2E_{12} + i(E_{11} - E_{22})](M_2 - iM_1)^2\}$,
$P_3 \cdot \text{Im}\{[2E_{12} + i(E_{11} - E_{22})(M_2 - iM_1)(E_{23} - iE_{31})\}$,
$P_3 \cdot \text{Im}\{[2E_{12} + i(E_{11} - E_{22})(M_2 - iM_1)(P_1 + iP_2)\}$.
Degree 5: $(P_1M_1 + P_2M_2) \cdot \text{Re}\{[2E_{12} + i(E_{11} - E_{22})]^3\}$,
$\text{Im}\{[2E_{12} - i(E_{11} - E_{22})](M_2 - iM_1)^3(P_1 + iP_2)\}$,
$\text{Im}\{[2E_{12} - i(E_{11} - E_{22})](M_2 - iM_1)(P_1 + iP_2)^3\}$,
$\text{Im}\{[2E_{12} - i(E_{11} - E_{22})](M_2 - iM_1)^2(P_1 + iP_2)^2\}$,
$\text{Im}\{[2E_{12} - i(E_{11} - E_{22})](M_2 - iM_1)^2(E_{23} - iE_{31})(P_1 + iP_2)\}$,
$\text{Im}\{[2E_{12} - i(E_{11} - E_{22})](E_{23} - iE_{31})^2(M_2 - iM_1)(P_1 + iP_2)\}$,
$\text{Im}\{[2E_{12} - i(E_{11} - E_{22})](M_2 - iM_1)(E_{23} - iE_{31})(P_1 + iP_2)^2\}$,
$P_3 \cdot \text{Im}\{[2E_{12} + i(E_{11} - E_{22})]^2(M_2 + iM_1)^2\}$,
$P_3 \cdot \text{Im}\{[2E_{12} + i(E_{11} - E_{22})]^2(M_2 + iM_1)(E_{23} + iE_{31})\}$,
$P_3 \cdot \text{Im}\{[2E_{12} + i(E_{11} - E_{22})]^2(M_2 + iM_1)(P_1 - iP_2)\}$.
Degree 6: $P_3 \cdot \text{Re}\{[2E_{12} - i(E_{11} - E_{22})](M_2 - iM_1)^4\}$,
$P_3 \cdot \text{Re}\{[2E_{12} - i(E_{11} - E_{22})](M_2 - iM_1)^3(E_{23} - iE_{31})\}$,
$P_3 \cdot \text{Re}\{[2E_{12} - i(E_{11} - E_{22})](M_2 - iM_1)^3(P_1 + iP_2)\}$,
$P_3 \cdot \text{Re}\{[2E_{12} - i(E_{11} - E_{22})](E_{23} - iE_{31})^3(M_2 - iM_1)\}$,
$P_3 \cdot \text{Re}\{[2E_{12} - i(E_{11} - E_{22})](P_1 + iP_2)^3(M_2 - iM_1)\}$,

$P_3 \cdot \mathrm{Re}\{[2E_{12} - i(E_{11} - E_{22})](M_2 - iM_1)^2(E_{23} - iE_{31})^2\}$,
$P_3 \cdot \mathrm{Re}\{[2E_{12} - i(E_{11} - E_{22})](M_2 - iM_1)^2(P_1 + iP_2)^2\}$,
$P_3 \cdot \mathrm{Re}\{[2E_{12} - i(E_{11} - E_{22})]$
$\times (M_2 - iM_1)^2(E_{23} - iE_{31})(P_1 + iP_2)\}$,
$P_3 \cdot \mathrm{Re}\{[2E_{12} - i(E_{11} - E_{22})]$
$\times (E_{23} - iE_{31})^2(M_2 - iM_1)(P_1 + iP_2)\}$,
$P_3 \cdot \mathrm{Re}\{[2E_{12} - i(E_{11} - E_{22})]$
$\times (P_1 + iP_2)^2(M_2 - iM_1)(E_{23} - iE_{31})\}$,
$\mathrm{Re}\{(M_2 - iM_1)^4(E_{23} - iE_{31})(P_1 + iP_2)\}$,
$\mathrm{Re}\{(M_2 - iM_1)^3(E_{23} - iE_{31})^2(P_1 + iP_2)\}$,
$\mathrm{Re}\{(M_2 - iM_1)^3(E_{23} - iE_{31})(P_1 + iP_2)^2\}$,
$\mathrm{Re}\{(M_2 - iM_1)^2(E_{23} - iE_{31})^3(P_1 + iP_2)\}$,
$\mathrm{Re}\{(M_2 - iM_1)^2(E_{23} - iE_{31})^2(P_1 + iP_2)^2\}$,
$\mathrm{Re}\{(M_2 - iM_1)^2(E_{23} - iE_{31})(P_1 + iP_2)^3\}$,
$\mathrm{Re}\{(M_2 - iM_1)(E_{23} - iE_{31})^4(P_1 + iP_2)\}$,
$\mathrm{Re}\{(M_2 - iM_1)(E_{23} - iE_{31})^3(P_1 + iP_2)^2\}$,
$\mathrm{Re}\{(M_2 - iM_1)(E_{23} - iE_{31})^2(P_1 + iP_2)^3\}$,
$\mathrm{Re}\{(M_2 - iM_1)(E_{23} - iE_{31})(P_1 + iP_2)^4\}$.

Degree 7: $P_3 \cdot \mathrm{Im}\{(M_2 - iM_1)^5(E_{23} - iE_{31})\}$,
$P_3 \cdot \mathrm{Im}\{(M_2 - iM_1)^4(E_{23} - iE_{31})^2\}$,
$P_3 \cdot \mathrm{Im}\{(M_2 - iM_1)^4(E_{23} - iE_{31})(P_1 + iP_2)\}$,
$P_3 \cdot \mathrm{Im}\{(M_2 - iM_1)^3(E_{23} - iE_{31})^3\}$,
$P_3 \cdot \mathrm{Im}\{(M_2 - iM_1)^3(E_{23} - iE_{31})^2(P_1 + iP_2)\}$,
$P_3 \cdot \mathrm{Im}\{(M_2 - iM_1)^3(E_{23} - iE_{31})(P_1 + iP_2)^2\}$,
$P_3 \cdot \mathrm{Im}\{(M_2 - iM_1)^2(E_{23} - iE_{31})^4\}$,
$P_3 \cdot \mathrm{Im}\{(M_2 - iM_1)^2(E_{23} - iE_{31})^3(P_1 + iP_2)\}$,
$P_3 \cdot \mathrm{Im}\{(M_2 - iM_1)^2(E_{23} - iE_{31})^2(P_1 + iP_2)^2\}$,
$P_3 \cdot \mathrm{Im}\{(M_2 - iM_1)^2(E_{23} - iE_{31})(P_1 + iP_2)^3\}$,
$P_3 \cdot \mathrm{Im}\{(M_2 - iM_1)(E_{23} - iE_{31})^5\}$,
$P_3 \cdot \mathrm{Im}\{(M_2 - iM_1)(E_{23} - iE_{31})^4(P_1 + iP_2)\}$,
$P_3 \cdot \mathrm{Im}\{(M_2 - iM_1)(E_{23} - iE_{31})^3(P_1 + iP_2)^2\}$,
$P_3 \cdot \mathrm{Im}\{(M_2 - iM_1)(E_{23} - iE_{31})^2(P_1 + iP_2)^3\}$,
$P_3 \cdot \mathrm{Im}\{(M_2 - iM_1)(E_{23} - iE_{31})(P_1 + iP_2)^4\}$.

7.42. Magnetic Crystal Class $\underline{6}2\underline{2}$ (see Table 3.3(13))

From Table 4.3(42) and Table B.12, we write

$$\phi, \phi' = E_{11} + E_{22}, E_{33},$$

$$\psi = P_3,$$

$$\eta = M_3,$$

$$\mathbf{A}, \mathbf{B} = \begin{pmatrix} E_{23} \\ -E_{31} \end{pmatrix}, \begin{pmatrix} P_1 \\ P_2 \end{pmatrix},$$

$$\mathbf{a}, \mathbf{b} = \begin{pmatrix} M_2 \\ -M_1 \end{pmatrix}, \begin{pmatrix} 2E_{12} \\ E_{11} - E_{22} \end{pmatrix}.$$

The TMEs of the integrity basis are given by (B.12). The actual elements are listed below.

Elements in E_{ij} only: (9)
Same elements as in the class $6\underline{\underline{22}}$.

Elements in P_i only: (4)
Same as in the class $6\underline{\underline{22}}$.

Elements in M_i only: (3)
Degree 2: M_3^2, Q_1.
Degree 3: Q_2.

Elements in E_{ij} and P_i only: (34)
Same elements as in the class $6\underline{\underline{22}}$.

Elements in E_{ij} and M_i only: (13)
Degree 2: T_1.
Degree 3: $M_3T_3, M_3R_4,$
$\mathrm{Im}\{[2E_{12} + i(E_{11} - E_{22})](M_2 - iM_1)^2\},$
$\mathrm{Im}\{[2E_{12} + i(E_{11} - E_{22}]^2(M_2 - iM_1)\},$
$\mathrm{Im}\{[(M_2 - iM_1)(E_{23} - iE_{31})^2\}.$
Degree 4: $M_3R_8,$
$M_3 \cdot \mathrm{Im}\{(E_{23} - iE_{31})(M_2 - iM_1)^2\},$
$M_3 \cdot \mathrm{Im}\{(E_{23} - iE_{31})(M_2 - iM_1)[2E_{12} + i(E_{11} - E_{22})]\},$
$M_3 \cdot \mathrm{Im}\{(E_{23} - iE_{31})[2E_{12} + i(E_{11} - E_{22})]^2\},$
$\mathrm{Re}\{(M_2 - iM_1)^2(E_{23} - iE_{31})^2\},$
$\mathrm{Re}\{(M_2 - iM_1)[2E_{12} + i(E_{11} - E_{22})](E_{23} - iE_{31})^2\}.$
Degree 5: $\mathrm{Im}\{(M_2 + iM_1)(E_{23} - iE_{31})^4\}.$

Elements in P_i and M_i only: (13)
Degree 2: None.
Degree 3: $\mathrm{Im}\{(M_2 - iM_1)(P_1 + iP_2)^2, M_3(P_1M_2 - P_2M_1)\}.$
Degree 4: $P_3Q_3, M_3S_3, (P_1M_1 + P_2M_2)P_3M_3,$
$M_3 \cdot \mathrm{Im}\{(P_1 + iP_2)(M_2 - iM_1)^2\},$
$P_3 \cdot \mathrm{Im}\{(M_2 - iM_1)(P_1 + iP_2)^2\},$
$\mathrm{Re}\{(M_2 - iM_1)^2(P_1 - iP_2)^2\}.$
Degree 5: $P_3M_3S_2,$
$P_3M_3 \cdot \mathrm{Re}\{(P_1 + iP_2)(M_2 - iM_1)^2\},$
$\mathrm{Im}\{(M_2 + iM_1)(P_1 + iP_2)^4\},$
$P_3 \cdot \mathrm{Im}\{(M_2 - iM_1)^2(P_1 - iP_2)^2\}.$
Degree 6: $P_3 \cdot \mathrm{Im}\{(M_2 + iM_1)(P_1 + iP_2)^4\}.$

Elements in E_{ij}, P_i and M_i: (41)
Degree 3: $P_3T_2, M_3N_2,$
$\mathrm{Im}(M_2 - iM_1)(E_{23} - iE_{31})(P_1 + iP_2).$
Degree 4: $T_2N_3, P_3M_3R_5, P_3M_3N_1, P_3M_3T_4,$
$P_3 \cdot \mathrm{Re}\{(M_2 - iM_1)(E_{23} - iE_{31})^2\},$

$$\begin{aligned}
&P_3 \cdot \operatorname{Re}\{(M_2 - iM_1)(E_{23} - iE_{31})(P_1 + iP_2)\},\\
&P_3 \cdot \operatorname{Re}\{(M_2 - iM_1)^2[2E_{12} + i(E_{11} - E_{22})]\},\\
&P_3 \cdot \operatorname{Re}\{(M_2 - iM_1)[2E_{12} + i(E_{11} - E_{22})]^2\},\\
&M_3 \cdot \operatorname{Im}\{(P_1 + iP_2)(M_2 - iM_1)[2E_{12} + i(E_{11} - E_{22})]\},\\
&M_3 \cdot \operatorname{Im}\{(P_1 + iP_2)[2E_{12} + i(E_{11} - E_{22})]^2\},\\
&M_3 \cdot \operatorname{Im}\{(P_1 + iP_2)(E_{23} - iE_{31})^2\},\\
&M_3 \cdot \operatorname{Im}\{(P_1 + iP_2)^2(E_{23} - iE_{31})\},\\
&\operatorname{Re}\{(M_2 - iM_1)^2(E_{23} + iE_{31})(P_1 + iP_2)\},\\
&\operatorname{Re}\{(M_2 - iM_1)[2E_{12} + i(E_{11} - E_{22})](E_{23} + iE_{31})(P_1 - iP_2)\},\\
&\operatorname{Re}\{(M_2 - iM_1)[2E_{12} + i(E_{11} - E_{22})](P_1 - iP_2)^2\}.
\end{aligned}$$

Degree 5:

$$\begin{aligned}
&Q_3N_3,\ P_3M_3R_9,\\
&N_3 \cdot \operatorname{Re}\{(M_2 - iM_1)^2[2E_{12} + i(E_{11} - E_{22})]\},\\
&N_3 \cdot \operatorname{Re}\{(M_2 - iM_1)[2E_{12} + i(E_{11} - E_{22})]^2\},\\
&\operatorname{Im}\{(M_2 + iM_1)(E_{23} - iE_{31})^3(P_1 + iP_2)\},\\
&\operatorname{Im}\{(M_2 + iM_1)(E_{23} - iE_{31})^2(P_1 + iP_2)^2\},\\
&\operatorname{Im}\{(M_2 + iM_1)(E_{23} - iE_{31})(P_1 + iP_2)^3\},\\
&P_3 \cdot \operatorname{Im}\{(M_2 - iM_1)^2(E_{23} + iE_{31})^2\},\\
&P_3 \cdot \operatorname{Im}\{(M_2 - iM_1)^2(E_{23} + iE_{31})(P_1 - iP_2)\},\\
&P_3 \cdot \operatorname{Im}\{(M_2 - iM_1)[2E_{12} + i(E_{11} - E_{22})](E_{23} + iE_{31})^2\},\\
&P_3 \cdot \operatorname{Im}\{(M_2 - iM_1)[2E_{12} + i(E_{11} - E_{22})]\\
&\times (E_{23} + iE_{31})(P_1 - iP_2)\},\\
&P_3 \cdot \operatorname{Im}\{(M_2 - iM_1)[2E_{12} + i(E_{11} - E_{22})](P_1 - iP_2)^2\},\\
&P_3M_3 \cdot \operatorname{Re}\{(E_{23} - iE_{31})^2(P_1 + iP_2)\},\\
&P_3M_3 \cdot \operatorname{Re}\{(E_{23} - iE_{31})(P_1 + iP_2)^2\},\\
&P_3M_3 \cdot \operatorname{Re}\{(E_{23} - iE_{31})(M_2 - iM_1)^2\},\\
&P_3M_3 \cdot \operatorname{Re}\{(E_{23} - iE_{31})(M_2 - iM_1)[2E_{12} + i(E_{11} - E_{22})]\},\\
&P_3M_3 \cdot \operatorname{Re}\{(E_{23} - iE_{31})[2E_{12} + i(E_{11} - E_{22})]^2\},\\
&P_3M_3 \cdot \operatorname{Re}\{(P_1 + iP_2)(M_2 - iM_1)[2E_{12} + i(E_{11} - E_{22})]\},\\
&P_3M_3 \cdot \operatorname{Re}\{(P_1 + iP_2)[2E_{12} + i(E_{11} - E_{22}]^2\}.
\end{aligned}$$

Degree 6:

$$\begin{aligned}
&P_3 \cdot \operatorname{Re}\{(M_2 + iM_1)(E_{23} - iE_{31})^4\},\\
&P_3 \cdot \operatorname{Re}\{(M_2 + iM_1)(E_{23} - iE_{31})^3(P_1 + iP_2)\},\\
&P_3 \cdot \operatorname{Re}\{(M_2 + iM_1)(E_{23} - iE_{31})^2(P_1 + iP_2)^2\},\\
&P_3 \cdot \operatorname{Re}\{(M_2 + iM_1)(E_{23} - iE_{31})(P_1 + iP_2)^3\}.
\end{aligned}$$

7.43. Magnetic Crystal Class $6\underline{m}\underline{m}$ (see Table 3.3(13))

From Table 4.3(43) and Table B.12, we have

$$\phi, \phi', \phi'', \phi''' = M_3,\ E_{11} + E_{22},\ E_{33},\ P_3,$$

$$\mathbf{A}, \mathbf{B}, \mathbf{C} = \begin{pmatrix} M_1 \\ M_2 \end{pmatrix}, \begin{pmatrix} E_{13} \\ E_{23} \end{pmatrix}, \begin{pmatrix} P_1 \\ P_2 \end{pmatrix},$$

$$\mathbf{a} = \begin{pmatrix} 2E_{12} \\ E_{11} - E_{22} \end{pmatrix}.$$

The TMEs of the integrity basis are given by (B.12). The actual elements are listed below.

Elements in E_{ij} only: (9)
Degree 1: R_0, R_1.
Degree 2: R_2, R_3.
Degree 3: R_6, R_{10}.
Degree 4: $R_4 R_4$.
Degree 5: $R_4 R_8$.
Degree 6: $R_8 R_8$.

Elements in P_i only: (3)
Degree 1: P_3.
Degree 2: S_1.
Degree 3: None.
Degree 4: None.
Degree 5: None.
Degree 6: $S_2 S_2$.

Elements in M_i only: (3)
Degree 1: M_3.
Degree 2: Q_1.
Degree 3: None.
Degree 4: None.
Degree 5: None.
Degree 6: $Q_2 Q_2$.

Elements in E_{ij} and P_i only: (15)
Degree 2: N_3.
Degree 3: N_9, N_{11}.
Degree 4: $R_5 N_2, N_1 N_1$.
Degree 5: $R_7 N_4, R_9 N_2, N_1 N_7, S_3 R_5, S_2 N_1$.
Degree 6: $R_8 N_7, N_7 N_7, N_7 N_{13}, S_2 N_7, S_2 N_{13}$.

Elements in E_{ij} and M_i only: (15)
Degree 2: T_3.
Degree 3: T_9, T_{11}.
Degree 4: $R_5 T_2, T_1 T_1$.
Degree 5: $R_7 T_4, R_9 T_2, T_1 T_7, Q_3 R_5, Q_2 T_1$.
Degree 6: $R_8 T_7, T_7 T_7, T_7 T_{13}, Q_2 T_7, Q_2 T_{13}$.

Elements in P_i and M_i only: (6)
Degree 2: $P_1 M_1 + P_2 M_2$.
Degree 3: None.
Degree 4: None.
Degree 5: None.

Degree 6: $S_2Q_2, S_2J_1(\mathbf{M}, \mathbf{P}), S_2J_1(\mathbf{P}, \mathbf{M})$,
$Q_2J_1(\mathbf{M}, \mathbf{P}), Q_2J_1(\mathbf{P}, \mathbf{M})$.

Elements in E_{ij}, P_i, and M_i: (19)
Degree 3: $\operatorname{Im}\{(M_1 + iM_2)(P_1 + iP_2)[2E_{12} + i(E_{11} - E_{22})]\}$.
Degree 4: $\operatorname{Im}\{(M_1 - iM_2)(P_1 - iP_2)[2E_{12} + i(E_{11} - E_{22})]^2\}$.
Degree 5: $(P_1M_2 - P_2M_1)R_7$,
$\operatorname{Im}\{[2E_{12} - i(E_{11} - E_{22})](M_1 + iM_2)^3(P_1 + iP_2)\}$,
$\operatorname{Im}\{[2E_{12} - i(E_{11} - E_{22})](P_1 + iP_2)^3(M_1 + iM_2)\}$,
$\operatorname{Im}\{[2E_{12} - i(E_{11} - E_{22})](M_1 + iM_2)^2(P_1 + iP_2)^2\}$,
$\operatorname{Im}\{[2E_{12} - i(E_{11} - E_{22})](M_1 + iM_2)^2(E_{13} + iE_{23})(P_1 + iP_2)\}$,
$\operatorname{Im}\{[2E_{12} - i(E_{11} - E_{22})](E_{13} + iE_{23})^2(M_1 + iM_2)(P_1 + iP_2)\}$,
$\operatorname{Im}\{[2E_{12} - i(E_{11} - E_{22})](P_1 + iP_2)^2(M_1 + iM_2)(E_{13} + iE_{23})\}$.
Degree 6: $\operatorname{Re}\{(M_1 + iM_2)^4(E_{13} + iE_{23})(P_1 + iP_2)\}$,
$\operatorname{Re}\{(M_1 + iM_2)^3(E_{13} + iE_{23})^2(P_1 + iP_2)\}$,
$\operatorname{Re}\{(M_1 + iM_2)^3(E_{13} + iE_{23})(P_1 + iP_2)^2\}$,
$\operatorname{Re}\{(M_1 + iM_2)^2(E_{13} + iE_{23})^3(P_1 + iP_2)\}$,
$\operatorname{Re}\{(M_1 + iM_2)^2(E_{13} + iE_{23})^2(P_1 + iP_2)^2\}$,
$\operatorname{Re}\{(M_1 + iM_2)^2(E_{13} + iE_{23})(P_1 + iP_2)^3\}$,
$\operatorname{Re}\{(M_1 + iM_2)(E_{13} + iE_{23})^4(P_1 + iP_2)\}$,
$\operatorname{Re}\{(M_1 + iM_2)(E_{13} + iE_{23})^3(P_1 + iP_2)^2\}$,
$\operatorname{Re}\{(M_1 + iM_2)(E_{13} + iE_{23})^2(P_1 + iP_2)^3\}$,
$\operatorname{Re}\{(M_1 + iM_2)(E_{13} + iE_{23})(P_1 + iP_2)^4\}$.

7.44. Magnetic Crystal Class $\underline{6}m\underline{m}$ (see Table 3.3(13))

From Table 4.3(44) and Table B.12, we write

$$\phi, \phi', \phi'' = E_{11} + E_{22}, E_{33}, P_3,$$
$$\xi = M_3,$$
$$\mathbf{A}, \mathbf{B} = \begin{pmatrix} E_{13} \\ E_{23} \end{pmatrix}, \begin{pmatrix} P_1 \\ P_2 \end{pmatrix},$$
$$\mathbf{a}, \mathbf{b} = \begin{pmatrix} M_2 \\ -M_1 \end{pmatrix}, \begin{pmatrix} 2E_{12} \\ E_{11} - E_{22} \end{pmatrix}.$$

The TMEs of the integrity basis are given by (B.12). The actual elements are listed below.

Elements in E_{ij} only: (9)
Same elements as in the class $6\underline{m}\underline{m}$.

Elements in P_i only: (3)
Same as in the class $6\underline{m}\underline{m}$.

Elements in M_i only: (3)
Degree 1: None.
Degree 2: M_3^2, Q_1.
Degree 3: Q_2.

Elements in E_{ij} and P_i only: (15)
Same as in the class $6\underline{m}\underline{m}$.

Elements in E_{ij} and M_i only: (13)
Degree 2: T_1.
Degree 3: $M_3 T_3$, $M_3 R_4$,
$\operatorname{Im}\{(M_2 - iM_1)^2[2E_{12} + i(E_{11} - E_{22})]\}$,
$\operatorname{Im}\{(M_2 - iM_1)[2E_{12} + i(E_{11} - E_{22})]^2\}$,
$\operatorname{Im}\{(M_2 - iM_1)(E_{13} + iE_{23})^2\}$.
Degree 4: $M_3 R_8$,
$\operatorname{Re}\{(M_2 - iM_1)^2(E_{13} - iE_{23})^2\}$,
$\operatorname{Re}\{(M_2 - iM_1)[2E_{12} + i(E_{11} - E_{22})](E_{13} - iE_{23})^2\}$,
$M_3 \cdot \operatorname{Re}\{(M_2 - iM_1)^2(E_{13} + iE_{23})\}$,
$M_3 \cdot \operatorname{Re}\{(M_2 - iM_1)[2E_{12} + i(E_{11} - E_{22})](E_{13} + E_{23})\}$,
$M_3 \cdot \operatorname{Re}\{[2E_{12} + i(E_{11} - E_{22})]^2(E_{13} + iE_{23})\}$.
Degree 5: $\operatorname{Im}\{(M_2 + iM_1)(E_{13} + iE_{23})^4\}$.

Elements in P_i and M_i only: (6)
Degree 2: None.
Degree 3: $M_3(P_1 M_1 + P_2 M_2)$,
$\operatorname{Im}\{(M_2 - iM_1)(P_1 + iP_2)^2\}$.
Degree 4: $M_3 \cdot \operatorname{Re}\{(M_2 - iM_1)^2(P_1 + iP_2)\}$, $M_3 S_2$,
$\operatorname{Re}\{(M_2 - iM_1)^2(P_1 - iP_2)^2\}$.
Degree 5: $\operatorname{Im}\{(M_2 + iM_1)(P_1 + iP_2)^4\}$.

Elements in E_{ij}, P_i and M_i: (16)
Degree 3: $M_3 N_1$,
$\operatorname{Im}\{(M_2 - iM_1)(E_{13} + iE_{23})(P_1 + iP_2)\}$.
Degree 4: $T_2 N_4$,
$\operatorname{Re}\{(M_2 - iM_1)^2(E_{13} - iE_{23})(P_1 - iP_2)\}$,
$\operatorname{Re}\{(M_2 - iM_1)[2E_{12} + i(E_{11} - E_{22})](E_{13} - iE_{23})(P_1 - iP_2)\}$,
$\operatorname{Re}\{(M_2 - iM_1)[2E_{12} + i(E_{11} - E_{22})](P_1 - iP_2)^2\}$,
$M_3 \cdot \operatorname{Re}\{(M_2 - iM_1)[2E_{12} + i(E_{11} - E_{22})](P_1 + iP_2)\}$,
$M_3 \cdot \operatorname{Re}\{[2E_{12} + i(E_{11} - E_{22})]^2(P_1 + iP_2)\}$,
$M_3 \cdot \operatorname{Re}\{(E_{13} + iE_{23})^2(P_1 + iP_2)\}$,
$M_3 \cdot \operatorname{Re}\{(E_{13} + iE_{23})(P_1 + iP_2)^2\}$.
Degree 5: $N_4 \cdot \operatorname{Re}\{(M_2 - iM_1)^3\}$,
$N_4 \cdot \operatorname{Re}\{(M_2 - iM_1)^2[2E_{12} + i(E_{11} - E_{22})]\}$,
$N_4 \cdot \operatorname{Re}\{(M_2 - iM_1)[2E_{12} + i(E_{11} - E_{22})]^2\}$,
$\operatorname{Im}\{(M_2 + iM_1)(E_{13} + iE_{23})^3(P_1 + iP_2)\}$,
$\operatorname{Im}\{(M_2 + iM_1)(E_{13} + iE_{23})^2(P_1 + iP_2)^2\}$,
$\operatorname{Im}\{(M_2 + iM_1)(E_{13} + iE_{23})(P_1 + iP_2)^3\}$.

7.45. Magnetic Crystal Class $6/\underline{m}$ (see Table 3.3(12))

From Table 4.3(45) and Table B.13, we write

$$\begin{aligned} \phi, \phi' &= E_{11} + E_{22}, E_{33}, \\ a &= E_{11} - E_{22} + i2E_{12}, \\ A &= E_{13} - iE_{23}, \\ \pi, \pi' &= M_3, P_3, \\ x, y &= M_1 - iM_2, P_1 - iP_2. \end{aligned} \tag{7.42}$$

The TMEs of the integrity basis are given by (B.14). The actual elements are listed below.

Elements in E_{ij} only: (14)
Degree 1: R_0, R_1.
Degree 2: R_2, R_3.
Degree 3: R_6, R_7, R_{10}, R_{13}.
Degree 4: R_4R_4, R_4R_5.
Degree 5: R_4R_8, R_5R_8.
Degree 6: R_8R_8, R_8R_9.

Elements in P_i only: (4)
Degree 1: None.
Degree 2: P_3^2, S_1.
Degree 3: None.
Degree 4: None.
Degree 5: None.
Degree 6: S_2S_2, S_2S_3.

Elements in M_i only: (4)
Degree 1: None.
Degree 2: M_3^2, Q_1.
Degree 3: None.
Degree 4: None.
Degree 5: None.
Degree 6: Q_2Q_2, Q_2Q_3.

Elements in E_{ij} and P_i only: (30)
Degree 2: None.
Degree 3: $P_3N_3, P_3N_4, N_{11}, N_{12}$.
Degree 4: $P_3N_9, P_3N_{10}, N_3N_3, N_3N_4, N_1N_1, N_1N_2$.
Degree 5: $P_3R_5N_1, P_3R_5N_2, N_1N_7, N_2N_7, S_2N_1, S_2N_2$.
Degree 6: $P_3R_9N_2, P_3R_9N_1, N_7N_7, N_7N_8, S_2N_7, S_2N_8, P_3S_3R_4, P_3S_3R_5$.
Degree 7: $P_3R_8N_7, P_3R_8N_8, P_3N_7N_{13}, P_3N_7N_{14}, P_3S_2N_{13}, P_3S_2N_{14}$.

Elements in E_{ij} and M_i only: (30)

Degree 2: None.

Degree 3: $M_3 T_3, M_3 T_4, T_{11}, T_{12}$.

Degree 4: $M_3 T_9, M_3 T_{10}, T_3 T_3, T_3 T_4, T_1 T_1, T_1 T_2$.

Degree 5: $M_3 R_5 T_1, M_3 R_5 T_2, T_1 T_7, T_2 T_7, Q_2 T_1, Q_2 T_2$.

Degree 6: $M_3 R_9 T_2, M_3 R_9 T_1, T_7 T_7, T_7 T_8, Q_2 T_7, Q_2 T_8, M_3 Q_3 R_4, M_3 Q_3 R_5$.

Degree 7: $M_3 R_8 T_7, M_3 R_8 T_8, M_3 T_7 T_{13}, M_3 T_7 T_{13}, M_3 Q_2 T_{14}, M_3 Q_2 T_{14}$.

Elements in P_i and M_i only: (13)

Degree 2: $P_3 M_3, P_1 M_1 + P_2 M_2, P_1 M_2 - P_2 M_1$.

Degree 3: None.

Degree 4: None.

Degree 5: None.

Degree 6: $S_2 J_1(\mathbf{M}, \mathbf{P}), S_2 J_2(\mathbf{M}, \mathbf{P})$,
$Q_2 J_1(\mathbf{P}, \mathbf{M}), Q_2 J_2(\mathbf{P}, \mathbf{M})$,
$S_2 J_1(\mathbf{P}, \mathbf{M}), S_2 J_2(\mathbf{P}, \mathbf{M})$,
$Q_2 J_1(\mathbf{M}, \mathbf{P}), Q_2 J_2(\mathbf{M}, \mathbf{P})$,
$S_2 Q_2, S_2 Q_3$,

where $J_1(\mathbf{x}, \mathbf{y})$ and $J_2(\mathbf{x}, \mathbf{y})$ are defined by (7.38) and (7.41).

Elements in E_{ij}, P_i and M_i: (56)

Degree 3: $M_3 N_3, M_3 N_4, P_3 T_3, P_3 T_4$,
$\mathrm{Re}\{(M_1 - iM_2)(P_1 - iP_2)[(E_{11} - E_{22}) + 2iE_{12})\}$,
$\mathrm{Im}\{(M_1 - iM_2)(P_1 - iP_2)[(E_{11} - E_{22}) + 2iE_{12}]\}$.

Degree 4: $\mathrm{Re}\{[(E_{11} - E_{22}) + i2E_{12}]^2(M_1 + iM_2)(P_1 + iP_2)\}$,
$\mathrm{Im}\{[(E_{11} - E_{22}) + i2E_{12}]^2(M_1 + iM_2)(P_1 + iP_2)\}$,
$\mathrm{Re}\{(E_{13} - iE_{23})^2(M_1 + iM_2)(P_1 + iP_2)\}$,
$\mathrm{Im}\{(E_{13} - iE_{23})^2(M_1 + iM_2)(P_1 + iP_2)\}$,
$M_3 \cdot \mathrm{Re}\{[(E_{11} - E_{22}) + i2E_{12}](E_{13} - iE_{23})(P_1 - iP_2)\}$,
$M_3 \cdot \mathrm{Im}\{[(E_{11} - E_{22}) + i2E_{12}](E_{13} - iE_{23})(P_1 - iP_2)\}$,
$P_3 \cdot \mathrm{Re}\{[(E_{11} - E_{22}) + i2E_{12}](E_{13} - iE_{23})(M_1 - iM_2)\}$,
$P_3 \cdot \mathrm{Im}\{[(E_{11} - E_{22}) + i2E_{12}](E_{13} - iE_{23})(M_1 - iM_2)\}$.

Degree 5: $\mathrm{Re}\{\bar{a}xxxy\}, \mathrm{Im}\{\bar{a}xxxy\}$,
$\mathrm{Re}\{\bar{a}xxyy\}, \mathrm{Im}\{\bar{a}xxyy\}$,
$\mathrm{Re}\{\bar{a}xyyy\}, \mathrm{Im}\{\bar{a}xyyy\}$,
$\mathrm{Re}\{\bar{a}AAxy\}, \mathrm{Im}\{\bar{a}AAxy\}$,
$P_3 \cdot \mathrm{Re}\{aa\bar{A}\bar{x}\}, P_3 \cdot \mathrm{Im}\{aa\bar{A}\bar{x}\}$,
$M_3 \cdot \mathrm{Re}\{aa\bar{A}\bar{y}\}, M_3 \cdot \mathrm{Im}\{aa\bar{A}\bar{y}\}$.

Degree 6: $\mathrm{Re}\{AAAAxy\}, \mathrm{Im}\{AAAAxy\}$,
$M_3 \cdot \mathrm{Re}\{\bar{a}AAAy\}, M_3 \cdot \mathrm{Im}\{\bar{a}AAAy\}$,
$P_3 \cdot \mathrm{Re}\{\bar{a}AAAx\}, P_3 \cdot \mathrm{Im}\{\bar{a}AAAx\}$,
$P_3 \cdot \mathrm{Re}\{\bar{a}Axxx\}, P_3 \cdot \mathrm{Im}\{\bar{a}Axxx\}$,
$P_3 \cdot \mathrm{Re}\{\bar{a}Axxy\}, P_3 \cdot \mathrm{Im}\{\bar{a}Axxy\}$,
$P_3 \cdot \mathrm{Re}\{\bar{a}Axyy\}, P_3 \cdot \mathrm{Im}\{\bar{a}Axyy\}$,
$M_3 \cdot \mathrm{Re}\{\bar{a}Axxy\}, M_3 \cdot \mathrm{Im}\{\bar{a}Axxy\}$,
$M_3 \cdot \mathrm{Re}\{\bar{a}Axyy\}, M_3 \cdot \mathrm{Im}\{\bar{a}Axyy\}$,
$M_3 \cdot \mathrm{Re}\{\bar{a}Ayyy\}, M_3 \cdot \mathrm{Im}\{\bar{a}Ayyy\}$,

$\text{Re}\{AAxxxy\}$, $\text{Im}\{AAxxxy\}$,
$\text{Re}\{AAxxyy\}$, $\text{Im}\{AAxxyy\}$,
$\text{Re}\{AAxyyy\}$, $\text{Im}\{AAxyyy\}$.

Degree 7: Real and imaginary parts of the following elements:

$M_3 AAAAAy, \; P_3 AAAAAx,$
$M_3 AAAxxy, \; P_3 AAAxxx,$
$M_3 AAAxyy, \; P_3 AAAxxy,$
$M_3 AAAyyy, \; P_3 AAAxyy,$
$M_3 Axxxxy, \; P_3 Axxxxx,$
$M_3 Axxyyy, \; P_3 Axxxyy,$
$M_3 Axyyyy, \; P_3 Axxyyy,$
$M_3 Ayyyyyy, \; P_3 Axyyyy,$
$M_3 Axxxyy, \; P_3 Axxxxy,$

where the complex quantities x, y, A, a are defined by (7.42).

7.46. Magnetic Crystal Class $\underline{6}/m$ (see Table 3.3(12))

From Table 4.3(46) and Table B.13, we write

$$\phi, \phi' = E_{11} + E_{22}, E_{33},$$

$$a = E_{11} - E_{22} + i2E_{12},$$

$$A = E_{13} - iE_{23},$$

$$\pi = P_3, \qquad\qquad (7.43)$$

$$X = M_1 - iM_2,$$

$$\delta = M_3,$$

$$x = P_1 - iP_2.$$

The TMEs of the integrity basis are given by (B.14). The actual elements are listed below.

Elements in E_{ij} only: (14)
Same as in the class $6/\underline{m}$.

Elements in P_i only: (4)
Same as in the class $6/\underline{m}$.

Elements in M_i only: (4)
Degree 1: None.
Degree 2: $M_3^2, M_1^2 + M_2^2.$
Degree 3: None.
Degree 4: None.
Degree 5: None.
Degree 6: $Q_2Q_2, Q_2Q_3.$

Elements in E_{ij} and P_i only: (30)
Same as in the class $6/\underline{m}$.

Elements in E_{ij} and M_i only: (30)

Degree 2: None.

Degree 3: $M_3T_3, M_3T_4, T_{11}, T_{12}$.

Degree 4: $T_3T_3, T_3T_4, T_1T_1, T_1T_2, M_3T_9, M_3T_{10}$.

Degree 5: $M_3R_5T_1, M_3R_5T_2, T_1T_7, T_2T_7, Q_2T_1, Q_2T_2$.

Degree 6: $M_3R_9T_2, M_3R_9T_1, T_7T_7, T_7T_8, Q_2T_7, Q_2T_8, M_3Q_3R_4, M_3Q_3R_5$.

Degree 7: $M_3R_8T_7, M_3R_8T_8, M_3T_7T_{13}, M_3T_7T_{14}, M_3Q_2T_{13}, M_3Q_2T_{14}$.

Elements in P_i and M_i only: (19)

Degree 2: None.

Degree 3: None.

Degree 4: $P_3Q_2, P_3Q_3, P_3M_3(P_1M_1 + P_2M_2)$,
$P_3M_3(P_1M_2 - P_2M_1), M_3S_2, M_3S_3$,
$(P_1M_1 + P_2M_2)(P_1M_2 - P_2M_1)$,
$(P_1M_1 + P_2M_2)^2 - (P_1M_2 - P_2M_1)^2$
$P_3J_1(\mathbf{M}, \mathbf{P}), P_3J_2(\mathbf{M}, \mathrm{P})$,
$M_3J_1(\mathbf{P}, \mathbf{M}), M_3J_2(\mathbf{P}, \mathbf{M})$.

Degree 5: None.

Degree 6: Both real and imaginary parts of the following elements,
$XXXXxx$,
$xxxxXX$,
$P_3xxxx\bar{X}$,
$M_3XXXX\bar{x}$.

Elements in E_{ij}, P_i, and M_i: (84)

Degree 3: $M_3N_1, M_3N_2, P_3T_1, P_3T_2$,
$\mathrm{Re}\{(E_{13} - iE_{23})(M_1 - iM_2)(P_1 - iP_2)\}$,
$\mathrm{Im}\{(E_{13} - iE_{23})(M_1 - iM_2)(P_1 - iP_2)\}$.

Degree 4: Both real and imaginary parts of the elements
$a\bar{A}X\bar{x}, a\bar{A}\bar{X}x, aA\bar{X}\bar{x}$,
P_3aaX, P_3AAX,
M_3aax, M_3AAx, and $P_3M_3R_4, P_3M_3R_5$.

Degree 5: Both real and imaginary parts of the following elements
$\bar{a}XXxx, P_3AA\bar{a}\bar{X}, P_3M_3aaA$,
$AAA\bar{X}x, P_3xx\bar{a}\bar{X}, P_3M_3AAA$,
$AAAX\bar{x}, P_3XX\bar{A}\bar{x}, P_3M_3XXA$,
$XXX\bar{A}x, M_3AA\bar{a}\bar{x}, P_3M_3xxA$,
$XXXA\bar{x}, M_3XX\bar{a}\bar{x}, P_3M_3aXx$,
$aaxA\bar{X}, M_3xx\bar{A}\bar{X}$,
$aax\bar{A}X$,
$aa\bar{x}AX$,
$xxxA\bar{X}$,
$xxx\bar{A}X$.

Degree 6: Both real and imaginary parts of the following elements
$aAx\bar{X}\bar{X}\bar{X}, P_3AAAA\bar{X}, M_3AAAA\bar{x}$,
$aAX\bar{x}\bar{x}\bar{x}, P_3AAxxX, M_3AAXX\bar{x}$,
$aXx\bar{A}\bar{A}\bar{A}$.

Degree 7: Both real and imaginary parts of the elements

$AAAAA\bar{X}\bar{x}$,

$XXXXX\bar{A}\bar{x}$,

$xxxxx\bar{A}\bar{X}$,

where the complex quantities a, A, x, X are defined in (7.43).

7.47. Magnetic Crystal Class $\underline{6}/\underline{m}$ (see Table 3.3(12))

From Table 4.3(47) and Table B.13, we write

$$\begin{aligned}
\phi, \phi' &= E_{11} + E_{22}, E_{33}, \\
a, b &= M_1 - iM_2, E_{11} - E_{22} + i2E_{12}, \\
\xi &= M_3, \\
A &= E_{13} - iE_{23}, \\
\pi &= P_3, \\
x &= P_1 - iP_2.
\end{aligned} \tag{7.44}$$

The TMEs of the integrity basis are given by (B.14). The actual elements are listed below.

Elements in E_{ij} only: (14)
Same as in the class $6/\underline{m}$.

Elements in P_i only: (4)
Same as in the class $6/\underline{m}$.

Elements in M_i only: (4)
Degree 1: None.
Degree 2: M_3^2, $M_1^2 + M_2^2$.
Degree 3: Q_2, Q_3.

Elements in E_{ij} and P_i only: (30)
Same as in the class $6/\underline{m}$.

Elements in E_{ij} and M_i only: (26)
Degree 2: T_1, T_2.
Degree 3: M_3T_3, M_3T_4, M_3R_4, M_3R_5,
and both real and imaginary parts of the following elements:

aab, abb, AAa.

Degree 4: $aa\bar{A}\bar{A}$, $ab\bar{A}\bar{A}$, M_3Aaa,

M_3Aab,

M_3Abb,

M_3AAA.

Degree 5: $AAAA\bar{a}$.

Elements in P_i and M_i only: (12)
Degree 2: None.
Degree 3: Both real and imaginary parts of the following elements:

xxa.

Degree 4: $aa\bar{x}\bar{x}$, $P_3 M_3 \bar{x}a$.
Degree 5: $P_3 M_3 xaa$, $P_3 M_3 xxx$, $xxxx\bar{a}$.

Elements in E_{ij}, P_i *and* M_i: (30)
Degree 3: None.
Degree 4: Both real and imaginary parts of the following elements:
$ab\bar{x}\bar{x}$, $P_3 Axa$, $M_3 xxA$, $P_3 M_3 \bar{x}b$.
Degree 5: $AAxx\bar{a}$, $P_3 \bar{A}\bar{x}aa$, $M_3 xx\bar{A}\bar{a}$, $P_3 \bar{A}\bar{x}ab$, $M_3 xx\bar{A}\bar{b}$,
$P_3 M_3 AAx$, $P_3 M_3 xab$, $P_3 M_3 xbb$,
Degree 6: $P_3 xAAA\bar{a}$, $M_3 xxxx\bar{A}$, $P_3 Axxx\bar{a}$,
where the complex quantities a, b, A, x are defined by (7.44).

7.48. Magnetic Crystal Class $6/\underline{mmm}$ (see Table 3.3(13))

From Table 4.3(49) and Table B.14, we write

$$\begin{aligned} \phi, \phi' &= E_{11} + E_{22}, E_{33}, \\ \mathbf{A} &= \begin{pmatrix} E_{23} \\ -E_{31} \end{pmatrix}, \\ \mathbf{a} &= \begin{pmatrix} 2E_{12} \\ E_{11} - E_{22} \end{pmatrix}, \\ \rho, \rho' &= M_3, P_3, \\ \mathbf{X}, \mathbf{Y} &= \begin{pmatrix} M_1 \\ M_2 \end{pmatrix}, \begin{pmatrix} P_1 \\ P_2 \end{pmatrix}. \end{aligned} \tag{7.45}$$

The TMEs of the integrity basis are given in Appendix B.

Elements in E_{ij} *only*: (Same as in the class $6\underline{mm}$) (9)
Degree 1: R_0, R_1.
Degree 2: R_2, R_3.
Degree 3: R_6, R_{10}.
Degree 4: $R_4 R_4$.
Degree 5: $R_4 R_8$.
Degree 6: $R_8 R_8$.

Elements in P_i *only*: (3)
Degree 1: None.
Degree 2: P_1^2, $P_2^2 + P_2^2 \equiv S_1$.
Degree 3: None.
Degree 4: None.
Degree 5: None.
Degree 6: $S_2 S_2$.

Elements in M_i *only*: (3)
Degree 1: None.
Degree 2: M_3^2, Q_1.

Degree 3: None.
Degree 4: None.
Degree 5: None.
Degree 6: Q_2Q_2.

Elements in E_{ij} and P_i only: (16)
Degree 2: None.
Degree 3: P_3N_3, N_{11}.
Degree 4: P_3N_9, N_4N_4, N_1N_1.
Degree 5: $P_3R_5N_2$, N_1N_7, S_2N_1.
Degree 6: $P_3R_9N_2$, $P_3R_7N_4$, N_7N_7, S_2N_7, $P_3S_3R_5$.
Degree 7: $P_3R_8N_7$, $P_3N_7N_{13}$, $P_3S_2N_{13}$.

Elements in E_{ij} and M_i only: (16)
Degree 2: None.
Degree 3: M_3T_3, T_{11}.
Degree 4: M_3T_9, T_4T_4, T_1T_1.
Degree 5: $M_3R_5T_2$, T_1T_7, Q_2T_1.
Degree 6: $M_3R_9T_2$, $M_3R_7T_4$, T_7T_7, Q_2T_7, $M_3Q_3R_5$.
Degree 7: $M_3R_8T_7$, $M_3T_7T_{13}$, $M_3Q_2T_{13}$.

Elements in P_i and M_i only: (7)
Degree 2: $P_1M_1 + P_2M_2$, P_3M_3.
Degree 3: None.
Degree 4: None.
Degree 5: None.
Degree 6: $S_2J_1(\mathbf{M}, \mathbf{P})$, $S_2J_1(\mathbf{P}, \mathbf{M})$,
$Q_2J_1(\mathbf{P}, \mathbf{M})$, $Q_2J_1(\mathbf{M}, \mathbf{P})$, S_2Q_2.

7.49. Magnetic Crystal Class $6/m\underline{m}\underline{m}$ (see Table 3.3(14))

From Table 4.3(50) and Table B.14, we write

$$\begin{aligned}
\phi, \phi', \phi'' &= M_3, E_{11} + E_{22}, E_{33}, \\
\mathbf{A}, \mathbf{B} &= \begin{pmatrix} M_2 \\ -M_1 \end{pmatrix}, \begin{pmatrix} E_{23} \\ -E_{31} \end{pmatrix}, \\
\mathbf{a} &= \begin{pmatrix} 2E_{12} \\ E_{11} - E_{22} \end{pmatrix}, \\
\rho &= P_3, \\
\mathbf{X} &= \begin{pmatrix} P_1 \\ P_2 \end{pmatrix}.
\end{aligned} \tag{7.46}$$

The TMEs of the integrity basis are given in Appendix B.

Elements in E_{ij} only: (9)
Same as in the class $6/\underline{m}\underline{m}\underline{m}$.

Elements in P_i only: (3)
Same as in the class $6/\underline{m}\underline{m}\underline{m}$.

Elements in M_i only: (3)
Degree 1: M_3.
Degree 2: $M_1^2 + M_2^2$.
Degree 3: None.
Degree 4: None.
Degree 5: None.
Degree 6: $Q_2 Q_2$.

Elements in E_{ij} and P_i only: (16)
Same elements as in the class $6/\underline{m}\underline{m}\underline{m}$.

Elements in E_{ij} and M_i only: (18)
Degree 2: $\operatorname{Re}\{A\bar{B}\}$.
Degree 3: $\operatorname{Im}\{AAa\}$, $\operatorname{Im}\{ABa\}$.
Degree 4: $\operatorname{Re}\{AA\bar{a}\bar{a}\}$, $\operatorname{Re}\{AB\bar{a}\bar{a}\}$,
$\operatorname{Re}\{A\bar{A}a\bar{a}\}$, $\operatorname{Re}\{A\bar{B}a\bar{a}\}$.
Degree 5: $\operatorname{Im}\{AAAA\bar{a}\}$, $\operatorname{Im}\{AAaaa\}$,
$\operatorname{Im}\{AAAB\bar{a}\}$, $\operatorname{Im}\{A\bar{B}aaa\}$,
$\operatorname{Im}\{AABB\bar{a}\}$,
$\operatorname{Im}\{ABBB\bar{a}\}$.
Degree 6: $\operatorname{Re}\{A^5B\}$, $\operatorname{Re}\{A^4B^2\}$, $\operatorname{Re}\{A^3B^3\}$,
$\operatorname{Re}\{A^2B^4\}$, $\operatorname{Re}\{AB^5\}$,
where A, B and a defined by (7.46) as $A = A_1 + iA_2$, $B = B_1 + iB_2$, and so on.

Elements in P_i and M_i only: (7)
Degree 2: None.
Degree 3: $P_3(P_1M_1 + P_2M_2)$.
Degree 4: $(P_1M_2 - P_2M_1)^2$.
Degree 5: None.
Degree 6: $\operatorname{Re}\{A^4X^2\}$, $\operatorname{Re}\{A^2X^4\}$.
Degree 7: $P_3 \cdot \operatorname{Im}\{A^5X\}$,
$P_3 \cdot \operatorname{Im}\{A^3X^3\}$,
$P_3 \cdot \operatorname{Im}\{AX^5\}$,
where A and X are defined by (7.46) as $A = A_1 + iA$ and $X = X_1 + iX_2$.

7.50. Magnetic Crystal Class $6/\underline{m}mm$ (see Table 3.3(14))

From Table 4.3(48) and Table B.14, we write

$$\phi, \phi' = E_{11} + E_{22}, E_{33},$$

$$\mathbf{A} = \begin{pmatrix} E_{23} \\ -E_{31} \end{pmatrix},$$

$$\mathbf{a} = \begin{pmatrix} 2E_{12} \\ E_{11} - E_{22} \end{pmatrix},$$

$$\pi = M_3,$$

$$\rho = P_3,$$

$$\mathbf{X}, \mathbf{Y} = \begin{pmatrix} M_2 \\ -M_1 \end{pmatrix}, \begin{pmatrix} P_1 \\ P_2 \end{pmatrix}. \tag{7.47}$$

The TMEs of the integrity basis are given in Appendix B.

Elements in E_{ij} only: (9)
Same as in the class $6/\underline{m}\underline{m}\underline{m}$.

Elements in P_i only: (3)
Same as in the class $6/\underline{m}\underline{m}\underline{m}$.

Elements in M_i only: (3)
Degree 1: None.
Degree 2: M_3^2, $M_1^2 + M_2^2$.
Degree 3: None.
Degree 4: $M_3 Q_2$.
Degree 5: None.
Degree 6: $Q_2 Q_2$.

Elements in E_{ij} and P_i only: (16)
Same as in the class $6/\underline{m}\underline{m}\underline{m}$.

7.51. Magnetic Crystal Class $\underline{6}/mm\underline{m}$ (see Table 3.3(14))

From Table 4.3(51) and Table B.14, we write

$$\phi, \phi' = E_{11} + E_{22}, E_{33},$$

$$\mathbf{A} = \begin{pmatrix} E_{23} \\ -E_{31} \end{pmatrix},$$

$$\mathbf{a} = \begin{pmatrix} 2E_{12} \\ E_{11} - E_{22} \end{pmatrix},$$

$$\rho = P_3, \tag{7.48}$$

$$\gamma = M_3,$$

$$X = \begin{pmatrix} P_1 \\ P_2 \end{pmatrix},$$

$$\mathbf{x} = \begin{pmatrix} M_2 \\ -M_1 \end{pmatrix}.$$

The TMEs of the integrity basis are given in Appendix B.

Elements in E_{ij} only: (9)
Same as in the class $6/\underline{m}\underline{m}\underline{m}$.

Elements in P_i only: (3)
Same as in the class $6/\underline{m}\underline{m}\underline{m}$.

Elements in M_i only: (3)
Degree 1: None.
Degree 2: M_3^2, Q_1.
Degree 3: None.
Degree 4: None.
Degree 5: None.
Degree 6: $Q_2 Q_2$.

Elements in E_{ij} and P_i only: (16)
Same as in the class $6/\underline{m}\underline{m}\underline{m}$.

7.52. Magnetic Crystal Class $\underline{6}/\underline{m}m\underline{m}$ (see Table 3.3(14))

From Table 4.3(52) and Table B.14, we write

$$\begin{aligned}
\phi, \phi' &= E_{11} + E_{22}, E_{33}, \\
\eta &= M_3, \\
\mathbf{A} &= \begin{pmatrix} E_{23} \\ -E_{31} \end{pmatrix}, \\
\mathbf{a}, \mathbf{b} &= \begin{pmatrix} M_2 \\ -M_1 \end{pmatrix}, \begin{pmatrix} 2E_{12} \\ E_{11} - E_{22} \end{pmatrix}, \\
\rho &= P_3, \\
\mathbf{X} &= \begin{pmatrix} P_1 \\ P_2 \end{pmatrix}.
\end{aligned} \tag{7.49}$$

The TMEs of the integrity basis are given in Appendix B.

Elements in E_{ij} only: (9)
Same as in the class $6/\underline{m}\underline{m}\underline{m}$.

Elements in P_i only: (3)
Same as in the class $6/\underline{m}\underline{m}\underline{m}$.

Elements in M_i, only: (Same as in the Class $6\underline{2}\underline{2}$)(3)
Degree 1: None.
Degree 2: M_3^2, Q_1.
Degree 3: Q_2.

Elements in E_{ij} and P_i only: (16)
Same as in the class $6/\underline{mmm}$.

Elements in E_{ij} and M_i only: (3)
Same as in the class $\underline{6}2\underline{2}$.

7.53. Magnetic Crystal Class $\underline{m}3$ (see Table 3.3(16))

From Table 4.3(53) we define

$$\phi = E_{11} + E_{22}E_{33} \quad : \quad \Gamma_1,$$
$$\psi = E_{11} + w^2E_{22} + wE_{33}: \quad \Gamma_2,$$
$$\mathbf{x} = \begin{pmatrix} E_{23} \\ E_{31} \\ E_{12} \end{pmatrix}, \tag{7.50}$$
$$\mathbf{X}, \mathbf{Y} = \begin{pmatrix} M_1 \\ M_2 \\ M_3 \end{pmatrix}, \begin{pmatrix} P_1 \\ P_2 \\ P_3 \end{pmatrix} \quad : \quad \Gamma_4'.$$

With the aid of the results given by Smith et al. [1963] and Smith and Rivlin [1964], we determine the elements of the integrity basis. They are listed below.

Elements in E_{ij} only: (14)
Degree 1: $\Sigma E_{11} \equiv I_1$. (7.51)
Degree 2: $\Sigma E_{11}E_{22} \equiv I_2, \Sigma E_{23}^2 \equiv I_4$. (7.52)
Degree 3: $E_{11}E_{22}E_{33} \equiv I_3, E_{23}E_{31}E_{12} \equiv I_6,$
$\Sigma E_{11}E_{23}^2 \equiv I_7, \Sigma E_{11}(E_{31}^2 - E_{12}^2) \equiv K_1,$
$\Sigma E_{11}E_{22}(E_{11} - E_{22}) \equiv K_2$. (7.53)
Degree 4: $\Sigma E_{23}^2E_{31}^2 \equiv I_5, \Sigma E_{11}E_{22}E_{12}^2 \equiv I_9,$
$\Sigma E_{11}E_{22}(E_{23}^2 - E_{31}^2) \equiv K_4$. (7.54)
Degree 5: $\Sigma E_{11}E_{31}^2E_{12}^2 \equiv I_8, \Sigma E_{23}^2E_{31}^2(E_{11} - E_{22}) \equiv K_3$. (7.55)
Degree 6: $\Sigma E_{23}^2E_{31}^2(E_{23}^2 - E_{31}^2) \equiv K_5$. (7.56)

Elements in P_i only: (4)
Degree 1: None.
Degree 2: $\Sigma P_1^2 \equiv I_{10}$. (7.57)
Degree 3: None.
Degree 4: $\Sigma P_1^2P_2^2 \equiv I_{11}$. (7.58)
Degree 5: None.
Degree 6: $(P_1P_2P_3)^2 \equiv I_{12}, \Sigma P_1^2P_2^2(P_1^2 - P_2^2) \equiv K_6$. (7.59)

Elements in M_i only: (4)
Degree 1: None.
Degree 2: ΣM_1^2.

Degree 3: None.
Degree 4: $\Sigma M_1^2 M_2^2$ (7.60)
Degree 5: None.
Degree 6: $(M_1 M_2 M_3)^2, \Sigma M_1^2 M_2^2 (M_1^2 - M_2^2)$.

Elements in E_{ij} and P_i only: (38)
Degree 2: None.
Degree 3: $\Sigma P_2 P_3 E_{23} \equiv I_{13}, \Sigma P_1^2 E_{11} \equiv I_{14},$
$\Sigma P_1^2 (E_{22} - E_{33}) \equiv K_7.$ (7.61)
Degree 4: $\Sigma P_1^2 E_{23}^2 \equiv I_{15}, \Sigma P_2 P_3 E_{31} E_{12} \equiv I_{16},$
$\Sigma P_1^2 E_{11}(E_{22} - E_{33}) \equiv K_{10}, \Sigma P_2 P_3 E_{11} E_{23} \equiv I_{17},$
$\Sigma P_1^2 E_{22} E_{33} \equiv I_{18}, \Sigma P_1^2 (E_{31}^2 - E_{12}^2) \equiv K_8,$
$\Sigma P_2 P_3 E_{23}(E_{22} - E_{33}) \equiv K_9.$ (7.62)
Degree 5: $\Sigma P_2 P_3 E_{23}^3 \equiv I_{19}, \Sigma P_1^2 E_{11} E_{23}^2 \equiv I_{20},$
$\Sigma P_2 P_3 E_{11} E_{31} E_{12} \equiv I_{21}, \Sigma P_2 P_3 E_{22} E_{33} E_{23} \equiv I_{22},$
$P_1 P_2 P_3 \Sigma P_1 E_{23} \equiv I_{26}, \Sigma P_1^2 P_2^2 E_{33} \equiv I_{27},$
$\Sigma P_2 P_3 E_{23}(E_{31}^2 - E_{12}^2) \equiv K_{11},$
$\Sigma P_1 E_{23}^2 (E_{22} - E_{33}) \equiv K_{12},$
$\Sigma P_2 P_3 E_{31} E_{12}(E_{22} - E_{33}) \equiv K_{13},$
$\Sigma P_2 P_3 E_{23} E_{11}(E_{22} - E_{33}) \equiv K_{14},$
$\Sigma P_1^3 (P_3 E_{31} - P_2 E_{12}) \equiv K_{20},$
$\Sigma P_1^2 P_2^2 (E_{11} - E_{22}) \equiv K_{21}.$ (7.63)
Degree 6: $\Sigma P_1^2 E_{31}^2 E_{12}^2 \equiv I_{23}, \Sigma P_2 P_3 E_{11} E_{23}^3 \equiv I_{24},$
$\Sigma P_2 P_3 E_{22} E_{33} E_{31} E_{12} \equiv I_{25}, \Sigma P_1^2 P_2^2 (E_{23}^2 - E_{31}^2) \equiv K_{22},$
$\Sigma P_2 P_3 E_{31} E_{12}(P_3^2 - P_2^2) \equiv K_{23},$
$P_1 P_2 P_3 \Sigma P_1 E_{23}(E_{22} - E_{33}) \equiv K_{24}, P_1 P_2 P_3 \Sigma P_1 E_{31} E_{21} \equiv I_{28},$
$P_1 P_2 P_3 \Sigma P_1 E_{11} E_{23} \equiv I_{29}, \Sigma P_1^2 E_{23}^2 (E_{31}^2 - E_{12}^2) \equiv K_{15},$
$\Sigma P_2 P_3 E_{31} E_{12}(E_{31}^2 - E_{12}^2) \equiv K_{16}, \Sigma P_2 P_3 E_{23}^3 (E_{22} - E_{33}) \equiv K_{17},$
$\Sigma P_2 P_3 E_{31} E_{12} E_{11}(E_{22} - E_{33}) \equiv K_{18}.$ (7.64)
Degree 7: $\Sigma P_1^3 P_2^3 E_{12} \equiv I_{30},$
$\Sigma P_2 P_3 E_{23}^3 (E_{31}^2 - E_{12}^2) \equiv K_{19},$
$P_1 P_2 P_3 \Sigma P_1 E_{23}(E_{31}^2 - E_{12}^2) \equiv K_{25},$
$P_1 P_2 P_3 \Sigma P_1^2 (P_3 E_{12} - P_2 E_{31}) \equiv K_{26}.$ (7.65)

Elements in E_{ij} and M_i only: (38)
Degree 2: None.
Degree 3: Replace **P** by **M** in
$I_{13}, I_{14}, K_7.$
Degree 4: Replace **P** by **M** in
$I_{15}, I_{16}, I_{17}, I_{18}, K_8, K_9, K_{10}.$
Degree 5: Replace **P** by **M** in
$I_{19}, I_{20}, I_{21}, I_{22}, I_{26}, I_{27},$
$K_{11}, K_{12}, K_{13}, K_{14}, K_{20}, K_{21}.$
Degree 6: Replace **P** by **M** in
$I_{23}, I_{24}, I_{25}, I_{28}, I_{29},$
$K_{22}, K_{23}, K_{24}, K_{15}, K_{16}, K_{17}. K_{18}.$

Degree 7: Replace **P** by **M** in
$I_{30}, K_{19}, K_{25}, K_{26}$.

Elements in P_i and M_i only: (17)
Degree 2: $\Sigma P_1 M_1$.
Degree 3: None.
Degree 4: $\Sigma P_1 P_2 M_1 M_2, \Sigma P_1^2(P_2 M_2 + P_3 M_3)$,
$\Sigma M_1^2(P_2 M_2 + P_3 M_3), \Sigma P_1^2(M_2^2 - M_3^2)$,
$\Sigma P_1^2(P_2 M_2 - P_3 M_3), \Sigma M_1^2(P_2 M_2 - P_3 M_3)$.
Degree 5: None.
Degree 6: $P_1 P_2 P_3 M_1 M_2 M_3, P_1 P_2 P_3 \Sigma P_1 P_2 M_3$,
$P_1 P_2 P_3 \Sigma M_1 M_2 P_3, M_1 M_2 M_3 \Sigma P_1 P_2 M_3$,
$M_1 M_2 M_3 \Sigma M_1 M_2 P_3, \Sigma P_1^2 M_2^2(P_2 M_2 - P_3 M_3)$,
$\Sigma P_1^2 P_2^2(M_1^2 - M_2^2), \Sigma M_1^2 M_2^2(P_1^2 - P_2^2)$,
$\Sigma P_1^2 P_2^2(P_1 M_1 - P_2 M_2), \Sigma M_1^2 M_2^2(P_1 M_1 - P_2 M_2)$.

7.54. Magnetic Crystal Class $\underline{4}3\underline{m}$ (see Table 3.3(15))

From Table 4.3(54) we define

$$\begin{aligned} \phi &= E_{11} + E_{22} + E_{33} &&: \Gamma_1, \\ \mathbf{a} &= \begin{pmatrix} 2E_{11} - E_{22} - E_{33} \\ \sqrt{3}(E_{22} - E_{33}) \end{pmatrix} &&: \Gamma_3, \\ \mathbf{x}, \mathbf{y}, \mathbf{z} &= \begin{pmatrix} M_1 \\ M_2 \\ M_3 \end{pmatrix}, \begin{pmatrix} P_1 \\ P_2 \\ P_3 \end{pmatrix}, \begin{pmatrix} E_{23} \\ E_{31} \\ E_{12} \end{pmatrix} &&: \Gamma_4. \end{aligned} \tag{7.66}$$

Let us introduce three symmetric tensors **A**, **B**, and **C** by

$$\begin{aligned} &A_{ij} = E_{ij} \quad (i, j = 1, 2, 3), \\ &B_{11} = B_{22} = B_{33} = 0, \quad B_{23} = P_1, \quad B_{31} = P_2, \quad B_{12} = P_3, \\ &C_{11} = C_{22} = C_{33} = 0, \quad C_{23} = M_1, \quad C_{31} = M_2, \quad C_{12} = M_3. \end{aligned} \tag{7.67}$$

Thus, the integrity basis for three tensors, A, B, and C are obtained as a special case of the results given by Smith and Kiral [1969, p. 16], where arbitrary numbers of the symmetric tensors are considered.

Elements in E_{ij} only: (9)
Degree 1: I_1.
Degree 2: I_2, I_4.
Degree 3: I_3, I_6, I_7.
Degree 4: I_5, I_9.
Degree 5: I_8.

Elements in P_i only: (3)
Degree 1: None.
Degree 2: I_{10}.
Degree 3: $P_1P_2P_3 \equiv J_0$.
Degree 4: $\Sigma P_1^2 P_2^2$. (7.68)

Elements in M_i only: (3)
Degree 1: None.
Degree 2: ΣM_1^2.
Degree 3: $M_1M_2M_3$.
Degree 4: $\Sigma M_1^2 M_2^2$.

Elements in E_{ij} and P_i only: (18)
Degree 2: $\Sigma P_1 E_{23} \equiv J_1$. (7.69)
Degree 3: $\Sigma P_1 E_{31} E_{12} \equiv J_2$, $\Sigma P_1 E_{23} E_{11} \equiv J_3$, I_{13}, I_{14}. (7.70)
Degree 4: $\Sigma P_1 E_{23}^3 \equiv J_4$, $\Sigma P_1 E_{31} E_{12} E_{11} \equiv J_5$,
$\Sigma P_1 E_{23} E_{22} E_{33} \equiv J_6$, I_{15}, I_{17}, I_{18},
$\Sigma P_1^3 E_{23} \equiv J_9$. (7.71)
Degree 5: $\Sigma P_1 E_{23}^3 E_{11} \equiv J_7$, $\Sigma P_1 E_{31} E_{12} E_{22} E_{23} \equiv J_8$,
I_{20}, I_{22}, I_{27}, $\Sigma P_1^3 E_{23} E_{11} \equiv J_{10}$. (7.72)

Elements in E_{ij} and M_i only: (18)
Degree 2: $\Sigma M_1 E_{23}$.
Degree 3: Replace **P** by **M** in J_2, J_3, I_{13}, I_{14}.
Degree 4: Replace **P** by **M** in J_4, J_5, J_6, J_9, I_{15}, I_{17}, I_{18}.
Degree 5: Replace **P** by **M** in J_7, J_8, J_{10}, I_{20}, I_{22}, I_{27}.

Elements in P_i and M_i only: (6)
Degree 2: $\Sigma P_1 M_1$.
Degree 3: $\Sigma P_1 P_2 M_3$, $\Sigma M_1 M_2 P_3$.
Degree 4: $\Sigma P_1 P_2 M_1 M_2$,
$\Sigma P_1^2(P_2M_2 + P_3M_3)$, $\Sigma M_1^2(P_2M_2 + P_3M_3)$.

Elements in E_{ij}, P_i, and *M_i*: (18)
Degree 3: $\Sigma E_{11} P_1 M_1$, $\Sigma E_{23}(P_2M_3 + P_3M_2)$,
$\Sigma P_1(E_{31}M_3 + E_{12}M_2)$, $\Sigma M_1(E_{31}P_3 + E_{12}P_2)$.
Degree 4: $\Sigma E_{11}^2 P_1 M_1$, $\Sigma E_{23}^2 P_1 M_1$, $\Sigma P_1^2 E_{23} M_1$,
$\Sigma M_1^2 E_{23} P_1$,
$\Sigma E_{11} E_{23}(P_2M_3 + P_3M_2) - \Sigma E_{11} P_1(E_{31}M_3 + E_{12}M_2)$,
$\Sigma E_{11} P_1(M_2E_{12} + M_3E_{31}) - \Sigma E_{11} M_1(P_2E_{12} + P_3E_{31})$.
Degree 5: $\Sigma E_{11} E_{23}^2 P_1 M_1$, $\Sigma E_{11} P_1^2 E_{23} M_1$, $\Sigma E_{11} M_1^2 E_{23} P_1$,
$\Sigma E_{11}^2 E_{23}(P_2M_3 + P_3M_2) - \Sigma E_{11}^2 P_1(E_{31}M_3 + E_{12}M_2)$,
$\Sigma E_{11}^2 P_1(M_2E_{12} + M_3E_{31}) - \Sigma E_{11}^2 M_1(P_2E_{13} + P_3E_{31})$.
Degree 6: $\Sigma E_{11}^2(E_{22} - E_{33}) \cdot \Sigma E_{23}(P_2M_3 - P_3M_2)$,
$\Sigma E_{11}^2(E_{22} - E_{33}) \cdot \Sigma P_1(P_2M_3 - P_3M_2)$,
$\Sigma E_{11}^2(E_{22} - E_{33}) \cdot \Sigma M_1(P_2M_3 - P_3M_2)$.

7.55. Magnetic Crystal Class $\underline{4}3\underline{2}$ (see Table 3.3(17))

From Table 4.3(55) we define

$$\begin{aligned}
\phi &= E_{11} + E_{22} + E_{33} &&: \quad \Gamma_1, \\
\mathbf{a} &= \begin{pmatrix} 2E_{11} - E_{22} - E_{23} \\ \sqrt{3}(E_{22} - E_{33}) \end{pmatrix} &&: \quad \Gamma_3, \\
\mathbf{x}, \mathbf{y} &= \begin{pmatrix} M_1 \\ M_2 \\ M_3 \end{pmatrix}, \begin{pmatrix} E_{23} \\ E_{31} \\ E_{12} \end{pmatrix} &&: \quad \Gamma_4, \\
\mathbf{X} &= \begin{pmatrix} P_1 \\ P_2 \\ P_3 \end{pmatrix} &&: \quad \Gamma_5.
\end{aligned} \tag{7.73}$$

With the aid of the results given by Smith et al. [1963, p. 115] and Smith and Kiral [1969, p. 15], we list partial results for the elements of the integrity basis below.

Elements in E_{ij} only: (9)
Degree 1: I_1.
Degree 2: I_2, I_3.
Degree 3: I_3, I_6, I_7.
Degree 4: I_5, I_9.
Degree 5: I_8.

Elements in P_i only: (4)
Degree 1: None.
Degree 2: I_{10}.
Degree 3: None.
Degree 4: I_{11}.
Degree 5: None.
Degree 6: I_{12}.
Degree 7: None.
Degree 8: None.
Degree 9: $J_0 K_6$.

Elements in M_i only: (3)
Degree 1: None.
Degree 2: ΣM_1^2.
Degree 3: $M_1 M_2 M_3$.
Degree 4: $\Sigma M_1^2 M_2^2$.

Elements in E_{ij} and P_i only: (47)
Degree 2: None.
Degree 3: I_{13}, I_{14}, L_1.

Degree 4: $I_{15}, I_{16}, I_{17}, I_{18}, L_2, L_3, L_4, L_9$.
Degree 5: $I_{19}, I_{20}, I_{21}, I_{22}, I_{26}, I_{27}, L_5, L_6, L_7, L_{10}, L_{11}, J_1K_1, J_1K_2, J_1K_7$.
Degree 6: $I_{23}, I_{24}, I_{28}, I_{29}, L_8, L_{12}, L_{13}, J_2K_1,$
$J_2K_2, J_1K_8, J_2K_7, J_0K_2, J_1K_4, J_0K_1, J_1K_{10}, J_0K_7$.
Degree 7: $I_{30}, J_0K_8, J_0K_{10}, J_1K_{21}$.
Degree 8: J_0K_{21}, J_1K_6,
where L_α ($\alpha = 1, \ldots, 13$) are defined by

$$\begin{aligned}
L_1 &\equiv \Sigma P_1 E_{23}(E_{22} - E_{33}),\\
L_2 &\equiv \Sigma P_1 E_{23}(E_{31}^2 - E_{12}^2),\\
L_3 &\equiv \Sigma P_1 E_{31}E_{12}(E_{22} - E_{33}),\\
L_4 &\equiv \Sigma P_1 E_{23}E_{11}(E_{22} - E_{33}).\\
L_5 &\equiv \Sigma P_1 E_{31}E_{12}(E_{31}^2 - E_{12}^2),\\
L_6 &\equiv \Sigma P_1 E_{23}^3(E_{22} - E_{33}),\\
L_7 &\equiv \Sigma P_1 E_{31}E_{12}E_{11}(E_{22} - E_{33}),\\
L_8 &\equiv \Sigma P_1 E_{23}^3(E_{31}^2 - E_{12}^2),\\
L_9 &\equiv \Sigma P_1^2(P_2E_{31} - P_3E_{12}),\\
L_{10} &\equiv \Sigma P_1^2 E_{23}(P_2E_{12} - P_3E_{31}),\\
L_{11} &\equiv \Sigma P_1^3 E_{23}(E_{22} - E_{33}),\\
L_{12} &\equiv \Sigma P_1^3 E_{23}(E_{31}^2 - E_{12}^2),\\
L_{13} &\equiv \Sigma P_1{}^3 E_{23}(P_2^2 - P_3^2).
\end{aligned} \tag{7.74}$$

Elements in E_{ij} and M_i only: (20)
Degree 2: $\Sigma E_{23}M_1$.
Degree 3: $\Sigma E_{11}E_{23}M_1, \Sigma E_{11}M_1^2, \Sigma E_{23}(E_{31}M_3 + E_{12}M_2), \Sigma E_{23}M_2M_3$.
Degree 4: $\Sigma E_{11}^2E_{23}M_1, \Sigma E_{11}^2M_1^2, \Sigma E_{23}^3M_1, \Sigma E_{23}^2M_1^2, \Sigma E_{23}M_1^3,$
$\Sigma E_{11}E_{23}(M_2E_{12} + M_3E_{31}) - \Sigma E_{11}M_1(E_{31}E_{12} + E_{12}E_{31}),$
$\Sigma E_{11}E_{23}(M_2M_3 + M_3M_2) - \Sigma E_{11}M_1(E_{31}M_3 + E_{12}M_2)$.
Degree 5: $\Sigma E_{11}E_{23}^3M_1, \Sigma E_{11}M_1^4, \Sigma E_{11}E_{23}M_1^3,$
$\Sigma E_{11}E_{23}^2M_1^2,$
$\Sigma E_{11}^2E_{23}(M_2E_{12} + M_3E_{31}) - \Sigma E_{11}^2M_1(E_{31}E_{12} + E_{12}E_{31}),$
$\Sigma E_{11}^2E_{23}(M_2M_3 + M_3M_2) - \Sigma E_{11}^2M_1(E_{31}M_3 + E_{12}M_2)$.
Degree 6: $\Sigma E_{11}^2(E_{22} - E_{33})\cdot\Sigma E_{23}(E_{31}M_3 - E_{12}M_2),$
$\Sigma E_{11}^2(E_{22} - E_{33})\cdot\Sigma M_1(E_{31}M_3 - E_{12}M_2)$.

Elements in P_i and M_i only: (18)
Degree 2: None.
Degree 3: $\Sigma P_2P_3M_1$.
Degree 4: $\Sigma P_1^2M_1^2, \Sigma P_2P_3M_2M_3, \Sigma P_1M_1(M_2^2 - M_3^2),$
$\Sigma P_1^2(P_2M_2 - P_3M_3)$.

Degree 5: $\Sigma P_2 P_3 M_1^3$, $P_1 P_2 P_3 \Sigma P_1 M_1$,
$\Sigma P_1 M_2 M_3 (M_2^2 - M_3^2)$, $\Sigma P_1^2 M_1 (P_2 M_3 - P_3 M_2)$.
Degree 6: $\Sigma P_1^2 M_2^2 M_3^2$, $P_1 P_2 P_3 \Sigma P_1 M_2 M_3$, $\Sigma P_1 M_1^3 (M_2^2 - M_3^2)$,
$\Sigma P_1 M_1 \cdot \Sigma P_1^2 (M_2^2 - M_3^2)$, $\Sigma P_1^3 M_1 (M_2^2 - M_3^2)$, $\Sigma P_1^3 M_1 (P_2^2 - P_3^2)$.
Degree 7: $\Sigma P_1^3 P_2^3 M_3$, $P_1 P_2 P_3 \Sigma P_1^2 (M_2^2 - M_3^2)$.
Degree 8: $\Sigma P_1 M_1 \cdot \Sigma P_1^2 P_2^2 (P_1^2 - P_2^2)$.

7.56. Magnetic Crystal Class $\underline{m}3\underline{m}$ (see Table 3.3(16))

From Table 4.3(56) we define

$$\phi = E_{11} + E_{22} + E_{33} \quad : \quad \Gamma_1,$$

$$\mathbf{a} = \begin{pmatrix} 2E_{11} - E_{22} - E_{33} \\ \sqrt{3}(E_{22} - E_{33}) \end{pmatrix} : \quad \Gamma_3,$$

$$\mathbf{x} = \begin{pmatrix} E_{23} \\ E_{31} \\ E_{12} \end{pmatrix} \quad : \quad \Gamma_4,$$

$$\mathbf{X}, \mathbf{Y} = \begin{pmatrix} M_1 \\ M_2 \\ M_3 \end{pmatrix}, \begin{pmatrix} P_1 \\ P_2 \\ P_3 \end{pmatrix} \quad : \quad \Gamma_5'.$$

With the aid of the results given by Smith et al. [1963, p. 110] and Smith and Rivlin [1964, p. 199], we list partial results for the elements of the integrity basis below.

Elements in E_{ij} only: (9)
Degree 1: I_1.
Degree 2: I_2, I_4.
Degree 3: I_3, I_6, I_7.
Degree 4: I_5, I_9.
Degree 5: I_8.

Elements in P_i only: (3)
Degree 1: None.
Degree 2: I_{10}.
Degree 3: None.
Degree 4: I_{11}.
Degree 5: None.
Degree 6: I_{12}.

Elements in M_i only: (3)
Degree 1: None.
Degree 2: ΣM_1^2.
Degree 3: None.

Degree 4: $\Sigma M_1^2 M_2^2$.
Degree 5: None.
Degree 6: $(M_1 M_2 M_3)^2$.

Elements in E_{ij} and P_i only: (18)
Degree 2: None.
Degree 3: I_{13}, I_{14}.
Degree 4: $I_{15}, I_{16}, I_{17}, I_{18}$.
Degree 5: $I_{19}, I_{20}, I_{21}, I_{22}, I_{26}, I_{27}$.
Degree 6: $I_{23}, I_{24}, I_{25}, I_{28}, I_{29}$.
Degree 7: I_{30}.

Elements in E_{ij} and M_i only: (18)
Degree 2: None.
Degree 3: $\Sigma M_2 M_3 E_{23}, \Sigma M_1^2 E_{11}$.
Degree 4: $\Sigma M_1^2 E_{23}^2, \Sigma M_2 M_3 E_{31} E_{12}, \Sigma M_2 M_3 E_{23} E_{11}, \Sigma M_1^2 E_{22} E_{33}$.
Degree 5: $\Sigma M_2 M_3 E_{23}^3, \Sigma M_1^2 E_{11} E_{23}^2, \Sigma M_2 M_3 E_{31} E_{12} E_{11}$,
$\Sigma M_2 M_3 E_{23} E_{22} E_{33}, M_1 M_2 M_3 \Sigma M_1 E_{23}$,
$\Sigma M_1^2 M_2^2 E_{33}$.
Degree 6: $\Sigma M_1^2 E_{31}^2 E_{12}^2, \Sigma M_2 M_3 E_{23}^3 E_{11}, \Sigma M_2 M_3 E_{31} E_{12} E_{22} E_{33}$,
$M_1 M_2 M_3 \Sigma M_1 E_{31} E_{12}, M_1 M_2 M_3 \Sigma M_1 E_{11} E_{23}$.
Degree 7: $\Sigma M_1^3 M_2^3 E_{12}$.

Elements in P_i and M_i only: (9)
Degree 2: $\Sigma P_1 M_1$.
Degree 3: None.
Degree 4: $\Sigma P_1 P_2 M_1 M_2, \Sigma P_1^2 (P_2 M_2 + P_3 M_3), \Sigma M_1^2 (P_2 M_2 + P_3 M_3)$.
Degree 5: None.
Degree 6: $P_1 P_2 P_3 M_1 M_2 M_3, P_1 P_2 P_3 \Sigma P_1 P_2 M_3$.
$P_1 P_2 P_3 \Sigma M_1 M_2 P_3, M_1 M_2 M_3 \Sigma P_1 P_2 M_3$,
$M_1 M_2 M_3 \Sigma M_1 M_2 P_3$.

7.57. Magnetic Crystal Class $m3\underline{m}$ (see Table 3.3(18))

From Table 4.3(57) we define

$$\phi = E_{11} + E_{22} + E_{33} \quad : \quad \Gamma_1,$$

$$\mathbf{a} = \begin{pmatrix} 2E_{11} - E_{22} - E_{33} \\ \sqrt{3}(E_{22} - E_{33}) \end{pmatrix} : \quad \Gamma_3,$$

$$\mathbf{x}, \mathbf{y} = \begin{pmatrix} E_{23} \\ E_{31} \\ E_{12} \end{pmatrix}, \begin{pmatrix} M_1 \\ M_2 \\ M_3 \end{pmatrix} \quad : \quad \Gamma_4,$$

$$\mathbf{X} = \begin{pmatrix} P_1 \\ P_2 \\ P_3 \end{pmatrix} \quad : \quad \Gamma_5'.$$

Referring to the results given by Smith et al. [1963, p. 110], and Smith and Kiral [1969, p. 15], we list partial results for the elements of the integrity basis below.

Elements in E_{ij} *only*: (9)
Same elements as in the class $\underline{m}3\underline{m}$.

Elements in P_i *only*: (3)
Same as in the class $\underline{m}3\underline{m}$.

Elements in M_i *only*: (3)
Degree 1: None.
Degree 2: ΣM_1^2.
Degree 3: $M_1 M_2 M_3$.
Degree 4: $\Sigma M_1^2 M_2^2$.

Elements in E_{ij} *and* P_i *only*: (18)
Same as in the class $\underline{m}3\underline{m}$.

Elements in E_{ij} *and* M_i *only*: (20)
Same as in the class $\underline{4}3\underline{2}$.

Elements in P_i *and* M_i *only*: (8)
Degree 2: None.
Degree 3: $\Sigma P_2 P_3 M_1$.
Degree 4: $\Sigma P_1^2 M_1^2$, $\Sigma P_2 P_3 M_2 M_3$.
Degree 5: $\Sigma P_2 P_3 M_1^3$, $P_1 P_2 P_3 \Sigma P_1 M_1$.
Degree 6: $\Sigma P_1^2 M_2^2 M_3^2$, $P_1 P_2 P_3 \Sigma P_1 M_2 M_3$.
Degree 7: $\Sigma P_1^3 P_2^3 M_3$.

7.58. Magnetic Crystal Class $\underline{m}3m$ (see Table 3.3(18))

From Table 4.3(58) we define

$$\phi = E_{11} + E_{22} + E_{33} \quad : \quad \Gamma_1,$$

$$\mathbf{a} = \begin{pmatrix} 2E_{11} - E_{22} - E_{33} \\ \sqrt{3}(E_{22} - E_{33}) \end{pmatrix} : \quad \Gamma_3,$$

$$\mathbf{x} = \begin{pmatrix} E_{23} \\ E_{31} \\ E_{12} \end{pmatrix} \quad : \quad \Gamma_4,$$

$$\mathbf{A} = \begin{pmatrix} M_1 \\ M_2 \\ M_3 \end{pmatrix} \quad : \quad \Gamma_4',$$

$$\mathbf{X} = \begin{pmatrix} P_1 \\ P_2 \\ P_3 \end{pmatrix} \quad : \quad \Gamma_5'.$$

With the aid of the results given by Smith et al. [1963], partial results for the elements of the integrity basis are listed below.

Elements in E_{ij} only: (9)
Same as in the class $\underline{m}3\underline{m}$.

Elements in P_i only: (3)
Same as in the class $\underline{m}3\underline{m}$.

Elements in M_i only: (3)
Degree 1: None.
Degree 2: ΣM_1^2.
Degree 3: None.
Degree 4: $\Sigma M_1^2 M_2^2$.
Degree 5: None.
Degree 6: $(M_1 M_2 M_3)^2$.

Elements in E_{ij} and P_i only: (18)
Same elements as in the class $\underline{m}3\underline{m}$.

7.59. Composite Symbols of Chapter 7

For convenience, we collect here composite quantities that appear in the elements of the integrity basis in various magnetic crystal classes. Numbers in parenthesis following symbols indicate the equation numbers in which they are first defined.

I_1 (7.51); I_2, I_4 (7.52); I_3, I_6, I_7 (7.53); I_9 (7.54); I_{10} (7.57); I_{11} (7.58); I_{12} (7.59); I_{13}, I_{14}, I_{15} (7.61); I_5, I_{16}, I_{17}, I_{18} (7.62); I_{19}, I_{20}, I_{21}, I_{22}, I_{26}, I_{27} (7.63); I_{23}, I_{24}, I_{25}, I_{28}, I_{29} (7.64); I_8, I_{30} (7.65).

J_0 (7.68); J_2, J_3 (7.70); J_4, J_5, J_6, J_9 (7.71); J_7, J_8, J_{10} (7.72).

K_1, K_2 (7.53); K_4 (7.54); K_3 (7.55); K_5 (7.56); K_6 (7.59); K_7 (7.61); K_8, K_9, K_{10} (7.62); K_{11}, K_{12}, K_{13}, K_{14}, K_{20}, K_{21}, K_{22}, K_{23} (7.63); K_{15}, K_{16}, K_{17}, K_{18}, K_{24} (7.64); K_{19}, K_{25}, K_{26} (7.65).

L_1–L_{13} (7.74).

N_1, N_6 (7.29); N_2 (7.15); N_3 (7.4); N_4 (7.30); N_5, N_{10} (7.32); N_7 (7.31); N_8 (7.17); N_9 (7.20); N_{10} (7.32); N_{11} (7.18); N_{12}, N_{13} (7.36); N_{14} (7.16).

Q_1 (7.12); Q_2 (7.37); Q_3 (7.13).

R_0, R_1, R_2, R_3 (7.9a); R_4 (7.34); R_5, R_6, R_9, R_{10} (7.9b); R_7, R_8, R_{11}, R_{13} (7.35); R_{12} (7.9c).

S_1 (7.10); S_2 (7.28); S_3 (7.11).

T_1, T_4, T_5, T_7, T_{10}, T_{12}, T_{13} (7.40); T_2 (7.22); T_3 (7.21); T_6 (7.26); T_8 (7.24); T_9 (7.27); T_{11} (7.25).

$J_1\left[\begin{pmatrix}x_1\\x_2\end{pmatrix},\begin{pmatrix}y_1\\y_2\end{pmatrix}\right]$ (7.38); $J_2\left[\begin{pmatrix}x_1\\x_2\end{pmatrix},\begin{pmatrix}y_1\\y_2\end{pmatrix}\right]$ (7.41).

$K_1\left[\begin{pmatrix}x_1\\x_2\end{pmatrix},\begin{pmatrix}y_1\\y_2\end{pmatrix},\begin{pmatrix}z_1\\z_2\end{pmatrix}\right]$ (7.39); $K_2(\mathbf{a}, \mathbf{b}, \mathbf{c})$ (7.9).

CHAPTER 8

Applications

Here we give a number of examples on the application of the results and tables obtained in the previous chapters.

EXAMPLE 1. In order to display the power of the method presented in Chapter 6, let us consider a less trivial example in which we find the number and location of the nonvanishing components of the material tensor C_{ijklm} in the relationship

$$E_{ij} = C_{ijklm}P_kM_lM_m, \tag{8.1}$$

where E_{ij} is the symmetric strain tensor and P_k and M_l are the polarization and magnetization vectors, respectively. Now introduce a third-order tensor $\mathbf{Z}$ by $Z_{klm} = P_kM_lM_m$, which is symmetric with respect to its last two indices. Note that since $\mathbf{M}$ occurs twice in the definition of $\mathbf{Z}$, it is not only a true tensor, it is also time-symmetric. Now, (8.1) becomes a linear relationship between $\mathbf{E}$ and $\mathbf{Z}$. To be specific, we consider the crystal class $C_{4v} = 4mm$.

The decomposition of E_{ij} is already given in Table 4.3(18). The decomposition of Z_{klm} can be obtained by employing (4.23), (4.24) with the aid of Table 3.3(7). We have

$C_{4v} = 4mm$	E_{ij}	$Z_{klm} = P_kM_lM_m$
Γ_1	$E_{11} + E_{22}, E_{33}$	$Z_{113} + Z_{223}, Z_{311} + Z_{322}, Z_{333}$
Γ_2	O	$Z_{123} - Z_{213}$
Γ_3	E_{12}	$Z_{123} + Z_{213}, Z_{312}$
Γ_4	$E_{11} - E_{22}$	$Z_{113} - Z_{223}, Z_{311} - Z_{322}$
		$\begin{pmatrix} Z_{111} \\ Z_{222} \end{pmatrix}, \begin{pmatrix} Z_{122} \\ Z_{211} \end{pmatrix},$
Γ_5	$\begin{pmatrix} E_{13} \\ E_{23} \end{pmatrix}$	$\begin{pmatrix} Z_{133} \\ Z_{233} \end{pmatrix}, \begin{pmatrix} Z_{221} \\ Z_{112} \end{pmatrix}, \begin{pmatrix} Z_{313} \\ Z_{323} \end{pmatrix}$

from which it follows that

$$
\begin{aligned}
E_{11} + E_{22} &= a_1(Z_{113} + Z_{223}) + a_2(Z_{311} + Z_{322}) + a_3 Z_{333}, \\
E_{33} &= a_4(Z_{113} + Z_{223}) + a_5(Z_{311} + Z_{322}) + a_6 Z_{333}, \\
E_{12} &= a_7(Z_{123} + Z_{213}) + a_8 Z_{312}, \\
E_{11} - E_{22} &= a_9(Z_{113} - Z_{223}) + a_{10}(Z_{311} - Z_{322}), \\
\begin{pmatrix} E_{13} \\ E_{23} \end{pmatrix} &= a_{11}\begin{pmatrix} Z_{111} \\ Z_{222} \end{pmatrix} + a_{12}\begin{pmatrix} Z_{122} \\ Z_{211} \end{pmatrix} + a_{13}\begin{pmatrix} Z_{133} \\ Z_{233} \end{pmatrix} + a_{14}\begin{pmatrix} Z_{221} \\ Z_{112} \end{pmatrix} \\
&\quad + a_{15}\begin{pmatrix} Z_{313} \\ Z_{323} \end{pmatrix}.
\end{aligned}
\tag{8.2}
$$

From (8.2), the fifteen nonvanishing independent components of C_{ijklm} are readily obtained as

$$
\begin{aligned}
&C_{11113} = C_{22223}, \quad C_{11311} = C_{22322}, \quad C_{11333} = C_{22333}, \\
&C_{11223} = C_{22113}, \quad C_{11322} = C_{22311}, \quad C_{33113} = C_{33223}, \\
&C_{33311} = C_{33322}, \quad C_{33333}, \quad C_{12123} = C_{12213}, C_{12312}, \\
&C_{13111} = C_{23222}, \quad C_{13122} = C_{23211}, \quad C_{13133} = C_{23233}, \\
&C_{13221} = C_{23112}, \quad C_{13313} = C_{23323}.
\end{aligned}
\tag{8.3}
$$

Note that the number of nonvanishing components may be alternatively obtained by observing the representations of E_{ij} and Z_{klm} (see Section 6.3). From Table 4.2(7) and Table 4.4(7), we write

$$
\begin{aligned}
\mathrm{Rep}(Z_{klm}) &= \mathrm{Rep}(P_k M_l M_m) = [1] \times [2] \\
&= (\Gamma_1 + \Gamma_5) \otimes (2\Gamma_1 + \Gamma_3 + \Gamma_4 + \Gamma_5) \\
&= 3\Gamma_1 + \Gamma_2 + 2\Gamma_3 + 2\Gamma_4 + 5\Gamma_5
\end{aligned}
\tag{8.4}
$$

and

$$
\mathrm{Rep}(E_{ij}) = 2\Gamma_1 + \Gamma_3 + \Gamma_4 + \Gamma_5. \tag{8.5}
$$

Combining (8.4) and (8.5) we find

$$
\begin{aligned}
\mathrm{Rep}(\mathbf{C}) &= \mathrm{Rep}(\mathbf{E}) \otimes \mathrm{Rep}(\mathbf{Z}) \\
&= 15\Gamma_1 + 11\Gamma_2 + 13\Gamma_3 + 13\Gamma_4 + 28\Gamma_5
\end{aligned}
\tag{8.6}
$$

from which we get $n_1 = 15$ which is the expected result. Scheme (8.3) will be the same for the magnetic crystal classes $4\underline{mm}$ and $\underline{4}m\underline{m}$ originated from $4mm$, since **E** and **Z** are both time-symmetric.

Let us now study the material tensor D_{ijklm} in the relationship

$$
E_{ij} = D_{ijklm} M_k P_l P_m. \tag{8.7}
$$

Let us again introduce a third-order tensor $\mathbf{Z}'$ by $Z'_{klm} = M_k P_l P_m$, which is still symmetric with respect to its last two indices, but, it is an axial and time-asymmetric tensor since $\mathbf{M}$ occurs once in the definition of $\mathbf{Z}'$. We then write

$$\text{Rep}(\mathbf{Z}')_{a,c} = \text{Rep}(\mathbf{Z})_{p,i} \otimes A \otimes T. \tag{8.8}$$

In order to be specific, we consider the magnetic class $4\underline{m}\underline{m}$. We have, from Table 3.3(7), that

	I	τR_2	τR_1	D_3	τT_3	$R_2 T_3$	$R_1 T_3$	$\tau D_3 T_3$
A	1	-1	-1	1	-1	1	1	-1
T	1	-1	-1	1	-1	1	1	-1
$A \otimes T$	1	1	1	1	1	1	1	$1 = \Gamma_1$

Hence, the nonvanishing independent components of D_{ijklm} will be identical to those of C_{ijklm}, since $\text{Rep}(\mathbf{Z}') = \text{Rep}(\mathbf{Z})$.

However, for the magnetic class $\underline{4}m\underline{m}$ the situation is different. For this case we have, from Table 3.3(7):

	I	R_2	R_1	D_3	τT_3	$\tau R_2 T_3$	$\tau R_1 T_3$	$\tau D_3 T_3$
A	1	-1	-1	1	-1	1	1	-1
T	1	1	1	1	-1	-1	-1	-1
$A \otimes T$	1	-1	-1	1	1	-1	-1	$1 = \Gamma_3$

Hence, we write

$$\text{Rep}(\mathbf{Z}') = \text{Rep}(\mathbf{Z}) \otimes A \otimes T = \text{Rep}(\mathbf{Z}) \otimes \Gamma_3 \tag{8.9}$$

and from (8.6) and Table 6.4(7) we arrive at

$$\text{Rep}(\mathbf{Z}') = 2\Gamma_1 + 2\Gamma_2 + 3\Gamma_3 + \Gamma_4 + 5\Gamma_5 \tag{8.10}$$

and

$$\begin{aligned} \text{Rep}(\mathbf{D}) &= \text{Rep}(\mathbf{E}) \otimes \text{Rep}(\mathbf{Z}') \\ &= 13\Gamma_1 + 13\Gamma_2 + 15\Gamma_3 + 11\Gamma_4 + 28\Gamma_5, \end{aligned}$$

which shows that there are 13 (not 15!) nonvanishing independent components of D_{ijklm} for this crystal class.

We note that the enumeration of C_{ijklm} can be obtained in an alternative manner with the aid of the integrity basis for $\mathbf{E}$, $\mathbf{P}$, $\mathbf{M}$, such that each term of total degree 5 in $W = W(E, P, M)$ must be linear in $\mathbf{E}$ and $\mathbf{P}$ and of second degree in $\mathbf{M}$. Observation of the integrity basis listed in Section 7.18 yields

$$\begin{aligned} W_{(E,P,M,M)} = {} & a_1(E_{31}P_1M_1^2 + E_{23}P_{23}P_2M_2^2) + a_2E_{12}P_3M_1M_2 \\ & + a_3(E_{12}P_2M_1M_3 + E_{12}P_1M_2M_3) + a_4E_{33}P_3M_3^2 \\ & + a_5(P_1M_1 + P_2M_2)(E_{31}M_1 + E_{23}M_2) \\ & + a_6P_3M_3^2(E_{11} + E_{22}) + a_7E_{33}P_3(M_1^2 + M_2^2) \end{aligned}$$

$$
\begin{aligned}
&+ a_8 P_3(E_{11} + E_{22})(M_1^2 + M_2^2) \\
&+ a_9(E_{11} - E_{22})(P_1 M_1 M_3 - P_2 M_2 M_3) \\
&+ a_{10}(E_{11} P_3 M_2^2 + E_{22} P_3 M_1^2) \\
&+ a_{11} M_3(E_{11} + E_{22})(P_1 M_1 + P_2 M_2) \\
&+ a_{12}(E_{31} P_3 M_1 M_3 + E_{23} P_3 M_2 M_3) \\
&+ a_{13}(E_{33} P_1 M_1 M_3 + E_{33} P_2 M_2 M_3) \\
&+ a_{14}(E_{31} P_1 + E_{23} P_2)(M_1^2 + M_2^2) \\
&+ a_{15}(E_{31} P_1 M_3^2 + E_{23} P_2 M_3^2), \qquad (8.11)
\end{aligned}
$$

where the first term is the fourth-degree element of the integrity basis, the remaining terms are obtained as products of the appropriate elements of the integrity basis of degree lower than four. Noting that

$$
W_{(E,P,M,M)} = C_{ijklm} E_{ij} P_k M_l M_m, \qquad (8.12)
$$

and comparing (8.11) and (8.12) we get the nonvanishing independent components of C_{ijklm}, which is identical to (8.3).

EXAMPLE 2. Piezomagnetoelectricity in Cr_2O_3 ($\bar{3}\underline{m}$). It has been pointed out by Rado and Folen [1962] that the free energy W may contain terms which are linear in the strain tensor E_{ij}, in the polarization P_k, and in the magnetization M_k, and that such terms lead to the possible existence of piezomagnetoelectric (PME) effects. Thus, the piezomagnetoelectric contribution to W is in the form:

$$
W_{(\mathrm{PME})} = C_{ijkl} E_{ij} P_k M_l \qquad (i, j, k, l = 1, 2, 3). \qquad (8.13)
$$

In the particular case of Cr_2O_3, the magnetic point group is $\underline{\bar{3}}\underline{m}$ and from the results given in Section 7.39:

$$
\begin{aligned}
W_{(\mathrm{PME})} = {} & a_1 M_3(E_{31} P_1 + E_{23} P_2) + a_2(2E_{12} P_1 + E_{11} P_2 - E_{22} P_2) \\
&+ a_3 P_3(E_{31} M_1 + E_{32} M_2) + a_4 P_3(2E_{12} M_1 + E_{11} M_2 - E_{22} M_2) \\
&+ a_5(E_{23} P_1 M_1 - E_{23} P_2 M_2 + E_{31} P_1 M_2 + E_{31} P_2 M_1) \\
&+ a_6[(E_{11} - E_{22})(P_1 M_1 - P_2 M_2) + 2E_{12}(P_1 M_2 + P_2 M_1)] \\
&+ a_7 E_{33} P_3 M_3 + a_8 E_{33}(P_1 M_1 + P_2 M_2) + a_9 P_3 M_3(E_{11} + E_{22}) \\
&+ a_{10}(E_{11} + E_{22})(P_1 M_1 + P_2 M_2). \qquad (8.14)
\end{aligned}
$$

We note that the first six terms in (8.14) are the third-degree elements of the integrity basis, and that the remaining four terms are the product of the appropriate integrity basis of lower degree.

Comparison of the two expressions (8.13) and (8.14) leads to the result that only 25 of the 54 coefficients C_{ijkl} do not vanish, and that only 10 of these are independent. The nonvanishing coefficients are

$$\begin{aligned}
&C_{1111} = C_{2222}, \qquad C_{1122} = C_{2211}, \qquad C_{1123} = C_{1213} = -C_{2223},\\
&C_{1132} = C_{1231} = -C_{2232}, \qquad C_{1133} = C_{2233},\\
&C_{1212} = C_{1221} = (C_{1111} - C_{1122})/2, \qquad C_{3333}, \qquad\qquad (8.15)\\
&C_{1312} = C_{1321} = C_{2311} = -C_{2322}, \qquad C_{1313} = C_{2323},\\
&C_{1331} = C_{2332}, \qquad C_{3311} = C_{3322}.
\end{aligned}$$

These results (8.15) are in agreement with Rado [1962]. Note that the number 10 of independent coefficients was misprinted as 9 in Rado and Folen [1962]. Lyubimov [1966] later listed only the number of independent coefficients of piezomagnetoelectricity for each magnetic crystal class.

Once $W_{(\mathrm{PME})}$ is constructed, the constitutive equations for the stress, **T**, the electric field, **E**, and the magnetic field, **B**, are obtained using the thermodynamic relations

$$\mathbf{T} = \frac{\partial W}{\partial \mathbf{E}}, \qquad \mathbf{E} = \frac{\partial W}{\partial \mathbf{P}}, \qquad \mathbf{B} = \frac{\partial W}{\partial \mathbf{M}}. \tag{8.16}$$

Alternatively, piezomagnetoelectricity can be studied in a very simple and direct manner by the method introduced in Chapter 6. Let us consider the constitutive equations for stress in the form

$$T_{kl} = C_{klmn} P_m M_n$$

under the crystal group $\underline{\bar{3}m}$. We have the basic quantities for T_{kl} and $P_m M_n$ listed below:

$\underline{\bar{3}m}$	T_{kl}	$P_m M_n$
Γ_1	$T_{11} + T_{22}, T_{33}$	$P_1 M_1 + P_2 M_2, P_3 M_3$
Γ_2		$P_1 M_2 - P_2 M_1$
Γ_6	$\begin{pmatrix} T_{13} \\ T_{23} \end{pmatrix}, \begin{pmatrix} 2T_{12} \\ T_{11} - T_{22} \end{pmatrix}$	$\begin{pmatrix} P_1 M_2 + P_2 M_1 \\ P_1 M_1 - P_2 M_2 \end{pmatrix}, \begin{pmatrix} P_1 M_3 \\ P_2 M_3 \end{pmatrix}, \begin{pmatrix} P_3 M_1 \\ P_3 M_2 \end{pmatrix}$

(8.17)

In (8.17) the basic quantities for T_{kl} are copied from Table 4.3(40); and for $P_m M_n$, (4.23), have been used. From (8.17) we readily write

$$\begin{aligned}
T_{11} + T_{22} &= b_1(P_1 M_1 + P_2 M_2) + b_2 P_3 M_3,\\
T_{33} &= b_3(P_1 M_1 + P_2 M_2) + b_4 P_3 M_3,\\
\begin{pmatrix} T_{13} \\ T_{23} \end{pmatrix} &= b_5 \begin{pmatrix} P_1 M_2 + P_2 M_1 \\ P_1 M_1 - P_2 M_2 \end{pmatrix} + b_6 \begin{pmatrix} P_1 M_3 \\ P_2 M_3 \end{pmatrix} + b_7 \begin{pmatrix} P_3 M_1 \\ P_3 M_2 \end{pmatrix}, \qquad (8.18)\\
\begin{pmatrix} 2T_{12} \\ T_{11} - T_{22} \end{pmatrix} &= b_8 \begin{pmatrix} P_1 M_2 + P_2 M_1 \\ P_1 M_1 - P_2 M_2 \end{pmatrix} + b_9 \begin{pmatrix} P_1 M_3 \\ P_2 M_3 \end{pmatrix} + b_{10} \begin{pmatrix} P_3 M_1 \\ P_3 M_2 \end{pmatrix}.
\end{aligned}$$

Example 3. The Magnetoelectric Effect in Cr_2O_3 ($\bar{3}\underline{m}$). In order to consider this effect, the free energy W is formed as a second-order function of **P** and **M** (polarization and magnetization, respectively), such that each term is linear in **P** and **M**. Since material symmetry belongs to $\bar{3}\underline{m}$, from Section 7.39 we write

$$\begin{aligned} W_{(\mathbf{P},\mathbf{M})} &= a_1 P_3^2 + a_2(P_1^2 + P_2^2) + b_1 M_3^2 + b_2(M_1^2 + M_2^2) \\ &\quad + C_1 P_3 M_3 + C_2(P_1 M_1 + P_2 M_2). \end{aligned} \tag{8.19}$$

With the aid of the thermodynamic relations

$$E_K = \frac{\partial W}{\partial P_K} \quad \text{and} \quad B_K = \frac{\partial W}{\partial M_K} \tag{8.20}$$

we derive the constitutive equations for E_K and B_K as

$$E_K = 2a_1 P_3 \delta_{3K} + 2a_2(P_1 \delta_{1K} + P_2 \delta_{2K}) + C_1 M_3 \delta_{3K} + C_2(M_1 \delta_{1K} + M_2 \delta_{2K}) \tag{8.21}$$

and for the magnetic induction as

$$B_K = 2b_1 M_3 \delta_{3K} + 2b_2(M_1 \delta_{1K} + M_2 \delta_{2K}) + C_1 P_3 \delta_{3K} + C_2(P_1 \delta_{1K} + P_2 \delta_{2K}). \tag{8.22}$$

Thus, a_1, a_2 and b_1, b_2 are the dielectric constant and susceptibilities, respectively. C_1 and C_2 are the material constants describing the *magnetoelectric* effects. These results are in agreement with Dzyaloshinskii [1960] and Rado and Folen [1962].

Example 4. Piezomagnetism in MnF_2, CoF_2, and FeF_2 ($\underline{4}/mm\underline{m}$). The terms in the potential W, containing both the strain tensor, E_{KL}, and the magnetization vector, M_K, linear in both E_{KL} and M_K, are responsible for the first-order piezomagnetism. From Section 7.29, we get

$$W = a_1(M_1 E_{23} + M_2 E_{31}) + a_2 M_3 E_{12}. \tag{8.23}$$

Now, using the thermodynamic relation $B_K = \partial W/\partial M_K$, from (8.23) we get

$$B_K = a_1(E_{23}\delta_{1K} + E_{31}\delta_{2K}) + a_2 E_{12}\delta_{3K}, \tag{8.24}$$

where a_1 and a_2 are the two material constants describing the piezomagnetic effects in materials having the magnetic symmetry $\underline{4}/mm\underline{m}$.

In Dyzyaloshinskii [1960] the second term $a_2 E_{12}\delta_{3K}$ is missing, which is responsible for the appearance of the piezomagnetic induction along the x_3-axis. This error has been pointed out by Borovik-Romanov [1960], where the experimental discovery of this longitudinal effect was first reported.

Example 5. Rigid and Nonconducting Electromagnetic Solids. For rigid materials, the strain tensor $E_{KL} = 0$, and if the material does not conduct heat we have $\theta_{,K} = 0$, so that the free energy W is independent of E_{KL} and $\theta_{,K}$, i.e.,

$$W = W(\Pi_K, M_K, \theta, X), \tag{8.25}$$

where $\Pi_K = P_K/\rho$ is the polarization per unit mass, and M_K is the magnetization vector per unit volume. Due to the C–D inequality we have

$$E_K = \frac{\partial W}{\partial \Pi_K} \quad \text{(electric field)}, \quad B_K = \frac{\partial W}{\delta M_K} \quad \text{(magnetic induction)}. \tag{8.26}$$

To be specific, we consider the antiferromagnetic fluorides CoF_2, MnF_2 whose magnetic symmetry is $\underline{4}/mm\underline{m}$.

The complete set of minimal integrity bases is listed in Section 7.29. We pick up those elements that are functions of $\Pi_K = P_K/\rho$ and M_K only. They are

$$\begin{aligned}
\mathbf{\Pi}:&\quad \mathrm{I}_1 = \Pi_3^2, \quad \mathrm{I}_2 = \Pi_1^2 + \Pi_2^2, \quad \mathrm{I}_3 = \Pi_1^2\Pi_2^2,\\
\mathbf{M}:&\quad \mathrm{I}_4 = M_3^2, \quad \mathrm{I}_5 = M_1^2 + M_2^2, \quad \mathrm{I}_6 = M_1M_2M_3, \quad \mathrm{I}_7 = M_1^2M_2^2,\\
(\mathbf{\Pi}, \mathbf{M}):&\quad \mathrm{I}_8 = \Pi_1\Pi_2M_3, \quad \mathrm{I}_9 = \Pi_3(\Pi_1M_2 + \Pi_2M_1),\\
&\quad \mathrm{I}_{10} = \Pi_1\Pi_2M_1M_2, \quad \mathrm{I}_{11} = \Pi_3M_3(\Pi_1M_1 + \Pi_2M_2),\\
&\quad \mathrm{I}_{12} = \Pi_1^2M_1^2 + \Pi_2^2M_2^2, \quad \mathrm{I}_{13} = \Pi_3(M_1^3\Pi_2 + M_2^3\Pi_1),\\
&\quad \mathrm{I}_{14} = \Pi_3(\Pi_1^3M_2 + \Pi_2^3M_1).
\end{aligned} \tag{8.27}$$

Hence, the free energy W becomes

$$W = W(\mathrm{I}_1, \ldots, \mathrm{I}_{14}, \theta, X). \tag{8.28}$$

Then the electric field, E_K, and the magnetic induction, B_K, take the canonical form

$$E_K = \frac{\partial W}{\partial \mathrm{I}_\alpha}\frac{\partial \mathrm{I}_\alpha}{\partial \Pi_K} \quad \text{and} \quad B_K = \frac{\partial W}{\partial \mathrm{I}_\alpha}\frac{\partial \mathrm{I}_\alpha}{\partial M_K} \quad (\alpha = 1, \ldots, 14). \tag{8.29}$$

Using (8.27), the quantities $\partial \mathrm{I}_\alpha/\partial \Pi_K$ and $\partial \mathrm{I}_\alpha/\partial M_K$ ($\alpha = 1, \ldots, 14$) are computed to be

$$\frac{\partial \mathrm{I}_1}{\partial \Pi_K} = 2\Pi_3\delta_{3K}, \quad \frac{\partial \mathrm{I}_2}{\partial \Pi_K} = 2\Pi_1\delta_{1K} + 2\Pi_2\delta_{2K},$$

$$\frac{\partial \mathrm{I}_3}{\partial \Pi_K} = 2\Pi_1\Pi_2^2\delta_{1K} + 2\Pi_1^2\Pi_2\delta_{2K},$$

$$\frac{\partial \mathrm{I}_4}{\partial \Pi_K} = \frac{\partial \mathrm{I}_5}{\partial \Pi_K} = \frac{\partial \mathrm{I}_6}{\partial \Pi_K} = \frac{\partial \mathrm{I}_7}{\partial \Pi_K} = 0,$$

$$\frac{\partial \mathrm{I}_8}{\partial \Pi_K} = \Pi_2M_3\delta_{1K} + \Pi_1M_3\delta_{2K},$$

$$\frac{\partial \mathrm{I}_9}{\partial \Pi_K} = (\Pi_1M_2 + \Pi_2M_1)\delta_{3K} + \Pi_3M_2\delta_{1K} + \Pi_3M_1\delta_{2K},$$

$$\frac{\partial \mathrm{I}_{10}}{\partial \Pi_K} = \Pi_2 M_1 M_2 \delta_{1K} + \Pi_1 M_1 M_2 \delta_{2K},$$

$$\frac{\partial \mathrm{I}_{11}}{\partial \Pi_K} = M_3(\Pi_1 M_1 + \Pi_2 M_2)\delta_{3K} + \Pi_3 M_1 M_3 \delta_{1K} + \Pi_3 M_2 M_3 \delta_{2K},$$

$$\frac{\partial \mathrm{I}_{12}}{\partial \Pi_K} = 2\Pi_1 M_1^2 \delta_{1K} + 2\Pi_2 M_2^2 \delta_{2K},$$

$$\frac{\partial \mathrm{I}_{13}}{\partial \Pi_K} = (\Pi_2 M_1^3 + \Pi_1 M_2^3)\delta_{3K} + \Pi_3 M_1^3 \delta_{2K} + \Pi_3 M_2^2 \delta_{1K},$$

$$\frac{\partial \mathrm{I}_{14}}{\partial \Pi_K} = (\Pi_1^3 M_2 + \Pi_2^3 M_1)\delta_{3K} + 3\Pi_1^2 \Pi_3 M_2 \delta_{1K} + 3\Pi_3 \Pi_2^2 M_1 \delta_{2K}, \quad (8.30)$$

and

$$\frac{\partial \mathrm{I}_1}{\partial M_K} = \frac{\partial \mathrm{I}_2}{\partial M_K} = \frac{\partial \mathrm{I}_3}{\partial M_K} = 0,$$

$$\frac{\partial \mathrm{I}_4}{\partial M_K} = 2M_3\delta_{3K}, \qquad \frac{\partial \mathrm{I}_5}{\partial M_K} = 2M_1\delta_{1K} + 2M_2\delta_{2K},$$

$$\frac{\partial \mathrm{I}_6}{\partial M_K} = M_2 M_3 \delta_{1K} + M_1 M_3 \delta_{2K} + M_1 M_2 \delta_{3K},$$

$$\frac{\partial \mathrm{I}_7}{\partial M_K} = 2M_1 M_2^2 \delta_{1K} + 2M_1^2 M_2 \delta_{2K},$$

$$\frac{\partial \mathrm{I}_8}{\partial M_K} = \Pi_1 \Pi_2 \delta_{3K}, \qquad \frac{\partial \mathrm{I}_9}{\partial M_K} = \Pi_3 \Pi_1 \delta_{2K} + \Pi_3 \Pi_2 \delta_{1K},$$

$$\frac{\partial \mathrm{I}_{10}}{\partial M_K} = \Pi_1 \Pi_2 M_2 \delta_{1K} + \Pi_1 \Pi_2 M_1 \delta_{2K}, \quad (8.31)$$

$$\frac{\partial \mathrm{I}_{11}}{\partial M_K} = \Pi_3(\Pi_1 M_1 + \Pi_2 M_2)\delta_{3K} + \Pi_1 \Pi_3 M_3 \delta_{1K} + \Pi_2 \Pi_3 M_3 \delta_{2K},$$

$$\frac{\partial \mathrm{I}_{12}}{\partial M_K} = 2\Pi_1^2 M_1 \delta_{1K} + 2\Pi_2^2 M_2 \delta_{2K},$$

$$\frac{\partial \mathrm{I}_{13}}{\partial M_K} = 3\Pi_2 \Pi_3 M_1^2 \delta_{1K} + 3\Pi_1 \Pi_3 M_2^2 \delta_{2K},$$

$$\frac{\partial \mathrm{I}_{14}}{\partial M_K} = \Pi_1^3 \Pi_3 \delta_{2K} + \Pi_3 \Pi_2^2 \delta_{1K}.$$

Observation of (8.30) and (8.31) reveals that the terms $\partial \mathrm{I}_1/\partial \Pi_K$ and $\partial \mathrm{I}_2/\partial \Pi_K$ are responsible for the electric polarization, and the terms $\partial \mathrm{I}_4/\partial M_K$ and $\partial \mathrm{I}_5/\partial M_K$ give rise to the magnetic polarization in the case of linear interactions. Thus, we have two dielectric constants and two magnetic susceptibilities. We also see that there is no linear magnetoelectric effect for the materials with magnetic symmetry $\underline{4}/mm\underline{m}$.

EXAMPLE 6. Nonlinear Magnetoelastic Nonconducting Solids. For nonpolarizable and non-heat-conducting materials, the free energy W does not depend on the polarization and temperature gradient. We have

$$W = W(E_{KL}, M_K, \theta, X). \tag{8.32}$$

Consider the magnetic crystal class $\underline{4}/mm\underline{m}$ for which the minimal integrity bases are listed in Section 7.29. The elements of the integrity bases that depend on E_{KL} and M_K only are repeated below for ease of reference.

$$\begin{aligned}
\mathbf{M}:\quad & \mathrm{I}_1 = M_3^2, \quad \mathrm{I}_2 = M_1^2 + M_2^2, \quad \mathrm{I}_3 = M_1 M_2 M_3, \quad \mathrm{I}_4 = M_1^2 M_2^2, \\
\mathbf{E}:\quad & \mathrm{I}_5 = E_{11} + E_{22}, \quad \mathrm{I}_6 = E_{33}, \quad \mathrm{I}_7 = E_{12}^2, \quad \mathrm{I}_8 = E_{11}E_{22}, \\
& \mathrm{I}_9 = E_{23}^2 + E_{31}^2, \quad \mathrm{I}_{10} = E_{23}E_{31}E_{12}, \\
& \mathrm{I}_{11} = E_{11}E_{23}^2 + E_{22}E_{31}^2, \\
& \mathrm{I}_{12} = E_{23}^2 E_{31}^2, \\
\mathbf{E}, \mathbf{M}:\quad & \mathrm{I}_{13} = M_1 E_{23} + M_2 E_{31}, \quad \mathrm{I}_{14} = M_3 E_{12}, \\
& \mathrm{I}_{15} = M_3(M_1 E_{31} + M_2 E_{23}), \\
& \mathrm{I}_{16} = M_1 M_2 E_{12}, \quad \mathrm{I}_{17} = M_3 E_{23} E_{31}, \\
& \mathrm{I}_{18} = E_{12}(M_1 E_{31} + M_2 E_{23}), \\
& \mathrm{I}_{19} = (E_{11} - E_{22})(M_1^2 - M_2^2), \\
& \mathrm{I}_{20} = (E_{11} - E_{22})(M_1 E_{23} - M_2 E_{31}), \\
& \mathrm{I}_{21} = M_1^3 E_{23} + M_2^3 E_{31}, \quad \mathrm{I}_{22} = M_1 E_{23}^3 + M_2 E_{31}^3, \\
& \mathrm{I}_{23} = M_1 M_2 E_{23} E_{31}, \quad \mathrm{I}_{24} = M_3(E_{11} - E_{22})(M_1 E_{31} - M_2 E_{23}), \\
& \mathrm{I}_{25} = E_{12}(E_{11} - E_{22})(M_1 E_{31} - M_2 E_{23}).
\end{aligned} \tag{8.33}$$

Due to the C–D inequality, the magnetic induction vector, B_K, and the stress tensor, T_{KL}, are derivable from the free energy W by

$$B_K = \frac{\partial W}{\partial M_K} \quad \text{and} \quad T_{KL} = \frac{\partial W}{\partial E_{KL}}. \tag{8.34}$$

Using (8.33), B_K takes the canonical form

$$B_K = \frac{\partial W}{\partial \mathrm{I}_\alpha}\frac{\partial \mathrm{I}_\alpha}{\partial M_K} \qquad (\alpha = 1, \ldots, 25), \tag{8.35}$$

where

$$\frac{\partial \mathrm{I}_1}{\partial M_K} = 2M_3\delta_{3K}, \qquad \frac{\partial \mathrm{I}_2}{\partial M_K} = 2M_1\delta_{1K} + 2M_2\delta_{2K},$$

$$\frac{\partial \mathrm{I}_3}{\partial M_K} = M_2 M_3 \delta_{1K} + M_1 M_3 \delta_{2K} + M_1 M_2 \delta_{3K},$$

$$\frac{\partial \mathrm{I}_4}{\partial M_K} = 2M_1 M_2^2 \delta_{1K} + 2M_1^2 M_2 \delta_{2K},$$

$$\frac{\partial \mathrm{I}_5}{\partial M_K} = \frac{\partial \mathrm{I}_6}{\partial M_K} = \cdots = \frac{\partial \mathrm{I}_{12}}{\partial M_K} = 0,$$

$$\frac{\partial \mathrm{I}_{13}}{\partial M_K} = E_{23}\delta_{1K} + E_{31}\delta_{2K}, \qquad \frac{\partial \mathrm{I}_{14}}{\partial M_K} = E_{12}\delta_{3K},$$

$$\frac{\partial \mathrm{I}_{15}}{\partial M_K} = (M_1 E_{31} + M_2 E_{23})\delta_{3K} + M_3(E_{31}\delta_{1K} + E_{23}\delta_{2K}),$$

$$\frac{\partial \mathrm{I}_{16}}{\partial M_K} = M_2 E_{12}\delta_{1K} + M_1 E_{12}\delta_{2K}, \qquad \frac{\partial \mathrm{I}_{17}}{\partial M_K} = E_{23}E_{31}\delta_{3K}, \tag{8.36}$$

$$\frac{\partial \mathrm{I}_{18}}{\partial M_K} = E_{12}(E_{31}\delta_{1K} + E_{23}\delta_{2K}),$$

$$\frac{\partial \mathrm{I}_{19}}{\partial M_K} = 2(E_{11} - E_{22})(M_1\delta_{1K} - M_2\delta_{2K}),$$

$$\frac{\partial \mathrm{I}_{20}}{\partial M_K} = (E_{11} - E_{22})(E_{23}\delta_{1K} - E_{31}\delta_{2K}),$$

$$\frac{\partial \mathrm{I}_{21}}{\partial M_K} = 3M_1^2 E_{23}\delta_{1K} + 3M_2^2 E_{31}\delta_{2K}, \qquad \frac{\partial \mathrm{I}_{22}}{\partial M_K} = E_{23}^3\delta_{1K} + E_{31}^3\delta_{2K},$$

$$\frac{\partial \mathrm{I}_{23}}{\partial M_K} = M_2 E_{23}E_{31}\delta_{1K} + M_1 E_{23}E_{31}\delta_{2K},$$

$$\frac{\partial \mathrm{I}_{24}}{\partial M_K} = (E_{11} - E_{22})(M_1 E_{31} - M_2 E_{23})\delta_{3K} + M_3(E_{11} - E_{22})(E_{31}\delta_{1K} - E_{23}\delta_{2K}),$$

$$\frac{\partial \mathrm{I}_{25}}{\partial M_K} = E_{12}(E_{11} - E_{22})(E_{31}\delta_{1K} - E_{23}\delta_{2K}).$$

Note that the terms $\partial \mathrm{I}_{13}/\partial M_K$ and $\partial \mathrm{I}_{14}/\partial M_K$ are responsible for linear piezomagnetism. This result, of course, checks with Table 6.2.

APPENDIX A

Review of Group Theory and Representation

Definition 1. Any collection of elements $\{A, B, \ldots, R, \ldots\}$ has the *group property* if an associative law of combination is defined such that for any ordered pair R, S there is a unique product, written RS, which is equivalent to some single element T which is also in the collection.

Natural numbers under addition, and also under multiplication, have the group property.

Definition 2. A collection where the law of combination is commutative is said to be "Abelian."

Natural numbers under addition and under multiplication are Abelian.

The example that follows provides a collection of elements whose combination is not Abelian.

EXAMPLE 1. Consider operations which bring the lamina of Figure A.1 into a position indistinguishable from that which is originally occupied. This operation is called a *symmetry operation.*

It is clear that this lamina is brought into self-coincidence by the six operations indicated in Figure A.1. C_3 and C_3^- stand for rotation through 120° anticlockwise and clockwise, while $\sigma^{(1)}$, $\sigma^{(2)}$, and $\sigma^{(3)}$ stand for reflection. We agree that RS indicates operation S followed by operation R. Since the combined effect of R and S is by definition, to send the lamina into self-coincidence we have $RS = T$, where T is one of the six possible symmetry operations.

Note that $\sigma^{(1)}C_3 = \sigma^{(2)}$ and $C_3\sigma^{(1)} = \sigma^{(3)}$, hence the operation is non-commutative.

Definition 3. A collection of elements $\{A, B, \ldots, R, \ldots\}$ is a group G if:

(i) it has the group property;
(ii) it contains a *unit element* E such that $RE = R$ (for all $R \in G$);

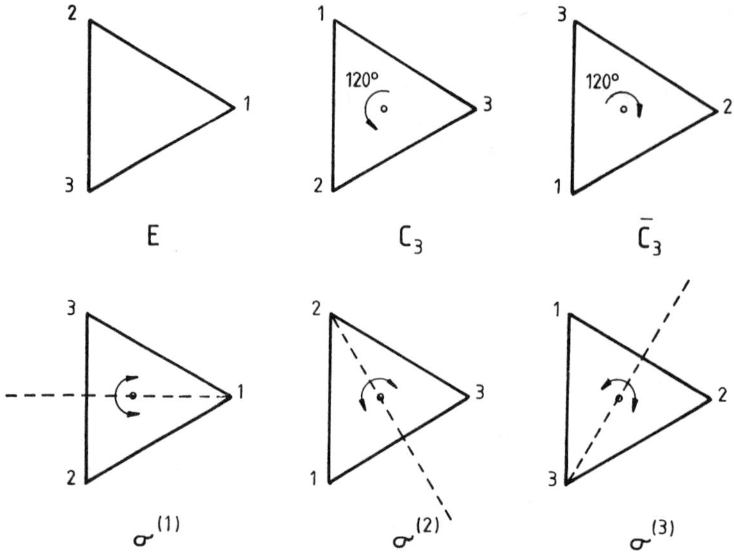

Figure A.1. Six possible symmetry operations applied in equilateral triangles.

(iii) it contains, for every element R, an *inverse* called R^{-1} such that $RR^{-1} = E$ (for all $R \in G$).

It can be readily shown that a right unit is also a left unit, and a right inverse is also a left inverse.

EXAMPLE 2. The collection of integers $\{\ldots, -n, \ldots, -2, -1, 1, 2, \ldots, n, \ldots\}$ forms a group under addition.

EXAMPLE 3. The collection of symmetry operations of an equilateral triangle

$$\{E, C_3, C_3^-, \sigma^{(1)}, \sigma^{(2)}, \sigma^{(3)}\} \tag{A.1}$$

obviously satisfies both (i) and (ii). It satisfies (iii) since for every element we can find an operation in the group which restores the lamina to its original position. The collection (A.1) is therefore a group. It is the symmetry group denoted by C_{3v}.

Definition 4. If a group G contains g elements, g is called its order. If g is finite it is said to be a finite group.

Thus, C_{3v} is a finite group of order 6.

In the case of finite groups, the properties of the group are conveniently indicated in a *multiplication table* which sets out systematically the products of all g^2 pairs of elements. The element in the table equivalent to RS is placed at the intersection of the R row and the S column.

EXAMPLE 4. By sequential performance of the operation defined in Figure A.1, we obtain the following multiplication table of C_{3v}:

	S: E	C_3	C_3^-	$\sigma^{(1)}$	$\sigma^{(2)}$	$\sigma^{(3)}$
R: E	E	C_3	C_3^-	$\sigma^{(1)}$	$\sigma^{(2)}$	$\sigma^{(3)}$
C_3	C_3	C_3^-	E	$\sigma^{(3)}$	$\sigma^{(1)}$	$\sigma^{(2)}$
C_3^-	C_3^-	E	C_3	$\sigma^{(2)}$	$\sigma^{(3)}$	$\sigma^{(1)}$
$\sigma^{(1)}$	$\sigma^{(1)}$	$\sigma^{(2)}$	$\sigma^{(3)}$	E	C_3	C_3^-
$\sigma^{(2)}$	$\sigma^{(2)}$	$\sigma^{(3)}$	$\sigma^{(1)}$	C_3^-	E	C_3
$\sigma^{(3)}$	$\sigma^{(3)}$	$\sigma^{(1)}$	$\sigma^{(2)}$	C_3	C_3^-	E

Theorem 1. *Each row (and column) of a multiplication table contains each element of the group once and only once. We write this result as $RG = G$ (for all $R \in G$). RG is used to denote*

$$RG = \{RA, RB, RC, \ldots\}.$$

With the help of the multiplication table any product of group elements can be reduced to a single element. For example, for the symmetry group C_{3v}, we have

$$C_3\sigma^{(1)}\sigma^{(3)} = C_3C_3^- = E.$$

Conversely, all the elements of a group of order g, may be expressed as products, whose factors are drawn from a limited number of elements.

EXAMPLE 5. In the group C_{3v}, we write

$$C_3^2 = C_3^-, \qquad C_3^3 = E, \qquad C_3^2\sigma^{(1)} = \sigma^{(2)}, \qquad C_3\sigma^{(1)} = \sigma^{(3)},$$

so that all elements of C_{3v} are expressed in terms of only two C_3 and $\sigma^{(1)}$.

Definition 5. A set P of elements of a group G is a *system of generators* of the group if every element of G can be written as the product of a finite number of factors, each of which is either an element of P or the inverse of such an element.

Now note that if R is an element of G, then R^2, $R^3, \ldots$ will also be in G. G will be infinite unless $R^n = E$ for some value of n.

Definition 6. A set of elements $\{R, R^2, \ldots, R^n(=E)\}$ is called a *cyclic group* of order n.

Definition 7. Any collection of elements of G, which by themselves form a group H, is called a *subgroup* of G. Two trivial subgroups, the unit element and the whole group itself, are called *improper* subgroups, other subgroups are said to be *proper* subgroups.

EXAMPLE 6. The set $\{E, C_3, C_3^-\}$ is a subgroup of C_{3v}. It is the cyclic group called C_3 and is defined by the single generator with the property $C_3^3 = E$. Note that $\{E, \sigma^{(1)}\}$, $\{E, \sigma^{(2)}\}$, and $\{E, \sigma^{(3)}\}$ are also subgroups and are also cyclic, since $(\sigma^{(1)})^2 = (\sigma^{(2)})^2 = (\sigma^{(3)})^2 = E$. Hence, the group C_{3v} contains four proper subgroups, each being cyclic.

Definition 8. Let $H = \{A_1, A_2, \ldots, A_p\}$ be a subgroup of G, and let $R_1, R_2, \ldots$ be the elements of G not contained in H. Then the collection

$$HR_k = \{A_1 R_k, A_2 R_k, \ldots, A_p R_k\}$$

is said to be the *right coset* of H with respect to R_k.

EXAMPLE 7. Consider $G = C_{3v}$ and its subgroup $H = \{E, C_3, C_3^-\}$. The right coset of H with respect to $\sigma^{(1)}$ is

$$H\sigma^{(1)} = \{\sigma^{(1)}, \sigma^{(3)}, \sigma^{(2)}\}.$$

Cosets have the following properties:

(i) Every element of a group appears either in the subgroup or in one of its cosets.
(ii) No element can be common to both a subgroup and one of its cosets.
(iii) No element can be common to two different cosets of the same subgroup.
(iv) No coset can contain the same element more than once.

Theorem 2. *The order in any subgroup H of G must be a divisor of the order of the group G.*

This result allows us to state, for example, that the group C_{3v} can have proper subgroups of order 2 and 3 only. See Example 6.

Definition 9. An element B is said to be *conjugate* to A with respect to R if $B = RAR^{-1}$. When we form all the elements conjugate to A as R runs through the whole group G and collect the distinct results, we get a *class*, that is, the class of all elements conjugate to A.

EXAMPLE 8. Consider the group C_{3v} and its multiplication table given in Example 4. The table of transformations RAR^{-1} is given below:

A \ R	E	C_3	C_3^-	$\sigma^{(1)}$	$\sigma^{(2)}$	$\sigma^{(3)}$	classes
E	E	E	E	E	E	E	$\{E\}$
C_3	C_3	C_3	C_3	C_3^-	C_3^-	C_3^-	$\{C_3, C_3^-\}$
C_3^-	C_3^-	C_3^-	C_3^-	C_3	C_3	C_3	$\{C_3, C_3^-\}$
$\sigma^{(1)}$	$\sigma^{(1)}$	$\sigma^{(2)}$	$\sigma^{(3)}$	$\sigma^{(1)}$	$\sigma^{(2)}$	$\sigma^{(3)}$	$\{\sigma^{(1)}, \sigma^{(2)}, \sigma^{(3)}\}$
$\sigma^{(2)}$	. . . . . .						$\{\sigma^{(1)}, \sigma^{(2)}, \sigma^{(3)}\}$
$\sigma^{(3)}$	. . . . . .						$\{\sigma^{(1)}, \sigma^{(2)}, \sigma^{(3)}\}$

Thus, the group C_{3v} is partitioned into three distinct classes $C_{3v} = \{E\} + \{C_3, C_3^-\} + \{\sigma^{(1)}, \sigma^{(2)}, \sigma^{(3)}\}$. The number of distinct classes is an important characteristic of a group.

Theorem 3. *Any two elements of a class are conjugate to each other with respect to some member of the group.*

Theorem 4. *If g is the order of a group, and the group has a class of α elements, then each element is transformed into itself, and every other element of the class, exactly g/α times. In the above example, C_3 and C_3^- are transformed into themselves and into each other $6/2 = 3$ times. Conjugate subgroups may be defined in much the same way as conjugate elements.*

Definition 10. If $H = \{A_1, \ldots, A_h\}$ is a subgroup of G and if we denote by RHR^{-1} the collection of conjugate elements of H with respect to R, then RHR^{-1} is called the *subgroup conjugate* to H with respect to R.

Theorem 5. *The collection RHR^{-1} is also a subgroup.*

Definition 11. When $RHR^{-1} = H$ for all R in G, then H is said to be a *normal (invariant, self-conjugate) subgroup.*

EXAMPLE 9. In the case of C_{3v}, only one of the four subgroups is normal, namely $H = \{E, C_3, C_3^-\}$, which can be checked from the multiplication table (Example 4): $EHE^{-1} = H$, $C_3HC_3^{-1} = H$, $\bar{C}_3H\bar{C}_3^{-1} = H$, so on. The subgroup $\{E, \sigma^{(1)}\}$ is not normal since $C_3\{E, \sigma^{(1)}\}C_3^{-1} = \{E, \sigma^{(2)}\} \neq \{E, \sigma^{(1)}\}$.

In group representations we are concerned with various collections of quantities which satisfy the same multiplication table as a given group, and which are thus similar in structure.

EXAMPLE 10. Consider the group C_{3v}, and associate with $\{E, C_3C_3^-\}$ plus 1 and with $\{\sigma^{(1)}, \sigma^{(2)}, \sigma^{(3)}\}$ minus 1, thus

E	C_3	C_3^-	$\sigma^{(1)}$	$\sigma^{(2)}$	$\sigma^{(3)}$
1	1	1	-1	-1	-1

Then, the associated quantities satisfy the same multiplication table, for example,

$C_3\sigma^{(1)} = \sigma^{(3)}$
$1(-1) = (-1)$

Let $\bar{G} = \{1, -1\}$, which forms a group itself.

Definition 12. A mapping of G into $\bar{G}$ is called *homomorphic* when the condition $(RS)' = R'S'$ ($R', S' \in \bar{G}$ and $R, S \in G$) is fulfilled for all pairs of elements

R, S in G. If, in addition, the association R to R' is unique (one-to-one), so that each element of G has a different image in $\bar{G}$, the mapping is called *isomorphic*.

EXAMPLE 11. Consider the set of matrices

$$\mathbf{A}_1 = \begin{Vmatrix} 1 & 0 \\ 0 & 1 \end{Vmatrix}, \quad \mathbf{A}_2 = \begin{Vmatrix} -1/2 & -\sqrt{3}/2 \\ \sqrt{3}/2 & -1/2 \end{Vmatrix}, \quad \mathbf{A}_3 = \begin{Vmatrix} -1/2 & \sqrt{3}/2 \\ -\sqrt{3}/2 & -1/2 \end{Vmatrix},$$
$$\mathbf{A}_4 = \begin{Vmatrix} 1 & 0 \\ 0 & -1 \end{Vmatrix}, \quad \mathbf{A}_5 = \begin{Vmatrix} -1/2 & -\sqrt{3}/2 \\ -\sqrt{3}/2 & 1/2 \end{Vmatrix}, \quad \mathbf{A}_6 = \begin{Vmatrix} -1/2 & \sqrt{3}/2 \\ \sqrt{3}/2 & 1/2 \end{Vmatrix}. \tag{A.2}$$

If we adopt matrix multiplication as the law of combination for this collection, it can be seen that the matrices have the same multiplication table as C_{3v} provided that we make the association

E	C_3	$\bar{C}_3$	$\sigma^{(1)}$	$\sigma^{(2)}$	$\sigma^{(3)}$
$\mathbf{A}_1$	$\mathbf{A}_2$	$\mathbf{A}_3$	$\mathbf{A}_4$	$\mathbf{A}_5$	$\mathbf{A}_6$

To illustrate, take $\sigma^{(1)}C_3 = \sigma^{(2)}$, then

$$\mathbf{A}_4\mathbf{A}_2 = \begin{Vmatrix} 1 & 0 \\ 0 & -1 \end{Vmatrix} \begin{Vmatrix} -1/2 & -\sqrt{3}/2 \\ \sqrt{3}/2 & -1/2 \end{Vmatrix} = \begin{Vmatrix} -1/2 & -\sqrt{3}/2 \\ -\sqrt{3}/2 & 1/2 \end{Vmatrix} = \mathbf{A}_5.$$

In this example the association is one-to-one, since the six matrices are all different. The mapping is isomorphic and $\bar{G} = \{A_1, \ldots, A_6\}$ forms a group under ordinary matrix multiplication.

Definition 13. If we can find g matrices $\{\mathbf{A}, \mathbf{B}, \mathbf{C}, \ldots, \mathbf{R}, \ldots\}$ and associate them with the g elements of a group $G = \{A, B, C, \ldots, R, \ldots\}$, in such a way that when $RS = T$ then $\mathbf{RS} = \mathbf{T}$ for all pairs (R, S), the matrices $\{\mathbf{A}, \mathbf{B}, \ldots\}$ are said to form a *representation* of G. If the representation of G is isomorphic it is said to be *faithful*.

The association of a matrix with a group element is often indicated by a functional notation. With this notation, we write $\mathbf{A} = \mathbf{D}(A)$, $\mathbf{B} = \mathbf{D}(B), \ldots$, $\mathbf{R} = \mathbf{D}(R), \ldots$ and, say, a representation D by matrices $\mathbf{D}(R)$.

The theory of representations is concerned with determining the number and type of certain special representations and with the explicit construction of the matrices.

Definition 14. Let $\mathbf{e}_1, \mathbf{e}_2, \ldots, \mathbf{e}_n$ be any n linearly independent vectors. Then any vector $\mathbf{r}$ is expressed as

$$\mathbf{r} = r_i \mathbf{e}_i, \tag{A.3}$$

where $r_1, \ldots, r_n$ are called the components of $\mathbf{r}$ relative to the bases $\mathbf{e}_1, \ldots, \mathbf{e}_n$.

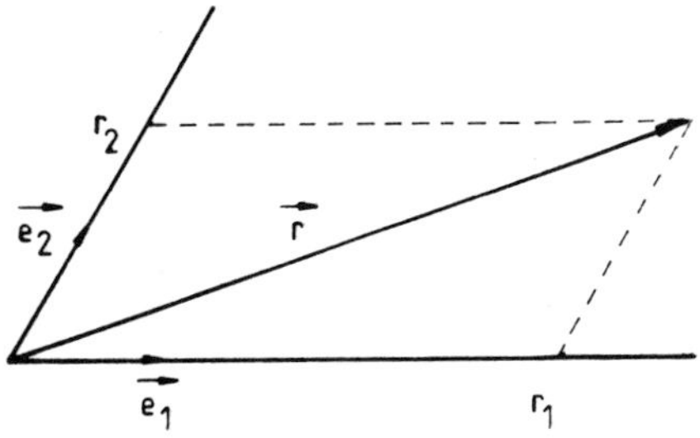

Figure A.2. Parallel projections of a vector on the base vectors.

Geometrically, the components may be regarded as parallel projections of a vector along the directions specified by the bases vectors (see Figure A.2).

If instead of $\mathbf{e}_1, \ldots, \mathbf{e}_n$, we choose a new bases $\bar{\mathbf{e}}_1, \ldots, \bar{\mathbf{e}}_n$, whose vectors are linear combinations of $\mathbf{e}_1, \ldots, \mathbf{e}_n$, we write

$$\bar{\mathbf{e}}_i = \mathbf{e}_j R_{ji} \qquad (i, j = 1, \ldots, n). \tag{A.4}$$

On the other hand, we have

$$\mathbf{r} = r_i \mathbf{e}_i = \bar{r}_i \bar{\mathbf{e}}_i. \tag{A.5}$$

Both (A.4) and (A.5) may be written in matrix form as

$$\vec{\mathbf{r}} = \mathbf{e}\mathbf{r} \qquad \text{and} \qquad \bar{\mathbf{e}} = \mathbf{e}\mathbf{R}, \tag{A.6}$$

where we placed an arrow to distinguish vector $\vec{\mathbf{r}}$ from the corresponding matrix $\mathbf{r}$, and defined the matrices

$$\underset{(n \times n)}{\mathbf{e} = (\mathbf{e}_1 \quad \mathbf{e}_2 \quad \ldots \quad \mathbf{e}_n)}, \qquad \underset{(n \times 1)}{\mathbf{r} = \begin{Vmatrix} r_1 \\ r_2 \\ \vdots \\ r_n \end{Vmatrix}}, \tag{A.7}$$

$$\underset{(n \times n)}{\mathbf{R} = \begin{Vmatrix} R_{11} & R_{12} & \ldots & R_{1n} \\ R_{21} & R_{22} & \ldots & R_{2n} \\ \ldots & \ldots & \ldots & \ldots \\ R_{n1} & R_{n2} & \ldots & R_{nn} \end{Vmatrix}}.$$

The first column of $\mathbf{e}$ is the component of $\mathbf{e}_1$, etc.

Example 12. Figure A.3 illustrates a three-dimensional situation where the bases vectors $\mathbf{e}_1, \mathbf{e}_2, \mathbf{e}_3$ are three perpendicular unit vectors and $\bar{\mathbf{e}}_1, \bar{\mathbf{e}}_2, \bar{\mathbf{e}}_3$ are obtained by rotating $\mathbf{e}_1, \mathbf{e}_2, \mathbf{e}_3$ through an angle θ about $\mathbf{e}_3$.

Note that

$$\overline{AC'} = \overline{AB'} = \cos\theta,$$

$$\overline{AC''} = \overline{AD'} = \sin\theta.$$

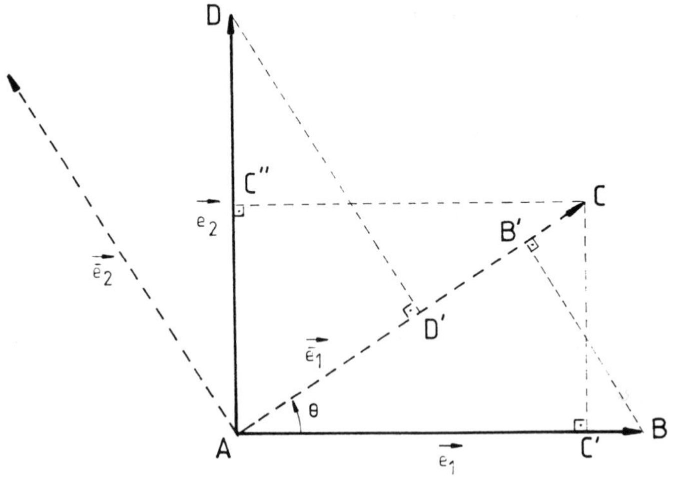

Figure A.3. Rotation of the base vectors through an angle θ about $\mathbf{e}_3$.

Hence,

$$\bar{\mathbf{e}}_1 = \overline{AC'}\,\mathbf{e}_1 + \overline{AC''}\,\mathbf{e}_2 = \mathbf{e}_1 \cos\theta + \mathbf{e}_2 \sin\theta,$$

and, similarly,

$$\bar{\mathbf{e}}_2 = -\mathbf{e}_1 \sin\theta + \mathbf{e}_2 \cos\theta,$$

$$\bar{\mathbf{e}}_3 = \mathbf{e}_3.$$

Hence, we get

$$(\bar{\mathbf{e}}_1 \quad \bar{\mathbf{e}}_2 \quad \bar{\mathbf{e}}_3) = (\mathbf{e}_1 \quad \mathbf{e}_2 \quad \mathbf{e}_3)\begin{Vmatrix} \cos\theta & -\sin\theta & 0 \\ \sin\theta & \cos\theta & 0 \\ 0 & 0 & 1 \end{Vmatrix}, \tag{A.8}$$

therefore $\mathbf{R}$ in (7) is

$$\mathbf{R} = \begin{Vmatrix} \cos\theta & -\sin\theta & 0 \\ \sin\theta & \cos\theta & 0 \\ 0 & 0 & 1 \end{Vmatrix}. \tag{A.9}$$

Now, in general, the n vectors $\bar{\mathbf{e}}_1, \ldots, \bar{\mathbf{e}}_n$ are linearly independent, hence they provide a basis in terms of which $\mathbf{e}_1, \ldots, \mathbf{e}_n$ may be expressed as

$$\mathbf{e} = \bar{\mathbf{e}}\mathbf{R}^{-1}. \tag{A.10}$$

From (A.6) and (A.10) we get

$$\vec{\mathbf{r}} = \mathbf{e}\mathbf{r} = \bar{\mathbf{e}}\mathbf{R}^{-1}\mathbf{r} = \overline{\mathbf{e}\mathbf{r}}$$

or

$$\bar{\mathbf{r}} = \mathbf{R}^{-1}\mathbf{r} \qquad \text{(fixed vector, basis rotated).} \tag{A.11}$$

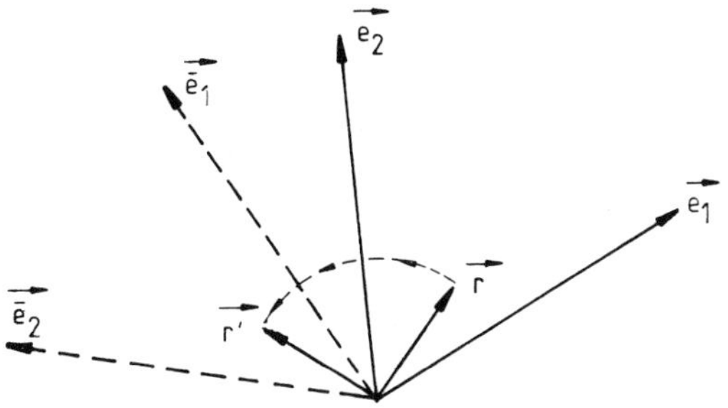

Figure A.4. The rotated vector $\mathbf{r}'$ has the same components in the new basis as $\mathbf{r}$ has in the old.

We shall now consider mappings of a vector space upon itself, by introducing operations which send every vector over into the *rotated* vector. Mapping of this kind forms the basis of the representation theory. We consider the transformations of an arbitrary vector $\mathbf{r}$ which is bounded to the basis. So that when the basis is rotated the vector $\mathbf{r}$ is carried with it. Thus the rotated $\mathbf{r}'$ will have the same components in the new basis as $\mathbf{r}$ had in the old (see Figure A.4), $\mathbf{r}'$ is called the image of $\mathbf{r}$.

Note that from (A.6)

$$\vec{\mathbf{r}}' = \bar{\mathbf{e}}\mathbf{r} = \mathbf{eRr} = \mathbf{er}'. \tag{A.12}$$

From which the components of $\mathbf{r}'$, relative to the original bases, are

$$\mathbf{r}' = \mathbf{Rr} \qquad \text{(rotated vector, basis fixed).} \tag{A.13}$$

It is convenient to regard a mapping $\vec{\mathbf{r}}$ to $\vec{\mathbf{r}}'$ as the result of an *operator* R which send every vector $\mathbf{r}$ into an image r'. That is,

$$\vec{\mathbf{r}}' = R\vec{\mathbf{r}} = R(\mathbf{er}) = R(\mathbf{e}_i r_i). \tag{A.14}$$

On the other hand

$$\mathbf{r}' = \bar{\mathbf{e}}\mathbf{r} = \vec{\bar{\mathbf{e}}}_i r_i.$$

Since $\vec{\mathbf{e}}_i = \bar{\mathbf{e}}_i' = R\mathbf{e}_i$, we write

$$\vec{\mathbf{r}}' = (R\mathbf{e}_i)r_i. \tag{A.15}$$

Comparison on (A.14) and (A.15) yields

$$R(\mathbf{e}_i r_i) = (R\mathbf{e}_i)r_i \tag{A.16}$$

or, explicitly,

$$R(\mathbf{e}_1 \quad \mathbf{e}_2 \quad \dots \quad \mathbf{e}_n)\begin{Vmatrix} r_1 \\ r_2 \\ \vdots \\ r_n \end{Vmatrix} = (R\mathbf{e}_1 \quad R\mathbf{e}_2 \quad \dots \quad R\mathbf{e}_n)\begin{Vmatrix} r_1 \\ r_2 \\ \vdots \\ r_n \end{Vmatrix}, \tag{A.17}$$

$$R(\mathbf{e}_1, \mathbf{e}_2, \dots, \mathbf{e}_n) = (R\mathbf{e}_1, R\mathbf{e}_2, \dots, R\mathbf{e}_n) = (\bar{\mathbf{e}}_1 \bar{\mathbf{e}}_2, \dots, \bar{\mathbf{e}}_n), \tag{A.18}$$

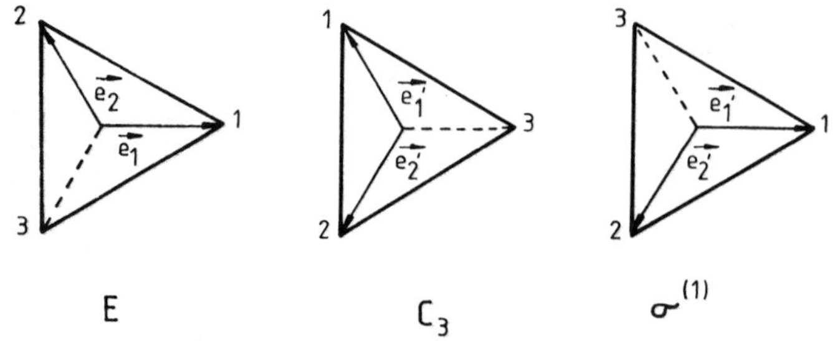

Figure A.5. The symmetry operation and the corresponding matrix representation is demonstrated on an equilateral lamina.

and, finally,

$$Re = \bar{\mathbf{e}} = \mathbf{eR}. \tag{A.19}$$

This relation makes a one-to-one association between a mapping R in a vector space and matrix $\mathbf{R}$. When a whole set of mappings $\{A, B, C, \ldots, R, \ldots\}$ is being considered, each will have its own matrix $\mathbf{A}, \mathbf{B}, \ldots, \mathbf{R}, \ldots$.

EXAMPLE 13. Consider the point group C_{3v}. If we embed two vectors $\mathbf{e}_1$ and $\mathbf{e}_2$ in the lamina, then any symmetry operations can be described alternatively, as a mapping R in which the two vectors are sent into images: $\mathbf{e}_1' = R\mathbf{e}_1$, $\mathbf{e}_2' = R\mathbf{e}_2$.

From Figure A.5, we have, for example,

$$\mathbf{e}_1' = C_3\mathbf{e}_1 = \mathbf{e}_2; \qquad \mathbf{e}_2' = C_3\mathbf{e}_2 = -\mathbf{e}_1 - \mathbf{e}_2.$$

Hence,

$$C_3(\mathbf{e}_1 \quad \mathbf{e}_2) = (\mathbf{e}_1 \quad \mathbf{e}_2)\begin{Vmatrix} 0 & -1 \\ 1 & -1 \end{Vmatrix} \quad \Rightarrow \quad \mathbf{D}(C_3) = \begin{Vmatrix} 0 & -1 \\ 1 & -1 \end{Vmatrix}.$$

Similarly,

$$\sigma^{(1)}\mathbf{e}_1 = \mathbf{e}_1, \qquad \sigma^{(1)}\mathbf{e}_2 = -\mathbf{e}_1 - \mathbf{e}_2.$$

Hence

$$\sigma^{(1)}(\mathbf{e}_1 \quad \mathbf{e}_2) = (\mathbf{e}_1 \quad \mathbf{e}_2)\begin{Vmatrix} 1 & -1 \\ 0 & -1 \end{Vmatrix} \quad \Rightarrow \quad \mathbf{D}(\sigma^{(1)}) = \begin{Vmatrix} 1 & -1 \\ 0 & -1 \end{Vmatrix}.$$

Finally, we have

	E	C_3	$\bar{C}_3$	$\sigma^{(1)}$	$\sigma^{(2)}$	$\sigma^{(3)}$
$\mathbf{D}(R)$:	$\begin{Vmatrix} 1 & 0 \\ 0 & 1 \end{Vmatrix}$	$\begin{Vmatrix} 0 & -1 \\ 1 & -1 \end{Vmatrix}$	$\begin{Vmatrix} -1 & 1 \\ -1 & 0 \end{Vmatrix}$	$\begin{Vmatrix} 1 & -1 \\ 0 & -1 \end{Vmatrix}$	$\begin{Vmatrix} -1 & 0 \\ -1 & 1 \end{Vmatrix}$	$\begin{Vmatrix} 0 & 1 \\ 1 & 0 \end{Vmatrix}$

We now show that the corresponding matrices obtained from $R\mathbf{e} = \mathbf{eR}$ provide the representation of the group. Consider the sequential performance of two operations, first S and then R. Let $RS = T$. Also let $\mathbf{r}$ be an arbitrary vector. Then from (A.6) and (A.19) we have

$$RS\mathbf{r} = RS\mathbf{er} = R\mathbf{eSr} = \mathbf{eRSr}. \tag{A.20}$$

Noting that

$$RS\mathbf{r} = T\mathbf{r} = T\mathbf{er} = \mathbf{eTr}, \tag{A.21}$$

and comparing (A.20) and (A.21) and making use of the independence of the basic vectors, we get

$$\mathbf{RS} = \mathbf{T}. \tag{A.22}$$

Hence, if $RS = T$ then (A.22) holds, which means that if the matrices $\mathbf{R}$ are associated with the mappings according to (A.19), they form a group representation.

Note that an arbitrary vector with components $\mathbf{r}$ is changed to one with components $\mathbf{r}'$, that is, $\mathbf{r}' = \mathbf{Rr}$. Whenever it is more convenient to work in terms of components, the matrices are obtained by this second relationship.

We also note that the matrices associated with a set of mappings (operations) are completely changed if we make a new choice of basis.

Example 14. Consider the same problem as in Example 13, but with a different choice of bases vectors (see Figure A.6).

From Example 12, with $\theta = 120°$, we write

$$\left.\begin{aligned} \mathbf{e}_1' &= -\tfrac{1}{2}\mathbf{e}_1 + \frac{\sqrt{3}}{2}\mathbf{e}_2, \\ \mathbf{e}_2' &= -\frac{\sqrt{3}}{2}\mathbf{e}_1 - \tfrac{1}{2}\mathbf{e}_2. \end{aligned}\right\} \tag{A.23}$$

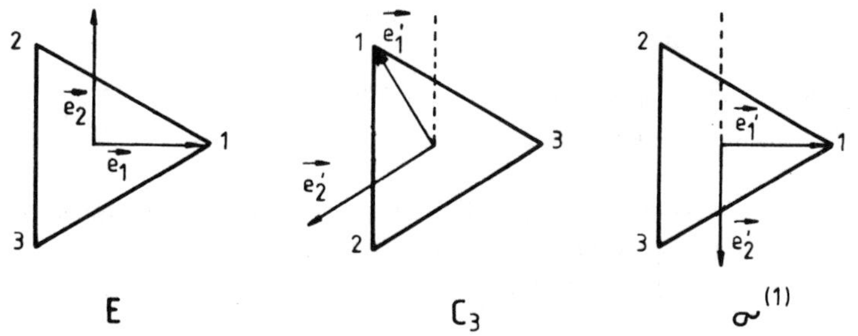

Figure A.6. A different choice of basis vectors produces a different matrix representation of the symmetry operation.

Hence

$$C_3(\mathbf{e}_1 \quad \mathbf{e}_2) = (\mathbf{e}_1 \quad \mathbf{e}_2)\begin{Vmatrix} -1/2 & -\sqrt{3}/2 \\ \sqrt{3}/2 & -1/2 \end{Vmatrix}$$

$$\Rightarrow \quad \mathbf{D}(C_3) = \begin{Vmatrix} -1/2 & -\sqrt{3}/2 \\ \sqrt{3}/2 & -1/2 \end{Vmatrix}.$$

In the case of reflection $\sigma^{(1)}$:

$$\left.\begin{aligned} \mathbf{e}_1' &= \sigma^{(1)}\mathbf{e}_1 = \mathbf{e}_1, \\ \mathbf{e}_2' &= \sigma^{(1)}\mathbf{e}_2 = -\mathbf{e}_2. \end{aligned}\right\} \tag{A.24}$$

Hence

$$\sigma^{(1)}(\mathbf{e}_1 \quad \mathbf{e}_2) = (\mathbf{e}_1 \quad \mathbf{e}_2)\begin{Vmatrix} 1 & 0 \\ 0 & -1 \end{Vmatrix} \quad \Rightarrow \quad \mathbf{D}(\sigma^{(1)}) = \begin{Vmatrix} 1 & 0 \\ 0 & -1 \end{Vmatrix}.$$

Similarly, we obtain

$$\mathbf{D}(E) = \begin{Vmatrix} 1 & 0 \\ 0 & 1 \end{Vmatrix}, \qquad \mathbf{D}(\bar{C}_3) = \begin{Vmatrix} -1/2 & \sqrt{3}/2 \\ -\sqrt{3}/2 & -1/2 \end{Vmatrix},$$

$$\mathbf{D}(\sigma^{(2)}) = \begin{Vmatrix} -1/2 & -\sqrt{3}/2 \\ -\sqrt{3}/2 & 1/2 \end{Vmatrix}, \qquad \mathbf{D}(\sigma^{(3)}) = \begin{Vmatrix} -1/2 & \sqrt{3}/2 \\ \sqrt{3}/2 & 1/2 \end{Vmatrix}.$$

Since the choice of basis is arbitrary, it appears that an infinite number of alternative representations can be obtained by simply changing the basis. However, the matrices in two representations are related. Suppose $\bar{\mathbf{e}} = \mathbf{eT}$. We write

$$R\mathbf{e} = \mathbf{eR} \qquad \text{(original basis)}, \tag{A.25}$$

$$R\bar{\mathbf{e}} = \bar{\mathbf{e}}\bar{\mathbf{R}} \qquad \text{(new basis)}. \tag{A.26}$$

It follows that

$$R\bar{\mathbf{e}} = R\mathbf{eT} = \mathbf{eRT} = \bar{\mathbf{e}}\mathbf{T}^{-1}\mathbf{RT},$$

and by comparison with (26) we get

$$\bar{\mathbf{R}} = \mathbf{T}^{-1}\mathbf{RT}.$$

Definition 15. The representations $\{\mathbf{R}\}$ and $\{\bar{\mathbf{R}}\}$, related by the similarity transformation, are called *equivalent*.

EXAMPLE 15. The representations of Examples 13 and 14, corresponding to the different choice of bases vectors $\mathbf{e}$ and $\bar{\mathbf{e}}$ must be equivalent.

From Figure A.7(a), (b), it follows that

$$\begin{cases} \mathbf{e}_1 = \bar{\mathbf{e}}_1, \\ \mathbf{e}_2 = -\tfrac{1}{2}\bar{\mathbf{e}}_1 + \dfrac{\sqrt{3}}{2}\bar{\mathbf{e}}_2, \end{cases} \qquad \text{or} \qquad \begin{cases} \bar{\mathbf{e}}_1 = \mathbf{e}_1, \\ \bar{\mathbf{e}}_2 = \dfrac{\sqrt{3}}{3}\mathbf{e}_1 + \dfrac{2\sqrt{3}}{2}\mathbf{e}_2. \end{cases}$$

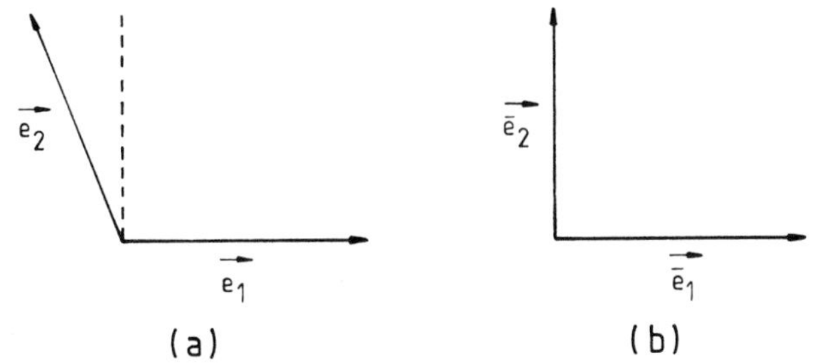

Figure A.7. Two different choices of base vectors.

Hence

$$(\mathbf{e}_1 \quad \mathbf{e}_2) = (\bar{\mathbf{e}}_1 \quad \bar{\mathbf{e}}_2)\begin{Vmatrix} 1 & -1/2 \\ 0 & \sqrt{3}/2 \end{Vmatrix} \quad \Rightarrow \quad \mathbf{T}^{-1} = \begin{Vmatrix} 1 & -1/2 \\ 0 & \sqrt{3}/2 \end{Vmatrix},$$

and

$$(\mathbf{e}_1 \quad \mathbf{e}_2) = (\bar{\mathbf{e}}_1 \quad \bar{\mathbf{e}}_2)\begin{Vmatrix} 1 & \sqrt{3}/3 \\ 0 & 2\sqrt{3}/3 \end{Vmatrix} \quad \Rightarrow \quad \mathbf{T} = \begin{Vmatrix} 1 & \sqrt{3}/3 \\ 0 & 2\sqrt{3}/3 \end{Vmatrix}.$$

Recalling that

$$\mathbf{D}(C_3) = \begin{Vmatrix} 0 & -1 \\ 1 & -1 \end{Vmatrix}$$

we obtain

$$\bar{\mathbf{D}}(C_3) = \mathbf{T}^{-1}\mathbf{D}(C_3)\mathbf{T} = \begin{Vmatrix} -1/2 & -\sqrt{3}/2 \\ \sqrt{3}/2 & -1/2 \end{Vmatrix},$$

which is the matrix found in Example 14.

Definition 16. The adjoint of $\mathbf{r}$ is denoted as $\mathbf{r}^*$ whose components are the complex conjugates of those of $\mathbf{r}$. In the n-dimensional complex vector space, $\mathbb{C}^n$, the adjoint of $\mathbf{r} = \mathbf{r}_i\mathbf{e}_i$ is

$$\mathbf{r} = \mathbf{r}_i^*\mathbf{e}_i^* \qquad (i = 1, \ldots, n). \tag{A.27}$$

Let us consider a Hermitian vector space, where the scalar product of any ordered pair of vectors $\mathbf{r}$ and $\mathbf{s}$ is defined by

$$\begin{aligned} (\mathbf{r}^*, \mathbf{s}) &= (\mathbf{s}^*, \mathbf{r})^*, \\ (\mathbf{r}^*, \mathbf{s} + \mathbf{t}) &= (\mathbf{r}^*, \mathbf{s}) + (\mathbf{r}^*, \mathbf{t}), \\ (\mathbf{r}^*, \alpha\mathbf{s}) &= \alpha(\mathbf{r}^*, \mathbf{s}), \\ (\mathbf{r}^*, \mathbf{r}) &> 0 \qquad \text{for any} \quad \mathbf{r} \neq \mathbf{0}. \end{aligned} \tag{A.28}$$

Definition 17. The matrix $\mathbf{G}$ whose elements G_{ij} are the scalar product of the basis vectors

$$G_{ij} = (\mathbf{e}_i^*, \mathbf{e}_j) \tag{A.29}$$

is called the *metrical matrix.*

The scalar product $(\mathbf{r}^*, \mathbf{s})$ may then take the form

$$(\mathbf{r}^*, \mathbf{s}) = r_i^* s_j (\mathbf{e}_1^*, \mathbf{e}_j) = r_i^* s_j G_{ij} = \mathbf{r}^\dagger \mathbf{G}\mathbf{s}, \tag{A.30}$$

where $\mathbf{r}$ and $\mathbf{s}$ are the usual columns of the components of $\mathbf{r}$ and $\mathbf{s}$, and the dagger (†) applied to a matrix means transposition accompanied by complex conjugation.

Note that $\mathbf{G}$ is a Hermitian matrix since

$$\mathbf{G}^\dagger = (\mathbf{e}^\dagger\mathbf{e})^\dagger = \mathbf{e}^\dagger\mathbf{e} = \mathbf{G}.$$

Definition 18. The two vectors $\mathbf{r}$ and $\mathbf{s}$ are said to be *orthogonal* if $(\mathbf{r}^*, \mathbf{s}) = 0$. If the vectors of a basis are mutually orthogonal and of unit length we have $(\mathbf{e}_i^*, \mathbf{e}_j) = \delta_{ij}$, the basis is then said to be unitary (orthonormal).

In an orthonormal basis, every component of a vector $\mathbf{r}$ can be expressed as a scalar product:

$$\begin{aligned}(\mathbf{e}_k^*, \mathbf{r}) &= (\mathbf{e}_k^*, \mathbf{e}_i r_i) = r_i(\mathbf{e}_k^*, \mathbf{e}_i)\\ &= r_i \delta_{ki} = r_k.\end{aligned} \tag{A.31}$$

Definition 19. A unitary basis, when all the quantities are real, is said to be a *rectangular Cartesian basis.*

Note that the elements of the materical matrix $\mathbf{G} = \|G_{ij}\|$ in the rectangular Cartesian basis are

$$G_{ij} = (\mathbf{e}_i^*, \mathbf{e}_j) = (\mathbf{e}_i, \mathbf{e}_j) = \delta_{ij}. \tag{A.32}$$

Definition 20. Nonsingular matrices which satisfy the condition $\mathbf{U}^\dagger\mathbf{U} = \mathbf{I}$ or $\mathbf{U}^\dagger = \mathbf{U}^{-1}$ are said to be *unitary*. When this is the case, the mapping $U\mathbf{e} = \mathbf{e}\mathbf{U}$ is called *unitary mapping.*

Theorem 6. *Unitary mapping has the property that lengths and angles remain invariant.*

Theorem 7. *Provided that the metric* $\mathbf{G}$ *is positive definite and that* $\mathbf{r}^\dagger\mathbf{G}\mathbf{r} > 0$ *for any* $\mathbf{r} \neq \mathbf{0}$, *a unitary basis can always be found.*

Now, the elements of the matrix $\mathbf{R}$ in the mapping $R\mathbf{e} = \mathbf{e}\mathbf{R}$ are metrically determined if the basis $\mathbf{e}$ is unitary. To show this, multiply both sides of $R\mathbf{e} = \mathbf{e}\mathbf{R}$ on the left by $\mathbf{e}^\dagger$

$$\mathbf{e}^\dagger R\mathbf{e} = (\mathbf{e}^\dagger\mathbf{e})\mathbf{R} = \mathbf{G}\mathbf{R} = \mathbf{R} \tag{A.33}$$

since $\mathbf{G} = \mathbf{I}_n$. This expresses $\mathbf{R}$ as an array of the scalar products

$$\mathbf{R} = \mathbf{e}^\dagger R\mathbf{e} = \begin{Vmatrix} \mathbf{e}_1^* \\ \mathbf{e}_2^* \\ \vdots \\ \mathbf{e}_n^* \end{Vmatrix} (R\mathbf{e}_1, R\mathbf{e}_2, \ldots, R\mathbf{e}_n)$$

$$= \begin{Vmatrix} (\mathbf{e}_1^*, R\mathbf{e}_1) & (\mathbf{e}_1^*, R\mathbf{e}_2) & \ldots & (\mathbf{e}_1^*, R\mathbf{e}_n) \\ \ldots & \ldots & \ldots & \ldots \\ (\mathbf{e}_n^*, R\mathbf{e}_1) & (\mathbf{e}_n^*, R\mathbf{e}_2) & \ldots & (\mathbf{e}_n^*, R\mathbf{e}_n) \end{Vmatrix},$$

hence, we get

$$R_{ij} = (\mathbf{e}_i^*, R\mathbf{e}_j). \tag{A.34}$$

Thus, if we have a space with a metric, it is not necessary to determine $\mathbf{R}$ by expressing $R\mathbf{e}_j$ in terms of $\mathbf{e}_1, \ldots, \mathbf{e}_n$ and picking out the appropriate components. We can simply evaluate the scalar product $\mathbf{e}_i^*$ and $\mathbf{e}_j' = R\mathbf{e}_j$ (image of $\mathbf{e}_j$).

Example 16. Consider the problem in Example 14. Let us find the matrix $\mathbf{D}(C_3)$ describing the mapping C_3. From Figure A.8 we have

$$C_3\mathbf{e} = \mathbf{e}\mathbf{D}(C_3)$$

and

$$D_{ij}(C_3) = (\mathbf{e}_i^*, C_3\mathbf{e}_j),$$

from which there follows

$$D_{11}(C_3) = (\mathbf{e}_1^*, C_3\mathbf{e}_1) = |\mathbf{e}_1^*|\cdot|C_3\mathbf{e}_1| \cos(\mathbf{e}_1^*, C_3\mathbf{e}_1) = \cos 120^\circ = -1/2,$$

$$D_{12}(C_3) = (\mathbf{e}_1^*, C_3\mathbf{e}_2) = |\mathbf{e}_1^*|\cdot|C_3\mathbf{e}_2| \cos(\mathbf{e}_1^*, C_3\mathbf{e}_2) = \cos 150^\circ = -\sqrt{3}/2,$$

$$D_{21}(C_3) = (\mathbf{e}_2^*, C_3\mathbf{e}_1) = |\mathbf{e}_2^*|\cdot|C_3\mathbf{e}_1| \cos(\mathbf{e}_2^*, C_3\mathbf{e}_1) = \cos 30^\circ = \sqrt{3}/2,$$

$$D_{22}(C_3) = (\mathbf{e}_2^*, C_3\mathbf{e}_2) = |\mathbf{e}_2^*|\cdot|C_3\mathbf{e}_2| \cos(\mathbf{e}_2^*, C_3\mathbf{e}_2) = \cos 120^\circ = -1/2,$$

which is in agreement with that of Example 14.

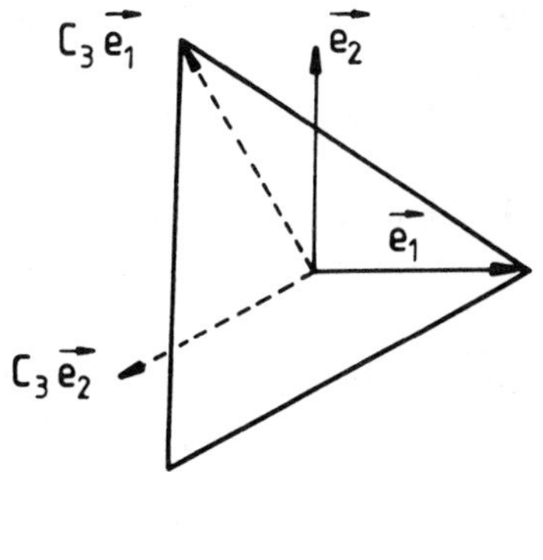

Figure A.8. Image of the base vectors under the mapping C_3.

Definition 21. Consider a group $G = \{R\}$ and its representation $\mathbf{D} = \{D(R)\}$. If the matrices $\mathbf{D}(R)$ are unitary, i.e.,

$$\mathbf{D}(R)\mathbf{D}^{\dagger}(R) = \mathbf{I} \qquad (\text{any } R \in G)$$

the representation D is then called a *unitary representation.*

Since unitary matrices have very useful properties, it is important to know whether a given representation $D = \{\mathbf{D}(R)\}$ is equivalent to a unitary representation.

Theorem 8. *For finite groups, every representation is equivalent to a unitary representation. That is, a new basis* $\bar{\mathbf{e}} = \mathbf{eT}$ *can be found such that* $\bar{\mathbf{R}} = \mathbf{T}^{-1}\mathbf{RT}$ *where* $\bar{\mathbf{R}}\bar{\mathbf{R}}^{\dagger} = \mathbf{I}$.

Recall that if the differences of two representations are due merely to a change of basis in a carrier space, the representations are equivalent and are not regarded as distinct.

Definition 22. Consider two inequivalent representations

$$D_1 = \{\mathbf{A}_1, \mathbf{B}_1, \ldots, \mathbf{R}_1, \ldots\},$$

$$D_2 = \{\mathbf{A}_2, \mathbf{B}_2, \ldots, \mathbf{R}_2, \ldots\},$$

of $G = \{A, B, \ldots, R, \ldots\}$. Form a third representation as block-form matrices

$$\mathbf{A} = \begin{Vmatrix} \mathbf{A}_1 & 0 \\ 0 & \mathbf{A}_2 \end{Vmatrix}, \qquad \mathbf{B} = \begin{Vmatrix} \mathbf{B}_1 & 0 \\ 0 & \mathbf{B}_2 \end{Vmatrix}, \ldots, \qquad \mathbf{R} = \begin{Vmatrix} \mathbf{R}_1 & 0 \\ 0 & \mathbf{R}_2 \end{Vmatrix}, \ldots.$$

The set $D = \{\mathbf{A}, \mathbf{B}, \ldots, \mathbf{R}, \ldots\}$ is called the *direct sum* of D_1, and D_2 and is denoted by $D = D_1 \dot{+} D_2$.

The representation $D = \{\mathbf{A}, \mathbf{B}, \ldots, \mathbf{R}\}$ may be replaced by an equivalent representation $\bar{D}$ of less simple appearance. By a similarity transformation, we have

$$\bar{\mathbf{R}} = \bar{\mathbf{D}}(R) = \mathbf{S}^{-1}\begin{Vmatrix} \mathbf{R}_1 & 0 \\ 0 & \mathbf{R}_2 \end{Vmatrix}, \qquad \mathbf{S} = \begin{Vmatrix} \bar{\mathbf{R}}_{11} & \bar{\mathbf{R}}_{12} \\ \bar{\mathbf{R}}_{21} & \bar{\mathbf{R}}_{22} \end{Vmatrix}, \tag{A.35}$$

which, in general, will not have a simple block form as $\mathbf{D}(R)$.

Now, in searching for *simple* representations, the above procedure is reversed. If we can find a similarity transformation, with a matrix $\mathbf{S}$ which brings *all* the matrices of the representation $\bar{D}$ to a similar block form D, the process is called *reduction.*

Definition 23. Suppose that the process of reduction is repeated until the various blocks cannot be reduced any further, then the set of matrices

$$D_1 = \{\mathbf{A}_1, \mathbf{B}_1, \ldots\},$$

$$D_2 = \{\mathbf{A}_2, \mathbf{B}_2, \ldots\},$$

$$D_3 = \{\mathbf{A}_3, \mathbf{B}_3, \ldots\},$$

$$\ldots\ldots\ldots\ldots\ldots\ldots\ldots$$

are called *irreducible representations* of the group G.

The number of inequivalent irreducible representations of any finite group G is determinable, and will be shown to be equal to the number of the distinct classes of G.

If we have the full set of the irreducible representations of G, we thus know that any other representation is equivalent either to one of them or to a direct sum of two or more.

Note that Theorem 8 implies that it is sufficient to prove all the theorems for representations with unitary matrices.

Theorem 9. *If two unitary representations*

$$D = \{\mathbf{A}, \mathbf{B}, \ldots, \mathbf{R}, \ldots\},$$
$$\bar{D} = \{\bar{\mathbf{A}}, \bar{\mathbf{B}}, \ldots, \bar{\mathbf{R}}, \ldots\},$$

are equivalent, i.e., $\bar{\mathbf{R}} = \mathbf{T}^{-1}\mathbf{R}\mathbf{T}$, *it is possible to find a unitary matrix* $\mathbf{U}$ *which relates them according to* $\bar{\mathbf{R}} = \mathbf{U}^{\dagger}\mathbf{R}\mathbf{U}$. *In fact,* $\mathbf{U} = (\mathbf{T}\mathbf{T}^{\dagger})^{-1/2}\mathbf{T}$.

EXAMPLE 17. In Examples 13, 14, and 15 we obtained two equivalent representations which we list below:

	E	C_3	$\bar{C}_3$
D:	$\begin{Vmatrix} 1 & 0 \\ 0 & 1 \end{Vmatrix}$	$\begin{Vmatrix} 0 & -1 \\ 1 & -1 \end{Vmatrix}$	$\begin{Vmatrix} -1 & 1 \\ -1 & 0 \end{Vmatrix}$
	$\sigma^{(1)}$	$\sigma^{(2)}$	$\sigma^{(3)}$
	$\begin{Vmatrix} 1 & -1 \\ 0 & -1 \end{Vmatrix}$	$\begin{Vmatrix} -1 & 0 \\ -1 & 1 \end{Vmatrix}$	$\begin{Vmatrix} 0 & 1 \\ 1 & 0 \end{Vmatrix}$
	E	C_3	$\bar{C}_3$
$\bar{D}$:	$\begin{Vmatrix} 1 & 0 \\ 0 & 1 \end{Vmatrix}$	$\begin{Vmatrix} -1/2 & -\sqrt{3}/2 \\ \sqrt{3}/2 & -1/2 \end{Vmatrix}$	$\begin{Vmatrix} -1/2 & \sqrt{3}/2 \\ -\sqrt{3}/2 & -1/2 \end{Vmatrix}$
	$\sigma^{(1)}$	$\sigma^{(2)}$	$\sigma^{(3)}$
	$\begin{Vmatrix} 1 & 0 \\ 0 & -1 \end{Vmatrix}$	$\begin{Vmatrix} -1/2 & -\sqrt{3}/2 \\ -\sqrt{3}/2 & 1/2 \end{Vmatrix}$	$\begin{Vmatrix} -1/2 & \sqrt{3}/2 \\ \sqrt{3}/2 & 1/2 \end{Vmatrix}$

with $\bar{\mathbf{D}}(R) = \mathbf{T}^{-1}\mathbf{D}(\mathbf{R})\mathbf{T}$, where

$$\mathbf{T} = \begin{Vmatrix} 1 & \sqrt{3}/3 \\ 0 & 2\sqrt{3}/3 \end{Vmatrix} \quad \text{and} \quad T^{-1} = \begin{Vmatrix} 1 & -1/2 \\ 0 & \sqrt{3}/2 \end{Vmatrix}.$$

Note that

$$\mathbf{T}\mathbf{T}^{\dagger} = \begin{Vmatrix} 1 & \sqrt{3}/3 \\ 0 & 2\sqrt{3}/3 \end{Vmatrix} \begin{Vmatrix} 1 & 0 \\ \sqrt{3}/3 & 2\sqrt{3}/3 \end{Vmatrix} = \frac{2}{3}\begin{Vmatrix} 2 & 1 \\ 1 & 2 \end{Vmatrix} \neq \mathbf{I}.$$

Hence $\mathbf{T}$ is not a unitary matrix. However, the matrix $\mathbf{U}$, given by

$$\mathbf{U} = (\mathbf{T}\mathbf{T}^{\dagger})^{-1/2}\mathbf{T} = \frac{1}{2}\begin{Vmatrix} \sqrt{2-\sqrt{3}} & -\sqrt{2+\sqrt{3}} \\ -\sqrt{2+\sqrt{3}} & \sqrt{2-\sqrt{3}} \end{Vmatrix} \cdot \begin{Vmatrix} 1 & \sqrt{3}/3 \\ 0 & 2/\sqrt{3} \end{Vmatrix}$$

is shown to be unitary.

EXAMPLE 18. Let us regard the symmetry operations of C_{3v} as mappings of a three-dimensional space upon itself, by putting the vertices of the equilateral triangle at the extremities of the three orthogonal unit vectors as shown in Figure A.9.

By the symmetry elements of C_{3v}, the basis vectors $\mathbf{e}_1$, $\mathbf{e}_2$, $\mathbf{e}_3$ are simply permuted. For example,

$$\mathbf{e}_1' = C_3\mathbf{e}_1 = \mathbf{e}_2,$$

$$\mathbf{e}_2' = C_3\mathbf{e}_2 = \mathbf{e}_3,$$

$$\mathbf{e}_3' = C_3\mathbf{e}_3 = \mathbf{e}_1,$$

or

$$C_3(\mathbf{e}_1 \quad \mathbf{e}_2 \quad \mathbf{e}_3) = (\mathbf{e}_1 \quad \mathbf{e}_2 \quad \mathbf{e}_3)\begin{Vmatrix} 0 & 0 & 1 \\ 1 & 0 & 0 \\ 0 & 1 & 0 \end{Vmatrix},$$

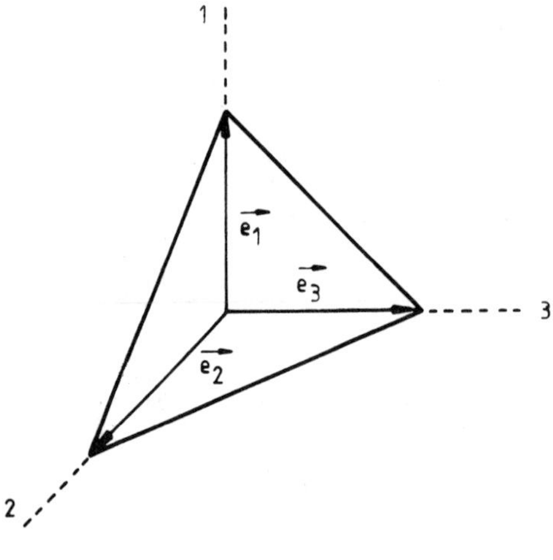

Figure A.9. Base vectors in three-dimensional space.

and for $\sigma^{(1)}$:

$$\sigma^{(1)}(\mathbf{e}_1 \quad \mathbf{e}_2 \quad \mathbf{e}_3) = (\mathbf{e}_1 \quad \mathbf{e}_2 \quad \mathbf{e}_3)\begin{Vmatrix} 1 & 0 & 0 \\ 0 & 0 & 1 \\ 0 & 1 & 0 \end{Vmatrix},$$

and so on. We list them below:

E	C_3	$\bar{C}_3$	$\sigma^{(1)}$	$\sigma^{(2)}$	$\sigma^{(3)}$
$\begin{Vmatrix} 1 & 0 & 0 \\ 0 & 1 & 0 \\ 0 & 0 & 1 \end{Vmatrix}$	$\begin{Vmatrix} 0 & 0 & 1 \\ 1 & 0 & 0 \\ 0 & 1 & 0 \end{Vmatrix}$	$\begin{Vmatrix} 0 & 1 & 0 \\ 0 & 0 & 1 \\ 1 & 0 & 0 \end{Vmatrix}$	$\begin{Vmatrix} 1 & 0 & 0 \\ 0 & 0 & 1 \\ 0 & 1 & 0 \end{Vmatrix}$	$\begin{Vmatrix} 0 & 0 & 1 \\ 0 & 1 & 0 \\ 1 & 0 & 0 \end{Vmatrix}$	$\begin{Vmatrix} 0 & 1 & 0 \\ 1 & 0 & 0 \\ 0 & 0 & 1 \end{Vmatrix}$

This gives a three-dimensional representation of C_{3v}, which is not in reduced form. To reduce this representation we must find a new set of basis vectors, some of which span an invariant subspace. For example, there is a vector perpendicular to the lamina through its centroid, which is invariant under all symmetry operations of C_{3v}. This vector spans an invariant one-dimensional subspace. It is not, however, an easy matter to pick invariant subspaces by inspection.

We now consider certain important relations between irreducible representations.

Theorem 10. *If $D^{(\alpha)}$ and $D^{(\beta)}$ are two irreducible representations of a group G, then the matrix elements satisfy the orthogonality relations*

$$\sum_{R \in G} D_{ij}^{(\alpha)}(R^{-1})D_{kl}^{(\beta)}(R) = \frac{g}{d_\alpha}\delta_{\alpha\beta}\delta_{kj}\delta_{il}, \tag{A.36}$$

where d_α and d_β denote the dimensions of $D^{(\alpha)}$ and $D^{(\beta)}$, respectively.

Example 19. In Example 13, we obtained the following nonunitary representations, which will shortly be shown to be irreducible.

	E	C_3	C_3^-	$\sigma^{(1)}$	$\sigma^{(2)}$	$\sigma^{(3)}$
$D(R)$	$\begin{Vmatrix} 1 & 0 \\ 0 & 1 \end{Vmatrix}$	$\begin{Vmatrix} 0 & -1 \\ 1 & -1 \end{Vmatrix}$	$\begin{Vmatrix} -1 & 1 \\ -1 & 0 \end{Vmatrix}$	$\begin{Vmatrix} 1 & -1 \\ 0 & -1 \end{Vmatrix}$	$\begin{Vmatrix} -1 & 0 \\ -1 & 1 \end{Vmatrix}$	$\begin{Vmatrix} 0 & 1 \\ 1 & 0 \end{Vmatrix}$
$D(R^{-1})$	$\begin{Vmatrix} 1 & 0 \\ 0 & 1 \end{Vmatrix}$	$\begin{Vmatrix} -1 & 1 \\ -1 & 0 \end{Vmatrix}$	$\begin{Vmatrix} 0 & -1 \\ 1 & -1 \end{Vmatrix}$	$\begin{Vmatrix} 1 & -1 \\ 0 & -1 \end{Vmatrix}$	$\begin{Vmatrix} -1 & 0 \\ -1 & 1 \end{Vmatrix}$	$\begin{Vmatrix} 0 & 1 \\ 1 & 0 \end{Vmatrix}$

We have $g = 6$, $d_\alpha = 2$, and $\alpha = \beta$. Take $i = j = k = l = 1$. Form

$$\sum_R D_{11}(R^{-1})D_{11}(R) = 1 + 0 + 0 + 1 + 1 + 0 = \tfrac{6}{2} \times 1 \times 1 \times 1 = 3,$$

and for $i = k = l = 1, j = 2$,

$$\sum_R D_{12}(R^{-1})D_{11}(R) = 0 + 0 + 1 - 1 + 0 + 0 = 0, \quad \text{and so on.}$$

EXAMPLE 20. We now consider all the irreducible representations of C_{3v}. They are collected in the following table. All these representations are unitary, or more specifically, orthogonal.

	E	C_3	C_3^-	$\sigma^{(1)}$	$\sigma^{(2)}$	$\sigma^{(3)}$
$D^{(1)}$	1	1	1	1	1	1
$D^{(2)}$	1	1	1	-1	-1	-1
$D^{(3)}$	$\begin{Vmatrix} 1 & 0 \\ 0 & 1 \end{Vmatrix}$	$\begin{Vmatrix} -c & -s \\ s & -c \end{Vmatrix}$	$\begin{Vmatrix} -c & s \\ -s & -c \end{Vmatrix}$	$\begin{Vmatrix} 1 & 0 \\ 0 & -1 \end{Vmatrix}$	$\begin{Vmatrix} -c & -s \\ -s & c \end{Vmatrix}$	$\begin{Vmatrix} -c & s \\ s & c \end{Vmatrix}$

where $s = \sqrt{3}/2$ and $c = \frac{1}{2}$. Then, for example,

$$\sum_R D_{11}^{(1)}(R^{-1})D_{11}^{(2)}(R) = 1 + 1 + 1 - 1 - 1 - 1 = 0,$$

$$\sum_R D_{11}^{(1)}(R^{-1})D_{11}^{(3)}(R) = 1 - c - c + 1 - c - c = 0,$$

$$\sum_R D_{11}^{(1)}(R^{-1})D_{12}^{(3)}(R) = 0 - s + s + 0 - s + s = 0, \qquad \text{and so on.}$$

The simplest criterion for reducibility depends on the notion of the *character*.

Definition 24. The character of an element R of a finite group $G = \{A, B, \ldots, R, \ldots\}$ in the representation

$$D = \{\mathbf{D}(A), \mathbf{D}(B), \ldots, \mathbf{D}(R), \ldots\}$$

is the *trace* of its representative matrix, i.e.,

$$\chi(R) = \text{trace } \mathbf{D}(R).$$

The whole set of characters

$$\chi(A), \chi(B), \ldots, \chi(R), \ldots,$$

is called the *character system* of the group G for the representation D.

Note that the character system is a property of the representation which is invariant under similarity transformations, that is, it does not depend on a particular choice of basis. Thus, the representations which are essentially distinct will have a distinct character system.

In giving the character system of a group it is only necessary to consider a typical element of each class, since any two elements in the same class are related by

$$A = RBR^{-1} \qquad (\text{any } R \text{ in } G).$$

Definition 25. A *character table* is the set of character systems associated with all the irreducible representations of the group considered.

EXAMPLE 21. The character table of C_{3v} is given by

χ \ class	$\{E\}$	$\{C_3, \bar{C}_3\}$	$\{\sigma^{(1)}, \sigma^{(2)}, \sigma^{(3)}\}$
$\chi^{(1)}$	1	1	1
$\chi^{(2)}$	1	1	-1
$\chi^{(3)}$	2	-1	0

where $\chi^{(i)} = \text{trace } D^{(i)}(R)$.

Theorem 11. *Character systems also form a set of orthogonal vectors, that is,*

$$\sum_{R \in G} \overset{*}{\chi}{}^{(\alpha)}(R)\chi^{(\beta)}(R) = g\delta_{\alpha\beta}. \tag{A.37}$$

EXAMPLE 22. Consider the character table of C_{3v} given in the previous example. We write, from (A.37),

$$\sum_R \overset{*}{\chi}{}^{(1)}(R)\chi^{(2)}(R) = 1 + 1 + 1 - 1 - 1 - 1 = 0,$$

$$\sum_R \overset{*}{\chi}{}^{(1)}(R)\chi^{(3)}(R) = 2 - 1 - 1 + 0 + 0 + 0 = 0,$$

$$\sum_R \overset{*}{\chi}{}^{(2)}(R)\chi^{(3)}(R) = 2 - 1 - 1 + 0 + 0 + 0 = 0,$$

$$\sum_R \overset{*}{\chi}{}^{(3)}(R)\chi^{(3)}(R) = 4 + 1 + 1 + 0 + 0 + 0 = 6,$$

The identity of the character systems is the necessary and sufficient condition for two representations to be equivalent. Verify by inspection that the two representations of C_{3v} obtained in Examples 13 and 14 are equivalent.

Furthermore, it is possible to find whether any given representation is reducible, and if so, how many times each irreducible representation occurs.

Theorem 12. *If* $D = \{D(R)\}$ *is an arbitrary representation of* $G = \{R\}$, *with character* $\chi(R)$ and *where* $D^{(1)}, D^{(2)}, \ldots, D^{(r)}$ *are the irreducible representations of* G, *then*

$$D = \sum_{l=1}^{r} n_i D^{(i)} = n_1 D^{(1)} \dotplus \cdots \dotplus n_r D^{(r)} \tag{A.38}$$

and n_i is given by

$$n_i = \frac{1}{g} \sum_{R \in G} \overset{*}{\chi}{}^{(i)}(R)\chi(R). \tag{A.39}$$

Thus, by means of a character table, we can determine the number of times each irreducible representation occurs in a given representation, knowing only the character of the given representation. However, finding a transformation matrix which will reduce a given representation is a much more difficult problem.

EXAMPLE 23. Consider the three-dimensional representation of C_{3v} obtained in Example 18. The character system is

	E	C_3	C_3^-	$\sigma^{(1)}$	$\sigma^{(2)}$	$\sigma^{(3)}$
χ	3	0	0	1	1	1

From (A.39) and the character table of C_{3v} (see Example 21), we have

$$n_1 = \tfrac{1}{6}(1 \times 3 + 1 \times 0 + 1 \times 0 + 1 \times 1 + 1 \times 1 + 1 \times 1) = 1,$$

$$n_2 = \tfrac{1}{6}(1 \times 3 + 1 \times 0 + 1 \times 0 - 1 \times 1 - 1 \times 1 - 1 \times 1) = 0,$$

$$n_3 = \tfrac{1}{6}(2 \times 3 - 1 \times 0 - 1 \times 0 + 0 \times 1 + 0 \times 1 + 0 \times 1) = 1.$$

Hence,

$$D = D^{(1)} \dotplus D^{(3)}. \tag{A.40}$$

Theorem 13. *The necessary and sufficient condition that a representation $D = \{\mathbf{D}(R)\}$ be irreducible is that*

$$\sum_{R \in G} \overset{*}{\chi}(R)\chi(R) = g. \tag{A.41}$$

EXAMPLE 24. The three representations of C_{3v} listed in Example 20 are each irreducible, since

$$\begin{aligned}\sum_{R \in G} \overset{*}{\chi}(R)\chi(R) &= 1 + 1 + 1 + 1 + 1 + 1 = 6 \\ &= 1 + 1 + 1 + 1 + 1 + 1 = 6 \\ &= 4 + 1 + 1 + 0 + 0 + 0 = 6,\end{aligned}$$

where 6 is the order of C_{3v}.

Theorem 14. *The number of distinct irreducible representations of a finite group is equal to the number of classes.*

Theorem 15. *The sum of the squares of the dimensions of the distinct irreducible representations $D^{(1)}, \ldots, D^{(r)}$ of a finite group G is equal to the order of the group*

$$\sum_{i=1}^{r} d_i^2 = g \qquad (r = \text{number of classes}), \tag{A.42}$$

where d_i = dimension $D^{(i)}$ and g = the number of elements in G.

Recall that the direct sum of two representations gives another representation. There is another way of combining two representations, in which we form the direct product.

Definition 26. Let $\mathbf{R}^{(\alpha)}$ and $\mathbf{R}^{(\beta)}$ be two square matrices of order m and n, respectively. The matrix $\mathbf{R}^{(\alpha\beta)}$ of order mn, defined by

$$R^{(\alpha\beta)}_{kl,ij} = R^{(\alpha)}_{ki} R^{(\beta)}_{lj} \tag{A.43}$$

is called the direct (Kronecker) product of matrices $\mathbf{R}^{(\alpha)}$ and $\mathbf{R}^{(\beta)}$, and is written as

$$\mathbf{R}^{(\alpha\beta)} = \mathbf{R}^{(\alpha)} \otimes \mathbf{R}^{(\beta)}. \tag{A.44}$$

Example 25. Suppose we have two 2×2 matrices $\mathbf{A}$ and $\mathbf{B}$; their 4×4 direct product is given by

$$\mathbf{A} \otimes \mathbf{E} = \begin{array}{c} \\ 11 \\ 12 \\ \\ 21 \\ 22 \end{array} \begin{array}{cc|cc} 11 & 12 & 21 & 22 \\ A_{11}B_{11} & A_{11}B_{12} & A_{12}B_{11} & A_{12}B_{12} \\ A_{11}B_{21} & A_{11}B_{22} & A_{12}B_{21} & A_{12}B_{22} \\ \hline A_{21}B_{11} & A_{21}B_{12} & A_{22}B_{11} & A_{22}B_{12} \\ A_{21}B_{21} & A_{21}B_{22} & A_{22}B_{21} & A_{22}B_{22} \end{array}$$

$$= \left\| \begin{array}{c|c} A_{11}\mathbf{B} & A_{12}\mathbf{B} \\ \hline A_{21}\mathbf{B} & A_{22}\mathbf{B} \end{array} \right\|, \tag{A.45}$$

provided that the index pairs (kl) and (ij) are arranged in dictionary order.

Theorem 16. *If* dim $\mathbf{A}$ = dim $\mathbf{B}$ *and* dim $\mathbf{C}$ = dim $\mathbf{D}$, *then*

$$(\mathbf{AB}) \otimes (\mathbf{CD}) = (\mathbf{A} \otimes \mathbf{C})(\mathbf{B} \otimes \mathbf{D}). \tag{A.46}$$

Note that if $\mathbf{A}$ *and* $\mathbf{B}$ *are nonsingular matrices of order* m *and* n, *respectively, it follows from* (A.46) *that*

$$(\mathbf{A} \otimes \mathbf{B})(\mathbf{A}^{-1} \otimes \mathbf{B}^{-1}) = \mathbf{I}_m \otimes \mathbf{I}_n = \mathbf{I}_{mn},$$

so that $(\mathbf{A} \otimes \mathbf{B})$ *is itself nonsingular. Also note that*

$$\mathbf{A} \otimes (\mathbf{B} \otimes \mathbf{C}) = (\mathbf{A} \otimes \mathbf{B}) \otimes \mathbf{C} \qquad (associative)$$

and

$$\mathbf{I}_m \otimes \mathbf{A} = \mathbf{A} \dotplus \mathbf{A} \dotplus \cdots \dotplus \mathbf{A} \qquad (m\ terms).$$

We also have the following properties:

$$\det(\mathbf{A} \otimes \mathbf{B}) = (\det \mathbf{A})^n (\det \mathbf{B})^m, \tag{A.47}$$

$$\operatorname{trace}(\mathbf{A} \otimes \mathbf{B}) = \operatorname{trace} \mathbf{A} \cdot \operatorname{trace} \mathbf{B}, \qquad \dim(\mathbf{A} \otimes \mathbf{B}) = \dim \mathbf{A} \cdot \dim \mathbf{B}. \tag{A.48}$$

Theorem 17. *The inner product* $D^{(\alpha\beta)} = D^{(\alpha)} \otimes D^{(\beta)}$ *of two representations* $D^{(\alpha)} = \{\mathbf{R}^{(\alpha)}\}$ *and* $D^{(\beta)} = \{\mathbf{R}^{(\beta)}\}$ *of a group* $G = \{R\}$ *form another representation.*

To show this, suppose $RS = T$, $(R, S, T \in G)$, *which implies*

$$\mathbf{R}^{(\alpha)}\mathbf{S}^{(\alpha)} = \mathbf{T}^{(\alpha)} \qquad \text{and} \qquad \mathbf{R}^{(\beta)}\mathbf{S}^{(\beta)} = \mathbf{T}^{(\beta)}.$$

Then, from (A.46), *it follows that*

$$\begin{aligned}(R^{(\alpha)} \otimes R^{(\beta)})(\mathbf{S}^{(\alpha)} \otimes \mathbf{S}^{(\beta)}) &= (\mathbf{R}^{(\alpha)}\mathbf{S}^{(\alpha)}) \otimes (\mathbf{R}^{(\beta)}\mathbf{S}^{(\beta)}) \\ &= \mathbf{T}^{(\alpha)} \otimes \mathbf{T}^{(\beta)},\end{aligned}$$

and consequently the inner direct (Kronecker) products also have the same multiplication table as the group elements.

EXAMPLE 26. Consider the group C_{3v} and its two-dimensional representation obtained in Example 13. Let us form the inner direct product representation $D^{(33)} = D^{(3)} \otimes D^{(3)}$

$$\mathbf{D}^{(33)}(E) = \begin{Vmatrix} 1 & 0 & 0 & 0 \\ 0 & 1 & 0 & 0 \\ 0 & 0 & 1 & 0 \\ 0 & 0 & 0 & 1 \end{Vmatrix}, \quad \mathbf{D}^{(33)}(C_3) = \begin{Vmatrix} 0 & 0 & 0 & 1 \\ 0 & 0 & -1 & 1 \\ 0 & -1 & 0 & 1 \\ 1 & -1 & -1 & 1 \end{Vmatrix},$$

$$\mathbf{D}^{(33)}(\bar{C}_3) = \begin{Vmatrix} 1 & -1 & -1 & 1 \\ 1 & 0 & -1 & 0 \\ 1 & -1 & 0 & 0 \\ 1 & 0 & 0 & 0 \end{Vmatrix}, \quad \mathbf{D}^{(33)}(\sigma^{(1)}) = \begin{Vmatrix} 1 & -1 & -1 & 1 \\ 0 & -1 & 0 & 1 \\ 0 & 0 & -1 & 1 \\ 0 & 0 & 0 & 1 \end{Vmatrix},$$

$$\mathbf{D}^{(33)}(\sigma^{(2)}) = \begin{Vmatrix} 1 & 0 & 0 & 0 \\ 1 & -1 & 0 & 0 \\ 1 & 0 & -1 & 0 \\ 1 & -1 & -1 & 1 \end{Vmatrix}, \quad \mathbf{D}^{(33)}(\sigma^{(3)}) = \begin{Vmatrix} 0 & 0 & 0 & 1 \\ 0 & 0 & 1 & 0 \\ 0 & 1 & 0 & 0 \\ 1 & 0 & 0 & 0 \end{Vmatrix}.$$

Take $\sigma^{(1)}C_3 = \sigma^{(2)}$,

$$\begin{aligned}\mathbf{D}^{(33)}(\sigma^{(1)})\mathbf{D}^{(33)}(C_3) &= \begin{Vmatrix} 1 & -1 & -1 & 1 \\ 0 & -1 & 0 & 1 \\ 0 & 0 & -1 & 1 \\ 0 & 0 & 0 & 1 \end{Vmatrix} \begin{Vmatrix} 0 & 0 & 0 & 1 \\ 0 & 0 & -1 & 1 \\ 0 & -1 & 0 & 1 \\ 1 & -1 & -1 & 1 \end{Vmatrix} \\ &= \begin{Vmatrix} 1 & 0 & 0 & 0 \\ 1 & -1 & 0 & 0 \\ 1 & 0 & -1 & 0 \\ 1 & -1 & -1 & 1 \end{Vmatrix} \\ &= \mathbf{D}^{(33)}(\sigma^{(2)}).\end{aligned}$$

We note that the inner direct product of two representations $\mathbf{D}^{(\alpha)}$ and $\mathbf{D}^{(\beta)}$ is, in general, reducible; that is,

$$D^{(\alpha\beta)} = D^{(\alpha)} \otimes D^{(\beta)} = \sum_{i=1}^{r} \gamma_{i,\alpha\beta} D^{(i)}. \tag{A.49}$$

The number of times each irreducible representation occurs in the product representation is given by

$$\gamma_{i,\alpha\beta} = \frac{1}{g} \sum_{R \in G} \overset{*}{\chi}^{(i)}(R)\chi^{(\alpha\beta)}(R) = \frac{1}{g} \sum_{R} \overset{*}{\chi}^{(i)}(R)\chi^{(\alpha)}(R)\chi^{(\beta)}(R). \tag{A.50}$$

in deriving this expression (A.39) and (A.48) are used.

EXAMPLE 27. Consider the character table of C_{3v}, obtained in Example 21.

	E	C_3	$\bar{C}_3$	$\sigma^{(1)}$	$\sigma^{(2)}$	$\sigma^{(3)}$
$\chi^{(1)}$	1	1	1	1	1	1
$\chi^{(2)}$	1	1	1	-1	-1	-1
$\chi^{(3)}$	2	-1	-1	0	0	0

Let us compose $D^{(33)}$:

for $i = 1$, $\quad \gamma_{1,33} = \frac{1}{6} \sum_R \overset{*}{\chi}^{(1)}(R)\chi^{(3)}(R)\chi^{(3)}(R) = \frac{1}{6}(4 + 1 + 1) = 1,$

for $i = 2$, $\quad \gamma_{2,33} = \frac{1}{6} \sum_R \overset{*}{\chi}^{(2)}(R)\chi^{(3)}(R)\chi^{(3)}(R) = \frac{1}{6}(4 + 1 + 1) = 1,$

for $i = 3$, $\quad \gamma_{3,33} = \frac{1}{6} \sum_R \overset{*}{\chi}^{(3)}(R)\chi^{(3)}(R)\chi^{(3)}(R) = \frac{1}{6}(8 - 1 - 1) = 1.$

Hence

$$D^{(33)} = D^{(1)} \dotplus D^{(2)} \dotplus D^{(3)}.$$

Considering all the irreducible representations of C_{3v} and using (A.50) we may form a table for the inner direct product of the irreducible representation for C_{3v}:

$\otimes$	$D^{(1)}$	$D^{(2)}$	$D^{(3)}$
$D^{(1)}$	$D^{(1)}$	$D^{(2)}$	$D^{(3)}$
$D^{(2)}$	$D^{(2)}$	$D^{(1)}$	$D^{(3)}$
$D^{(3)}$	$D^{(3)}$	$D^{(3)}$	$D^{(1)} \dotplus D^{(2)} \dotplus D^{(3)}$

Equation (A.50) can be rewritten as

$$\gamma_{k,ij} = \frac{1}{g} \sum_{m=1}^{r} r_m \overset{*}{\chi}{}_m^{(k)} \chi_m^{(i)} \chi_m^{(j)}, \tag{A.51}$$

where r_m is the number of elements of the mth class of G, and $\chi_m^{(i)}$ is the character of the irreducible representation $D^{(i)}$ associated with the mth class. On the other hand, from (A.49), it immediately follows that

$$\chi^{(i)}\chi^{(j)} = \sum_{k=1}^{r} \gamma_{k,ij}\chi^{(k)} \tag{A.52}$$

from which we get $\gamma_{k,ij} = \gamma_{k,ji}$.

It frequently happens that a physical system is invariant under two independent groups of symmetry operations.

Definition 27. Consider two groups of operators $G_1 = \{A_i\}$ and $G_2 = \{B_j\}$ where $A_iB_j = B_jA_i$ for all pairs of operators A_i, B_j, and suppose that G_1 and G_2 have no operator (element) in common, apart from the identity. Then the set $G = \{A_iB_j\}$, of all mn product operators A_iB_j, is said to be the outer product of G_1 and G_2, and is expressed as $G = G_1 \times G_2 = \{A_iB_j\}$.

Theorem 18. *The outer product $G = G_1 \times G_2$ forms a new group. The properties of a direct product are solely determined by those of the separate constituent groups, and it is therefore advantageous, wherever possible, to regard a group as a direct product of simpler groups.*

EXAMPLE 28. Consider the group $G_2 = C_{3v}$ operating on an equilateral triangle. Let us admit a new symmetry operation σ_h which reflects a point on the upper surface of the lamina to one directly beneath it on the lower surface

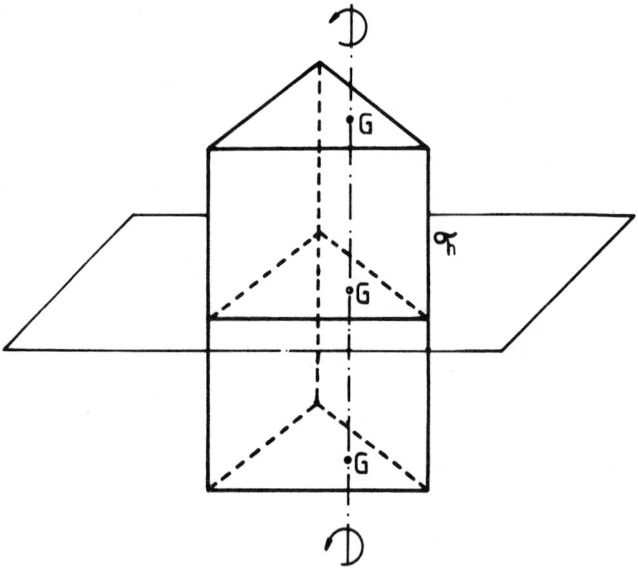

Figure A.10. A symmetry operation reflecting a point on the upper surface of a lamina from the one directly beneath it.

(see Figure A.10). Then $G_1 = \{E, \sigma_h\}$ forms a group where $\sigma_h\sigma_h = E$. The operations of G_1 clearly commute with those of $C_{3v} = G_2$. The direct product group is

$$G = G_1 \times G_2 = \{E, \sigma_h\} \times \{E, C_3, \bar{C}_3, \sigma^{(1)}, \sigma^{(2)}, \sigma^{(3)}\}$$
$$= \{E, C_3, \bar{C}_3, \sigma^{(1)}, \sigma^{(2)}, \sigma^{(3)}, \sigma_h, \sigma_h C_3, \sigma_h \bar{C}_3, \sigma_h\sigma^{(1)}, \sigma_h\sigma^{(2)}, \sigma_h\sigma^{(3)}\}.$$

Definition 28. If $D^{(\alpha)} = \{\mathbf{A}_i\}$ $(i = 1, 2, \ldots, M)$ is a set of d_α-dimensional matrices and $D^{(\beta)} = \{\mathbf{B}_j\}$ $(j = 1, 2, \ldots, N)$ is a set of d_β-dimensional matrices, then $D^{(\alpha)} \times D^{(\beta)} = \{A_i \times B_j\}$ is a set of $d_\alpha d_\beta$-dimensional MN matrices which is said to be the *outer* direct (Kronecker) product of the two sets.

Theorem 19. *The outer direct product of the representations of G_1 and G_2 forms a representation of $G = G_1 \times G_2$. Moreover, the outer direct product of any two irreducible representations of G_1 and G_2 gives an* irreducible *representation of $G = G_1 \times G_2$.*

Theorem 20. *The number of classes in $G = G_1 \times G_2$ is just the product of the number of classes in G_1 and G_2.*

APPENDIX B

Integrity Bases of Crystallographic Groups

Here we list, for ease of reference, the typical multilinear elements of the integrity basis for each of the crystal classes of the triclinic, monoclinic, rhombic, tetragonal, trigonal, and hexagonal crystal systems. These results are copied from Kiral and Smith [1974].

For each crystal class, there is listed a table of the form:

Name	$\mathbf{S}^1$	$\mathbf{S}^2$	...	$\mathbf{S}^g$	Basic quantities
$\boldsymbol{\Gamma}_1$	$\mathbf{D}^{(1)}(\mathbf{S}^1)$	$\mathbf{D}^{(1)}(\mathbf{S}^2)$	...	$\mathbf{D}^{(1)}(\mathbf{S}^g)$	$\varphi, \varphi', \ldots$
$\boldsymbol{\Gamma}_2$	$\mathbf{D}^{(2)}(\mathbf{S}^1)$	$\mathbf{D}^{(2)}(\mathbf{S}^2)$	...	$\mathbf{D}^{(2)}(\mathbf{S}^g)$	$a, b, \ldots$
$\vdots$	$\vdots$	$\vdots$	$\vdots$	$\vdots$	$\vdots$
$\boldsymbol{\Gamma}_r$	$\mathbf{D}^{(r)}(\mathbf{S}^1)$	$\mathbf{D}^{(r)}(\mathbf{S}^2)$	...	$\mathbf{D}^{(r)}(\mathbf{S}^g)$	$\mathbf{A}, \mathbf{B}, \ldots$

NAME identifies the classes by Schönflies symbols. The entries φ, φ', ..., a, b, ..., $\mathbf{A}$, $\mathbf{B}$, ... in the table indicate the basic quantities which form the carrier spaces for the irreducible representations $\boldsymbol{\Gamma}_1, \boldsymbol{\Gamma}_2, \ldots, \boldsymbol{\Gamma}_r$. Their explicit forms are determined in Section 4.4, and listed in Table 4.3(1–58) for the electromechanical tensors **M**, **J**, ..., **E**.

Table B.1. Pedial class, $C_1 = 1 = 1$: No symmetry. Hence all independent components of the vectors and tensors constitute the elements of the integrity basis.

Table B.2. Pinacoidal class, $C_i = \bar{1} = \bar{2}$. Domatic class, $C_s = m = m$. Sphenoidal class, $C_2 = 2 = 2$.

C_1 C_s C_2	**I** **I** **I**	**C** $\mathbf{R}_3$ $\mathbf{D}_3$	Basic quantities
$\boldsymbol{\Gamma}_1$	1	1	$a, a', \ldots$
$\boldsymbol{\Gamma}_2$	1	-1	$b, b', \ldots$

The typical multilinear elements of the integrity bases for C_i, C_s, and C_2 are given by

$$\begin{array}{l} 1.\ a; \\ 2.\ bb'. \end{array} \tag{B.1}$$

Table B.3. Prismatic class, $C_{2h} = 2/m = 2{:}m$.
Rhombic–pyramidal class, $C_{2v} = mm2 = 2 \cdot m$.
Rhombic–disphenoidal class, $D_2 = 222 = 2{:}2$.

C_{2h} C_{2v} D_2	**I** **I** **I**	$\mathbf{D}_3$ $\mathbf{D}_3$ $\mathbf{D}_1$	$\mathbf{R}_3$ $\mathbf{R}_1$ $\mathbf{D}_2$	**C** $\mathbf{R}_2$ $\mathbf{D}_3$	Basic quantities
Γ_1	1	1	1	1	$a, a', \ldots$
Γ_2	1	1	−1	−1	$b, b', \ldots$
Γ_3	1	−1	1	−1	$c, c', \ldots$
Γ_4	1	−1	−1	1	$d, d', \ldots$

The typical multilinear elements of the integrity bases for C_{2h}, C_{2v}, and D_2 are given by

$$\begin{array}{l} 1.\ a; \\ 2.\ bb',\ cc',\ dd'; \\ 3.\ bcd. \end{array} \tag{B.2}$$

Table B.4. Rhombic–dipyramidal class, $D_{2h} = mmm = m \cdot 2{:}m$.

D_{2h}	**I**	$\mathbf{D}_1$	$\mathbf{D}_2$	$\mathbf{D}_3$	**C**	$\mathbf{R}_1$	$\mathbf{R}_2$	$\mathbf{R}_3$	Basic quantities
Γ_1	1	1	1	1	1	1	1	1	$a, a', \ldots$
Γ_2	1	1	−1	−1	1	1	−1	−1	$b, b', \ldots$
Γ_3	1	−1	1	−1	1	−1	1	−1	$c, c', \ldots$
Γ_4	1	−1	−1	1	1	−1	−1	1	$d, d', \ldots$
Γ_1'	1	1	1	1	−1	−1	−1	−1	$A, A', \ldots$
Γ_2'	1	1	−1	−1	−1	−1	1	1	$B, B', \ldots$
Γ_3'	1	−1	1	−1	−1	1	−1	1	$C, C', \ldots$
Γ_4'	1	−1	−1	1	−1	1	1	−1	$D, D', \ldots$

The typical multilinear elements of the integrity basis for D_{2h} are given by

$$\begin{array}{l} 1.\ a; \\ 2.\ bb',\ cc',\ dd',\ AA',\ BB',\ CC',\ DD'; \\ 3.\ bcd,\ bAB,\ bCD,\ cAC,\ cBD,\ dAD,\ dBC; \\ 4.\ bcBC,\ bcAD,\ bdBD,\ bdAC,\ cdCD,\ cdAB,\ ABCD. \end{array} \tag{B.3}$$

Table B.5. Tetragonal–disphenoidal class, $\bar{C}_2 = \bar{4} = \bar{4}$. Tetragonal–pyramidal class, $C_4 = 4 = 4$.

$\bar{C}_2$ C_4	**I** **I**	$\mathbf{D}_3$ $\mathbf{D}_3$	$\mathbf{D}_1\mathbf{T}_3$ $\mathbf{R}_1\mathbf{T}_3$	$\mathbf{D}_2\mathbf{T}_3$ $\mathbf{R}_2\mathbf{T}_3$	Basic quantities
Γ_1	1	1	1	1	$\varphi, \varphi', \ldots$
Γ_2	1	1	-1	-1	$\psi, \psi', \ldots$
Γ_3	1	-1	i	$-i$	$a, b, \ldots$
Γ_4	1	-1	$-i$	i	$\bar{a}, \bar{b}, \ldots$

where the quantities $\bar{a}, \bar{b}, \ldots$ denote the complex conjugates of the quantities $a, b, \ldots$, respectively. The typical multilinear elements of the integrity basis for $\bar{C}_2$ and C_4 are given by

$$\begin{aligned} &1.\ \varphi; \\ &2.\ a\bar{b},\ \psi\psi'; \\ &3.\ \psi ab; \\ &4.\ abcd. \end{aligned} \tag{B.4}$$

We note that the presence of the complex invariants $a\bar{b}$, ψab, $abcd$ in the above list indicates that both the real and imaginary parts $a\bar{b} \pm \bar{a}b$, $\psi ab \pm \bar{a}\bar{b}$, and so on, are typical multilinear elements of the integrity basis.

Table B.6. Tetragonal–dipyramidal class, $C_{4h} = 4/m = 4{:}m$.

C_{4h}	**I**	$\mathbf{D}_3$	$\mathbf{R}_1\mathbf{T}_3$	$\mathbf{R}_2\mathbf{T}_3$	**C**	$\mathbf{R}_3$	$\mathbf{D}_1\mathbf{T}_3$	$\mathbf{D}_2\mathbf{T}_3$	Basic quantities
Γ_1	1	1	1	1	1	1	1	1	$\varphi, \varphi', \ldots$
Γ_2	1	1	-1	-1	1	1	-1	-1	$\Psi, \Psi', \ldots$
Γ_3	1	-1	i	$-i$	1	-1	i	$-i$	$a, b, \ldots$
Γ_4	1	-1	$-i$	i	1	-1	$-i$	i	$\bar{a}, \bar{b}, \ldots$
Γ_1'	1	1	1	1	-1	-1	-1	-1	$\xi, \xi', \ldots$
Γ_2'	1	1	-1	-1	-1	-1	1	1	$\eta, \eta', \ldots$
Γ_3'	1	-1	i	$-i$	-1	1	$-i$	i	$A, B, \ldots$
Γ_4'	1	-1	$-i$	i	-1	1	i	$-i$	$\bar{A}, \bar{B}, \ldots$

where the quantities $\bar{a}, \bar{b}, \ldots, \bar{A}, \bar{B}, \ldots$ denote the complex conjugates of $a, b, \ldots, A, B, \ldots$, respectively. The multilinear elements of the integrity basis for C_{4h} are given by

$$\begin{aligned} &1.\ \varphi; \\ &2.\ a\bar{b},\ A\bar{B},\ \Psi\Psi',\ \xi\xi',\ \eta\eta'; \\ &3.\ \Psi ab,\ \Psi AB,\ \xi a\bar{A},\ \eta aA,\ \Psi\xi\eta; \\ &4.\ abcd,\ abAB,\ ab\bar{A}\bar{B},\ ABCD,\ \Psi\xi aA,\ \Psi\eta a\bar{A},\ \xi\eta ab,\ \xi\eta AB; \\ &5.\ \xi aABC,\ \xi Aabc,\ \eta\bar{a}ABC,\ \eta\bar{A}abc. \end{aligned} \tag{B.5}$$

The presence of the complex invariants above indicates that both the real and imaginary parts of these invariants are typical multilinear elements of the integrity basis.

Table B.7. Ditetragonal–pyramidal class, $C_{4v} = 4mm = 4 \cdot m$. Tetragonal–trapezohedral class, $D_4 = 422 = 4:2$. Tetragonal–scalenohedral class, $D_{2v} = \bar{4}2m = \bar{4} \cdot m$.

C_{4v}	$\mathbf{I}$	$\mathbf{R}_2$	$\mathbf{R}_1$	$\mathbf{D}_3$	$\mathbf{T}_3$	$\mathbf{R}_2\mathbf{T}_3$	$\mathbf{R}_1\mathbf{T}_3$	$\mathbf{D}_3\mathbf{T}_3$	
D_4	$\mathbf{I}$	$\mathbf{D}_1$	$\mathbf{D}_2$	$\mathbf{D}_3$	$\mathbf{R}_2\mathbf{T}_3$	$\mathbf{R}_2\mathbf{T}_3$	$\mathbf{R}_1\mathbf{T}_3$	$\mathbf{C}\mathbf{T}_3$	Basic
D_{2v}	$\mathbf{I}$	$\mathbf{D}_1$	$\mathbf{D}_2$	$\mathbf{D}_3$	$\mathbf{T}_3$	$\mathbf{D}_1\mathbf{T}_3$	$\mathbf{D}_2\mathbf{T}_3$	$\mathbf{D}_3\mathbf{T}_3$	quantities
Γ_1	1	1	1	1	1	1	1	1	$\varphi, \varphi', \ldots$
Γ_2	1	−1	−1	1	−1	1	1	−1	$\psi, \psi', \ldots$
Γ_3	1	−1	−1	1	1	−1	−1	1	$\nu, \nu', \ldots$
Γ_4	1	1	1	1	−1	−1	−1	−1	$\tau, \tau', \ldots$
Γ_5	$\mathbf{E}$	$\mathbf{F}$	$-\mathbf{F}$	$-\mathbf{E}$	$\mathbf{K}$	$\mathbf{L}$	$-\mathbf{L}$	$-\mathbf{K}$	$\mathbf{a}, \mathbf{b}, \ldots$

The matrices $\mathbf{E}$, $\mathbf{F}$, $\mathbf{K}$, $\mathbf{L}$ in $\mathbf{\Gamma}_5$ are defined earlier in Chapter 3. The typical multilinear elements of the integrity basis for C_{4v}, D_4, and D_{2v} are given by

$$
\begin{aligned}
&1.\ \varphi;\\
&2.\ a_1b_1 + a_2b_2,\ \Psi\Psi',\ \nu\nu',\ \tau\tau';\\
&3.\ \Psi(a_1b_2 - a_2b_1),\ \nu(a_1b_2 + a_2b_1),\ \tau(a_1b_1 - a_2b_2),\ \Psi\nu\tau;\\
&4.\ a_1b_1c_1d_1 + a_2b_2c_2d_2,\ \Psi\nu(a_1b_1 - a_2b_2),\ \Psi\tau(a_1b_2 + a_2b_1),\\
&\quad\ \nu\tau(a_1b_2 + a_2b_1);\\
&5.\ \Psi(a_1b_1c_1d_2 + a_1b_1d_1c_2 + a_1c_1d_1b_2 + b_1c_1d_1a_2 - a_2b_2c_2d_1 - a_2b_2d_2c_1\\
&\quad\ - a_2c_2d_2b_1 - b_2c_2d_2a_1).
\end{aligned}
\tag{B.6}
$$

Table B.8. Ditetragonal–dipyramidal class, $D_{4h} = 4/mmm = m \cdot 4:m$.

D_{4h}	$\mathbf{I}$	$\mathbf{D}_1$	$\mathbf{D}_2$	$\mathbf{D}_3$	$\mathbf{C}\mathbf{T}_3$	$\mathbf{R}_1\mathbf{T}_3$	$\mathbf{R}_2\mathbf{T}_3$	$\mathbf{R}_3\mathbf{T}_3$	Basic quantities
Γ_1	1	1	1	1	1	1	1	1	$\varphi, \varphi', \ldots$
Γ_2	1	−1	−1	1	−1	1	1	−1	$\Psi, \Psi', \ldots$
Γ_3	1	−1	−1	1	1	−1	−1	1	$\nu, \nu', \ldots$
Γ_4	1	1	1	1	−1	−1	−1	−1	$\tau, \tau', \ldots$
Γ_5	$\mathbf{E}$	$\mathbf{F}$	$-\mathbf{F}$	$-\mathbf{E}$	$-\mathbf{K}$	$-\mathbf{L}$	$\mathbf{L}$	$\mathbf{K}$	$\mathbf{a}, \mathbf{b}, \ldots$
Γ_1'	1	1	1	1	1	1	1	1	$\xi, \xi', \ldots$
Γ_2'	1	−1	−1	1	−1	1	1	−1	$\eta, \eta', \ldots$
Γ_3'	1	−1	−1	1	1	−1	−1	1	$\theta, \theta', \ldots$
Γ_4'	1	1	1	1	−1	−1	−1	−1	$\gamma, \gamma', \ldots$
Γ_5'	$\mathbf{E}$	$\mathbf{F}$	$-\mathbf{F}$	$-\mathbf{E}$	$-\mathbf{K}$	$-\mathbf{L}$	$\mathbf{L}$	$\mathbf{K}$	$\mathbf{A}, \mathbf{B}, \ldots$

D_{4h}	$\mathbf{C}$	$\mathbf{R}_1$	$\mathbf{R}_2$	$\mathbf{R}_3$	$\mathbf{T}_3$	$\mathbf{D}_1\mathbf{T}_3$	$\mathbf{D}_2\mathbf{T}_3$	$\mathbf{D}_3\mathbf{T}_3$	Basic quantities
Γ_1	1	1	1	1	1	1	1	1	$\varphi, \varphi', \ldots$
Γ_2	1	−1	−1	1	−1	1	1	−1	$\Psi, \Psi', \ldots$
Γ_3	1	−1	−1	1	1	−1	−1	1	$\nu, \nu', \ldots$
Γ_4	1	1	1	1	−1	−1	−1	−1	$\tau, \tau', \ldots$
Γ_5	$\mathbf{E}$	$\mathbf{F}$	$-\mathbf{F}$	$-\mathbf{E}$	$-\mathbf{K}$	$-\mathbf{L}$	$\mathbf{L}$	$\mathbf{K}$	$\mathbf{a}, \mathbf{b}, \ldots$
Γ_1'	−1	−1	−1	−1	−1	−1	−1	−1	$\xi, \xi', \ldots$
Γ_2'	−1	1	1	−1	1	−1	−1	1	$\eta, \eta', \ldots$
Γ_3'	−1	1	1	−1	−1	1	1	−1	$\theta, \theta', \ldots$
Γ_4'	−1	−1	−1	−1	1	1	1	1	$\gamma, \gamma', \ldots$
Γ_5'	$-\mathbf{E}$	$-\mathbf{F}$	$\mathbf{F}$	$\mathbf{E}$	$\mathbf{K}$	$\mathbf{L}$	$-\mathbf{L}$	$-\mathbf{K}$	$\mathbf{A}, \mathbf{B}, \ldots$

The matrices **E**, **F**, **K**, **L** in Γ_5 and Γ_5' are defined earlier in Chapter 3. The typical multilinear elements of the integrity basis for D_{4h} are given by

$$
\begin{aligned}
&1.\ \varphi;\\
&2.\ a_1b_1 + a_2b_2,\ A_1B_1 + A_2B_2,\ \Psi\Psi',\ \nu\nu',\ \tau\tau'\ \xi\xi',\ \eta\eta',\ \theta\theta',\ \gamma\gamma';\\
&3.\ \Psi(a_1b_2 - a_2b_1),\ \Psi(A_1B_2 - A_2B_1),\ \nu(a_1b_2 - a_2b_1),\ \nu(A_1B_2 + A_2B_1),\\
&\quad \tau(a_1b_1 - a_2b_2),\ \tau(A_1B_1 - A_2B_2),\ \xi(a_1A_1 + a_2A_2),\ \eta(a_1A_2 - a_2A_1),\\
&\quad \theta(a_1A_2 + a_2A_1),\ \gamma(a_1A_1 - a_2A_2),\ \Psi\nu\tau,\ \Psi\theta\gamma,\ \Psi\xi\eta,\ \nu\eta\gamma,\ \tau\xi\gamma,\ \tau\eta\theta,\ \nu\xi\theta;\\
&4.\ a_1b_1c_1d_1 + a_2b_2c_2d_2,\ A_1B_1C_1D_1 + A_2B_2C_2D_2,\\
&\quad (a_1b_2 + a_2b_1)(A_1B_2 + A_2B_1),\ (a_1b_2 - a_2b_1)(A_1B_2 - A_2B_1),\\
&\quad (a_1b_1 - a_2b_2)(A_1B_1 - A_2B_2),\ (\Psi\nu,\ \xi\gamma,\ \eta\theta)(a_1b_1 - a_2b_2),\\
&\quad (\Psi\tau,\ \xi\theta,\ \eta\gamma)(a_1b_2 + a_2b_1),\ (\nu\tau,\ \xi\eta,\ \theta\gamma)(a_1b_2 - a_2b_1),\\
&\quad (\Psi\nu,\ \xi\gamma,\ \eta\theta)(A_1B_1 - A_2B_2),\ (\Psi\tau,\ \xi\theta,\ \eta\gamma)(A_1B_2 + A_2B_1),\\
&\quad (\nu\tau,\ \xi\eta,\ \theta\gamma)(A_1B_2 - A_2B_1),\ (\Psi\eta,\ \nu\theta,\ \tau\gamma)(a_1A_1 + a_2A_2),\\
&\quad (\Psi\theta,\ \nu\eta,\ \tau\xi)(a_1A_1 - a_2A_2),\ (\Psi\gamma,\ \nu\xi,\ \tau\eta)(a_1A_2 + a_2A_1),\\
&\quad (\Psi\xi,\ \nu\gamma,\ \tau\theta)(a_1A_2 - a_2A_1),\ \Psi\nu\xi\gamma,\ \Psi\nu\eta\theta,\ \Psi\tau\xi\theta,\ \Psi\tau\eta\gamma,\ \nu\tau\xi\eta,\ \nu\tau\theta\gamma,\ \xi\eta\theta\gamma;\\
&5.\ \Psi(a_1b_1c_1d_2 + a_1b_1d_1c_2 + a_1c_1d_1b_2 + b_1c_1d_1a_2 - a_2b_2c_2d_1 - a_2b_2d_2c_1\\
&\qquad\qquad - a_2c_2d_2b_1 - b_2c_2d_2a_1),\\
&\quad \Psi(A_1B_1C_1D_2 + A_1B_1D_1C_2 + A_1C_1D_1B_2 + B_1C_1D_1A_2 - A_2B_2C_2D_1\\
&\qquad\qquad - A_2B_2D_2C_1 - A_2C_2D_2B_1 - B_2C_2D_2A_1),\\
&\quad \Psi(a_1b_2 + a_2b_1)(A_1B_1 - A_2B_2),\ \Psi(a_1b_1 - a_2b_2)(A_1B_2 + A_2B_1),\\
&\quad \nu(a_1b_2 - a_2b_1)(A_1B_1 - A_2B_2),\ \nu(a_1b_1 - a_2b_2)(A_1B_2 - A_2B_1),\\
&\quad \tau(a_1b_2 - a_2b_1)(A_1B_2 + A_2B_1),\ \tau(a_1b_2 + a_2b_1)(A_1B_2 - A_2B_1),\\
&\quad \xi(a_1b_1c_1A_1 + a_2b_2c_2A_2),\ \xi(A_1B_1C_1a_1 + A_2B_2C_2a_2),\\
&\quad \eta(a_1b_1c_1A_2 - a_2b_2c_2A_2),\ \eta(A_1B_1C_1a_2 - A_2B_2C_2a_1),\\
&\quad \theta(a_1b_1c_1A_2 + a_2b_2c_2A_1),\ \theta(A_1B_1C_1a_2 + A_2B_2C_2a_1),\\
&\quad \gamma(a_1b_1c_1A_1 - a_2b_2c_2A_2),\ \gamma(A_1B_1C_1a_1 - A_2B_2C_2a_2),\\
&\quad (\Psi\xi\theta,\ \Psi\eta\gamma,\ \nu\xi\eta,\ \nu\theta\gamma)(a_1b_1 - a_2b_2),\ (a_1b_2 + a_2b_1)(\Psi\xi\gamma,\ \Psi\eta\theta,\ \tau\xi\eta,\ \tau\theta\gamma),\\
&\quad (\nu\xi\gamma,\ \nu\eta\theta,\ \tau\xi\theta,\ \tau\eta\gamma)(a_1b_2 - a_2b_1),\ (A_1B_1 - A_2B_2)(\Psi\xi\theta,\ \Psi\eta\gamma,\ \nu\xi\eta,\ \nu\theta\gamma),\\
&\quad (\Psi\xi\gamma,\ \Psi\eta\theta,\ \tau\xi\eta,\ \tau\theta\gamma)(A_1B_2 + A_2B_1),\ (A_1B_2 - A_2B_1)(\nu\xi\gamma,\ \nu\eta\theta,\ \tau\xi\theta,\ \tau\eta\gamma),\\
&\quad (\Psi\nu\gamma,\ \Psi\tau\theta,\ \nu\tau\eta,\ \eta\theta\gamma)(a_1A_1 + a_2A_2),\ (a_1A_1 - a_2A_2)(\Psi\eta\tau,\ \Psi\nu\xi,\ \nu\tau\theta,\ \xi\eta\theta),\\
&\quad (\Psi\nu\eta,\ \Psi\tau\xi,\ \nu\tau\gamma,\ \xi\eta\gamma)(a_1A_2 + a_2A_1),\ (a_1A_2 - a_2A_1)(\Psi\nu\theta,\ \Psi\tau\gamma,\ \nu\tau\xi,\ \xi\theta\gamma);\\
&6.\ \xi\eta(a_1b_1 - a_2b_2)(A_1B_2 + A_2B_1),\ \theta\gamma(a_1b_1 - a_2b_2)(A_1B_2 + A_2B_1),\\
&\quad \Psi\xi(a_1b_1c_1A_2 - a_2b_2c_2A_1),\ \Psi\xi(A_1B_1C_1a_2 - A_2B_2C_2a_1),\\
&\quad \Psi\eta(a_1b_1c_1A_1 + a_2b_2c_2A_2),\ \Psi\eta(A_1B_1C_1a_1 + A_2B_2C_2a_2),\\
&\quad \Psi\theta(a_1b_1c_1A_1 - a_2b_2c_2A_2),\ \Psi\theta(A_1B_1C_1a_1 - A_2B_2C_2a_2),\\
&\quad \Psi\gamma(a_1b_1c_1A_2 + a_2b_2c_2A_1),\ \Psi\gamma(A_1B_1C_1a_2 + A_2B_2C_2a_1),\\
&\quad (\xi\eta,\ \theta\gamma)(a_1b_1c_1d_2 + a_1b_1d_1c_2 + a_1c_1d_1b_2 + b_1c_1d_1a_2 - a_2b_2c_2d_1\\
&\qquad\qquad - a_2b_2d_2c_1 - a_2c_2d_2b_1 - b_2c_2d_2a_1),\\
&\quad (\xi\eta,\ \theta\gamma)(A_1B_1C_1D_2 + A_1B_1D_1C_2 + A_1C_1D_1B_2 + B_1C_1D_1A_2\\
&\qquad\qquad - A_2B_2C_2D_1 - A_2B_2D_2C_1 - A_2C_2D_2B_1 - B_2C_2D_2A_1),\\
&\quad (a_1b_2 - a_2b_1)(A_1B_1C_1D_2 + A_1B_1D_1C_2 + A_1C_1D_1B_2 + B_1C_1D_1A_2\\
&\qquad\qquad - A_2B_2C_2D_1 - A_2B_2D_2C_1 - A_2C_2D_2B_1 - B_2C_2D_2A_1),\\
&\quad (A_1B_2 - A_2B_1)(a_1b_1c_1d_2 + a_1b_1d_1c_2 + a_1c_1d_1b_2 + b_1c_1d_1a_2\\
&\qquad\qquad - a_2b_2c_2d_1 - a_2b_2d_2c_1 - a_2c_2d_2b_1 - b_2c_2d_2a_1).
\end{aligned}
\tag{B.7}
$$

Table B.9. Trigonal–pyramidal class, $C_3 = 3 = 3$.

C_3	**I**	$\mathbf{S}_1$	$\mathbf{S}_2$	Basic quantities
$\mathbf{\Gamma}_1$	1	1	1	$\varphi, \varphi', \ldots$
$\mathbf{\Gamma}_2$	1	ω	ω^2	$a, b, \ldots$
$\mathbf{\Gamma}_3$	1	ω^2	ω	$\bar{a}, \bar{b}, \ldots$

where ω and ω^2 are defined by

$$\omega = -1/2 + i\sqrt{3}/2, \qquad \omega^2 = -1/2 - i\sqrt{3}/2 \tag{B.8}$$

and that $\bar{a}, \bar{b}, \ldots$ denote the complex conjugates of the quantities $a, b, \ldots$, respectively. The typical multilinear elements of the integrity basis for C_3 are given by

$$\begin{aligned} &1.\ \varphi; \\ &2.\ a\bar{b}; \\ &3.\ abc. \end{aligned} \tag{B.9}$$

The presence of the complex invariants $a\bar{b}$ and abc in the above list indicates that both the real and imaginary parts $a\bar{b} \pm \bar{a}b$ and $abc \pm \bar{a}\bar{b}\bar{c}$ of these invariants are to be considered.

Table B.10. Ditrigonal–pyramidal class, $C_{3v} = 3m = 3 \cdot m$. Trigonal–trapezohedral class, $D_3 = 32$.

C_{3v} D_3	**I** **I**	$\mathbf{S}_1$ $\mathbf{S}_1$	$\mathbf{S}_2$ $\mathbf{S}_2$	$\mathbf{R}_1$ $\mathbf{D}_1$	$\mathbf{R}_1\mathbf{S}_1$ $\mathbf{D}_1\mathbf{S}_1$	$\mathbf{R}_1\mathbf{S}_2$ $\mathbf{D}_1\mathbf{S}_2$	Basic quantities
$\mathbf{\Gamma}_1$	1	1	1	1	1	1	$\varphi, \varphi', \ldots$
$\mathbf{\Gamma}_2$	1	1	1	-1	-1	-1	Ψ, Ψ'
$\mathbf{\Gamma}_3$	**E**	**A**	**B**	$-$**F**	$-$**G**	$-$**H**	**a**, **b**, ...

The matrices in $\mathbf{\Gamma}_3$ are defined earlier in Chapter 3. The typical multilinear elements of the integrity basis for C_{3v} and D_3 are given by

$$\begin{aligned} &1.\ \varphi; \\ &2.\ a_1b_1 + a_2b_2,\ \Psi\Psi'; \\ &3.\ a_2b_2c_2 - a_1b_1c_2 - b_1c_1a_2 - c_1a_1b_2,\ \Psi(a_1b_2 - a_2b_1); \\ &4.\ \Psi(a_1b_1c_1 - a_2b_2c_1 - b_2c_2a_1 - c_2a_2b_1). \end{aligned} \tag{B.10}$$

Table B.11. Rhombohedral class, $\bar{C}_3 = \bar{3} = \bar{6}$. Trigonal–dipyramidal class, $C_{3h} = \bar{6} = 3{:}m$. Hexagonal–pyramidal class, $C_6 = 6 = 6$.

$\bar{C}_3$	**I**	$\mathbf{S}_1$	$\mathbf{S}_2$	**C**	$\mathbf{CS}_1$	$\mathbf{CS}_2$	
C_{3h}	**I**	$\mathbf{S}_1$	$\mathbf{S}_2$	$\mathbf{R}_3$	$\mathbf{R}_3\mathbf{S}_1$	$\mathbf{R}_3\mathbf{S}_2$	Basic
C_6	**I**	$\mathbf{S}_1$	$\mathbf{S}_2$	$\mathbf{D}_3$	$\mathbf{D}_3\mathbf{S}_1$	$\mathbf{D}_3\mathbf{S}_2$	quantities
$\mathbf{\Gamma}_1$	1	1	1	1	1	1	$\varphi, \varphi', \ldots$
$\mathbf{\Gamma}_2$	1	ω	ω^2	1	ω	ω^2	$a, b, \ldots$
$\mathbf{\Gamma}_3$	1	ω^2	ω	1	ω^2	ω	$\bar{a}, \bar{b}, \ldots$
$\mathbf{\Gamma}_4$	1	1	1	-1	-1	-1	$\xi, \xi', \ldots$
$\mathbf{\Gamma}_5$	1	ω	ω^2	-1	$-\omega$	$-\omega^2$	$A, B, \ldots$
$\mathbf{\Gamma}_6$	1	ω^2	ω	-1	$-\omega^2$	$-\omega$	$\bar{A}, \bar{B}, \ldots$

The quantities ω and ω^2 appearing in the above table are defined by (B.8). The quantities $\bar{a}, \bar{b}, \ldots, \bar{A}, \bar{B}, \ldots$ denote the complex conjugates of $a, b, \ldots, A, B, \ldots$, respectively. The typical multilinear elements of the integrity basis for $\bar{C}_3$, C_{3h}, and C_6 are given by

$$
\begin{aligned}
&1.\ \varphi; \\
&2.\ a\bar{b},\ A\bar{B},\ \xi\xi'; \\
&3.\ abc,\ aAB,\ \xi a\bar{A}; \\
&4.\ ab\bar{A}\bar{B},\ \xi abA,\ \xi ABC; \\
&5.\ \bar{a}ABCD; \\
&6.\ ABCDEF.
\end{aligned}
\tag{B.11}
$$

The presence of the complex invariants $a\bar{b}$, $A\bar{B}$, ..., $ABCDEF$ in (B.11) indicates that both the real and imaginary parts of these invariants are typical multilinear elements of the integrity basis.

Table B.12. Ditrigonal–dipyramidal class, $D_{3h} = \overline{6}m2 = m\cdot 2:m$. Hexagonal–scalenohedral class, $D_{3v} = \overline{3}m = \overline{6}\cdot m$. Hexagonal–trapezohedral class, $D_6 = 622 = 6:2$.

D_{3h}	**I**	$\mathbf{S}_1$	$\mathbf{S}_2$	$\mathbf{R}_3$	$\mathbf{R}_3\mathbf{S}_1$	$\mathbf{R}_3\mathbf{S}_2$	
D_{3v}	**I**	$\mathbf{S}_1$	$\mathbf{S}_2$	**C**	$\mathbf{C}\mathbf{S}_1$	$\mathbf{C}\mathbf{S}_2$	
D_6	**I**	$\mathbf{S}_1$	$\mathbf{S}_2$	$\mathbf{D}_3$	$\mathbf{D}_3\mathbf{S}_1$	$\mathbf{D}_3\mathbf{S}_2$	Basic
C_{6v}	**I**	$\mathbf{S}_1$	$\mathbf{S}_2$	$\mathbf{D}_3$	$\mathbf{D}_3\mathbf{S}_1$	$\mathbf{D}_3\mathbf{S}_2$	quantities
Γ_1	1	1	1	1	1	1	$\varphi, \varphi', \ldots$
Γ_2	1	1	1	1	1	1	$\Psi, \Psi', \ldots$
Γ_3	1	1	1	−1	−1	−1	$\xi, \xi', \ldots$
Γ_4	1	1	1	−1	−1	−1	$\eta, \eta', \ldots$
Γ_5	**E**	**A**	**B**	−**E**	−**A**	−**B**	**A**, **B**, ...
Γ_6	**E**	**A**	**B**	**E**	**A**	**B**	**a**, **b**, ...
D_{3h}	$\mathbf{R}_1$	$\mathbf{R}_1\mathbf{S}_1$	$\mathbf{R}_1\mathbf{S}_2$	$\mathbf{D}_2$	$\mathbf{D}_2\mathbf{S}_1$	$\mathbf{D}_2\mathbf{S}_2$	
D_{3v}	$\mathbf{D}_1$	$\mathbf{D}_1\mathbf{S}_1$	$\mathbf{D}_1\mathbf{S}_2$	$\mathbf{R}_1$	$\mathbf{R}_1\mathbf{S}_1$	$\mathbf{R}_1\mathbf{S}_2$	
D_6	$\mathbf{D}_1$	$\mathbf{D}_1\mathbf{S}_1$	$\mathbf{D}_1\mathbf{S}_2$	$\mathbf{D}_2$	$\mathbf{D}_2\mathbf{S}_1$	$\mathbf{D}_2\mathbf{S}_2$	Basic
C_{6v}	$\mathbf{R}_2$	$\mathbf{R}_2\mathbf{S}_1$	$\mathbf{R}_2\mathbf{S}_2$	$\mathbf{R}_1$	$\mathbf{R}_1\mathbf{S}_1$	$\mathbf{R}_1\mathbf{S}_2$	quantities
Γ_1	1	1	1	1	1	1	$\varphi, \varphi', \ldots$
Γ_2	−1	−1	−1	−1	−1	−1	$\Psi, \Psi', \ldots$
Γ_3	1	1	1	−1	−1	−1	$\xi, \xi', \ldots$
Γ_4	−1	−1	−1	1	1	1	$\eta, \eta', \ldots$
Γ_5	**F**	**G**	**H**	−**F**	−**G**	−**H**	**A**, **B**, ...
Γ_6	−**F**	−**G**	−**H**	−**F**	−**G**	−**H**	**a**, **b**, ...

The typical multilinear elements of the integrity basis for D_{3v}, D_6, and C_{6v} are given by

$$\begin{aligned}
&1.\ \varphi;\\
&2.\ a\bar{b} + \bar{a}b,\ A\bar{B} + \bar{A}B,\ \Psi\Psi',\ \xi\xi',\ \eta\eta';\\
&3.\ abc - \bar{a}\bar{b}\bar{c},\ aAB - \bar{a}\bar{A}\bar{B},\\
&\quad \Psi(a\bar{b} - \bar{a}b),\ \Psi(A\bar{B} - \bar{A}B),\ \xi(a\bar{A} - \bar{a}A),\ \eta(a\bar{A} + \bar{a}A),\ \Psi\xi\eta;\\
&4.\ ab\bar{A}\bar{B} + \bar{a}\bar{b}AB,\ (a\bar{b} - \bar{a}b)(A\bar{B} - \bar{A}B),\ \Psi(abc + \bar{a}\bar{b}\bar{c}),\\
&\quad \Psi(aAB + \bar{a}\bar{A}\bar{B}),\ \xi(abA + \bar{a}\bar{b}\bar{A}),\ \xi(ABC + \bar{A}\bar{B}\bar{C}),\\
&\quad \eta(abA - \bar{a}\bar{b}\bar{A}),\ \eta(ABC - \bar{A}\bar{B}\bar{C}),\ \Psi\xi(a\bar{A} + \bar{a}A),\ \Psi\eta(a\bar{A} - \bar{a}A),\\
&\quad \xi\eta(a\bar{b} - \bar{a}b),\ \xi\eta(A\bar{B} - \bar{A}B);\\
&5.\ (abc + \bar{a}\bar{b}\bar{c})(A\bar{B} - \bar{A}B),\ a\bar{A}\bar{B}\bar{C}\bar{D} - \bar{a}ABCD,\\
&\quad \Psi(ab\bar{A}\bar{B} - \bar{a}\bar{b}AB),\ \Psi\xi(abA - \bar{a}\bar{b}\bar{A}),\ \Psi\xi(ABC - \bar{A}\bar{B}\bar{C}),\\
&\quad \Psi\eta(ABC + \bar{A}\bar{B}\bar{C}),\ \Psi\eta(abA + \bar{a}\bar{b}\bar{A}),\ \xi\eta(abc + \bar{a}\bar{b}\bar{c}),\ \xi\eta(aAB + \bar{a}\bar{A}\bar{B});\\
&6.\ ABCDEF + \bar{A}\bar{B}\bar{C}\bar{D}\bar{E}\bar{F},\ \Psi(a\bar{A}\bar{B}\bar{C}\bar{D} + \bar{a}ABCD);\\
&7.\ \Psi(ABCDEF - \bar{A}\bar{B}\bar{C}\bar{D}\bar{E}\bar{F}).
\end{aligned} \tag{B.12}$$

The complex quantities $A, \ldots, F, a, \ldots, c$ appearing in (B.12) are defined as

$$\begin{aligned}
A &= A_1 + iA_2, \ldots, & F &= F_1 + iF_2,\\
a &= a_1 + ia_2, \ldots, & c &= c_1 + ic_2,
\end{aligned} \tag{B.13}$$

where $A_1, A_2, \ldots, F_1, F_2$ are the components of the vectors $\mathbf{A}, \ldots, \mathbf{F}$ and a_1,

$a_2, \ldots, c_1, c_2$ are the components of the vectors **a**, ..., **c**. Again a superposed bar indicates complex conjugation.

Note that the matrices forming the irreducible representations $\boldsymbol{\Gamma}_5$ and $\boldsymbol{\Gamma}_6$ in Table B.12 are defined as in Table 3.3(3) in Chapter 3.

Table B.13. Hexagonal–dipyromidal class, $C_{6h} = 6/m = 6{:}m$.

C_{6h}	**I**	$\mathbf{S}_1$	$\mathbf{S}_2$	$\mathbf{D}_3$	$\mathbf{D}_3\mathbf{S}_1$	$\mathbf{D}_3\mathbf{S}_2$	Basic quantities
Γ_1	1	1	1	1	1	1	$\varphi, \varphi', \ldots$
Γ_2	1	ω	ω^2	1	ω	ω^2	$a, b, \ldots$
Γ_3	1	ω^2	ω	1	ω^2	ω	$\bar{a}, \bar{b}, \ldots$
Γ_4	1	1	1	-1	-1	-1	$\xi, \xi', \ldots$
Γ_5	1	ω	ω^2	-1	$-\omega$	$-\omega^2$	$A, B, \ldots$
Γ_6	1	ω^2	ω	-1	$-\omega^2$	$-\omega$	$\bar{A}, \bar{B}, \ldots$
Γ_1'	1	1	1	1	1	1	$\pi, \pi', \ldots$
Γ_2'	1	ω	ω^2	1	ω	ω^2	$X, Y, \ldots$
Γ_3'	1	ω^2	ω	1	ω^2	ω	$\bar{X}, \bar{Y}, \ldots$
Γ_4'	1	1	1	-1	-1	-1	$\delta, \delta', \ldots$
Γ_5'	1	ω	ω^2	-1	$-\omega$	$-\omega^2$	$x, y, \ldots$
Γ_6'	1	ω^2	ω	-1	$-\omega^2$	$-\omega$	$\bar{x}, \bar{y}, \ldots$
C_{6h}	**C**	$\mathbf{CS}_1$	$\mathbf{CS}_2$	$\mathbf{R}_3$	$\mathbf{R}_3\mathbf{S}_1$	$\mathbf{R}_3\mathbf{S}_2$	Basic quantities
Γ_1	1	1	1	1	1	1	$\varphi, \varphi', \ldots$
Γ_2	1	ω	ω^2	1	ω	ω^2	$\bar{a}, \bar{b}, \ldots$
Γ_3	1	ω^2	ω	1	ω^2	ω	$a, b, \ldots$
Γ_4	1	1	1	-1	-1	-1	$\xi, \xi', \ldots$
Γ_5	1	ω	ω^2	-1	$-\omega$	$-\omega^2$	$A, B, \ldots$
Γ_6	1	ω^2	ω	-1	$-\omega^2$	$-\omega$	$\bar{A}, \bar{B}, \ldots$
Γ_1'	-1	-1	-1	-1	-1	-1	$\pi, \pi', \ldots$
Γ_2'	-1	$-\omega$	$-\omega^2$	-1	$-\omega$	$-\omega^2$	$X, Y, \ldots$
Γ_3'	-1	$-\omega^2$	$-\omega$	-1	$-\omega^2$	$-\omega$	$\bar{X}, \bar{Y}, \ldots$
Γ_4'	-1	-1	-1	1	1	1	$\delta, \delta', \ldots$
Γ_5'	-1	$-\omega$	$-\omega^2$	1	ω	ω^2	$x, y, \ldots$
Γ_6'	-1	$-\omega^2$	$-\omega$	1	ω^2	ω	$\bar{x}, \bar{y}, \ldots$

The quantities ω and ω^2 appearing in Table B.13 are defined by (B.8). The typical multilinear elements of the integrity basis for the crystal class 6_{6h} are given by

1. φ;
2. $a\bar{b}, A\bar{B}, X\bar{Y}, x\bar{y}, \pi\pi', \xi\xi', \delta\delta'$;
3. $abc, ABa, XYa, xya, AXx, \xi a\bar{A}, \xi x\bar{X}$,
   $\delta a\bar{x}, \delta A\bar{X}, \pi a\bar{X}, \pi A\bar{x}, \pi\xi\delta$;
4. $ab\bar{A}\bar{B}, ab\bar{X}\bar{Y}, ab\bar{x}\bar{y}, AB\bar{X}\bar{Y}, AB\bar{x}\bar{y}, XY\bar{x}\bar{y}$,
   $a\bar{A}X\bar{x}, a\bar{A}\bar{X}x, aA\bar{X}\bar{x}$,
   $\pi abX, \pi ABX, \pi XYZ, \pi xyX, \pi aAx$,
   $\xi abA, \xi ABC, \xi XYA, \xi xyA, \xi aXx$,

$\delta abx, \delta ABx, \delta XYx, \delta xyz, \delta aAX,$
$\pi\xi a\bar{x}, \pi\xi A\bar{X}, \pi\delta a\bar{A}, \xi\delta x\bar{X}, \xi\delta a\bar{X}, \xi\delta A\bar{x};$

5. $\bar{a}ABCD, \bar{a}ABXY, \bar{a}ABxy, \bar{a}XYZU, \bar{a}XYxy, \bar{a}xyzu,$
$ABC\bar{X}x, ABCX\bar{x}, XYZ\bar{A}x, XYZA\bar{x}, abxA\bar{X},$
$abx\bar{A}X, ab\bar{x}AX, xyzA\bar{X}, \mathrm{xyz}\bar{A}X,$
$\pi ab\bar{A}\bar{x}, \pi AB\bar{a}\bar{X}, \pi xy\bar{a}\bar{X}, \pi XY\bar{A}\bar{x},$
$\xi ab\bar{X}\bar{x}, \xi AB\bar{X}\bar{x}, \xi XY\bar{A}\bar{a}, \xi xy\bar{A}\bar{a},$ (B.14)
$\delta ab\bar{A}\bar{X}, \delta AB\bar{a}\bar{x}, \delta XY\bar{a}\bar{x}, \delta xy\bar{A}\bar{X},$
$\pi\xi abx, \pi\xi ABx, \pi\xi XYx, \pi\xi xyz, \pi\xi aAX,$
$\pi\delta abA, \pi\delta ABC, \pi\delta XYA, \pi\delta xyA, \pi\delta aXx,$
$\xi\delta abX, \xi\delta ABX, \xi\delta XYZ, \xi\delta xyX, \xi\delta aAx;$

6. $ABCDEF, ABCDXY, ABXYZU, XYZUVW, ABCDxy,$
$ABxyzu, xyzuvw, XYZUxy, XYxyzu, aAx\bar{X}\bar{Y}\bar{Z},$
$aAX\bar{x}\bar{y}\bar{z}, aXx\bar{A}\bar{B}\bar{C},$
$\pi\bar{a}ABCx, \pi\bar{a}xyzA, \pi ABCD\bar{X}, \pi ABxy\bar{X}, \pi xyzu\bar{X},$
$\xi\bar{a}XYZx, \xi\bar{a}xyzX, \xi XYZU\bar{A}, \xi XYxy\bar{A}, \xi xyzu\bar{A},$
$\delta\bar{a}ABCX, \delta\bar{a}XYZA, \delta XYZU\bar{x}, \delta XYAB\bar{x}, \delta ABCD\bar{x};$

7. $ABCDE\bar{X}\bar{x}, XYZUV\bar{A}\bar{x}, xyzuv\bar{A}\bar{X},$
$\pi ABCDEx, \pi ABCxyz, \pi Axyzuv,$
$\xi XYZUVx, \xi XYZxyz, \xi Xxyzuv,$
$\delta ABCDEX, \delta ABCXYZ, \delta AXYZUV.$

The presence of the complex invariants $a\bar{b}$, $A\bar{B}$, ..., $\delta AXYZUV$ in (B.14) indicates that both the real and imaginary parts of these invariants are typical multilinear elements of the integrity basis.

APPENDIX C

Magnetic Point Symmetry of Certain Materials

In order to make any computations about constitutive equations for a particular material, it is necessary to know its magnetic symmetry. In over two decades, several hundred magnetic structures, ferromagnetic as well as antiferromagnetic, have been determined by neutron diffraction techniques. Some of the materials and their magnetic symmetry are listed below. The results are based on two works, by Oles et al. [1970] and Cox [1969], where further information on magnetic structures can be found. The temperatures in the second column of the following table indicate a transition from one magnetically ordered state to another as described in the last column.

References

[1] A. Oles, A. Bombik, M. Kubac, W. Sikora, and F. Kajzer. Tables of Magnetic Structures Determined by Neutron Diffraction, Reports 1/PS, 4/PS, 7/PS, 8/PS, 11/PS, 12/PS, 24/PS, Institute of Nuclear Techniques, Cracow, Poland (1970–1972).

[2] D.E. Cox. Table of Antiferromagnetic Materials Studied by Neutron Diffraction, Report BNL 13822, Brookhaven National Laboratory, Upton, New York (1969).

Material	T_N(K°)	Magnetic point group
Trirutile type compounds		
Cr_2TeO_6	105	$\underline{m}mm$
Fe_2TeO_6	219	$4/\underline{m}\underline{m}\underline{m}$
Cr_2WO_6	69	$\underline{m}mm$
Perovskite type compounds		
$DyAlO_3$	3.5	$\underline{m}\underline{m}\underline{m}$
$GdAlO_3$	4.0	$\underline{m}\underline{m}\underline{m}$
$TbAlO_3$	4.0	$\underline{m}\underline{m}\underline{m}$
$DyCoO_3$	8.8	$\underline{m}\underline{m}\underline{m}$
$HoCoO_3$	2.4	$\underline{m}\underline{m}\underline{m}$
$TbRhO_3$	1.9	$\underline{m}\underline{m}\underline{m}$
$TbCoO_3$	3.3	$mm\underline{m}$
Spinel type compounds		
$MnGa_2O_4$	33	$\underline{\bar{3}}\underline{m}$
Hausmannite type compounds		
$CuCr_2O_4$	135	$\underline{2}m\underline{m}$
$NiCr_2O_3$	65	$\underline{2}$
$CoMn_2O_4$	100	$\underline{2}m\underline{m}$
$CrMn_2O_4$	65	$\underline{2}m\underline{m}$
Corundum type compounds		
Cr_2O_3	318	$\underline{\bar{3}}\underline{m}$
$MnTiO_3$	64	$\underline{\bar{3}}$
$Nb_2Co_4O_9$	27	$\underline{\bar{3}}\underline{m}$
$Nb_2Mn_4O_9$	110	$\underline{\bar{3}}\underline{m}$
$Ta_2Co_4O_9$	21	$\underline{\bar{3}}\underline{m}$
$Ta_2Mn_4O_9$	104	$\underline{\bar{3}}\underline{m}$
Triphylite type compounds		
$LiCoPO_4$	23	$\underline{m}mm$
$LiFePO_4$	50	$\underline{m}mm$
$LiMnPO_4$	35	$\underline{m}\underline{m}\underline{m}$
$LiNiPO_4$	23	$mm\underline{m}$
Zircon type compounds		
$DyPO_4$	3.4	$\underline{4}/\underline{m}m\underline{m}$
$GdVO_4$	2.4	$\underline{4}/\underline{m}m\underline{m}$
$HoPO_4$	1.4	$\underline{4}/\underline{m}m\underline{m}$
$TbPO_4$	2.2	$2/\underline{m}$

Material	T_N(K°)	Magnetic point group
Miscellaneous ferromagnetoelectric compounds		
$Ni_3B_7O_{13}I$	60 T_c(K°)	$\underline{2}m\underline{m}$
$Ni_3B_7O_{13}Cl$	20 T_c(K°)	$\underline{2}m\underline{m}$
β-$NaFeO_2$	723 T_c(K°)	$\underline{2}m\underline{m}$
$MnGeN_2$	448 T_c(K°)	$\underline{2}m\underline{m}$
$Co_3B_7O_{13}Cl$	22 T_c(K°)	m
$MnNb_3S_6$	33 T_N(K°)	$2\underline{2}\underline{2}$
Miscellaneous ferrimagnetic compounds		
$FeGaO_3$	305 T_c(K°)	$\underline{2}m\underline{m}$
Na_2NiFeF_7	88 T_c(K°)	$\underline{2}m\underline{m}$
$Li_{0.5}Fe_{2.5}O_4$	923 T_N(K°)	$3\underline{2}$
Miscellaneous antiferromagnetoelectric compounds		
$CoGeO_3$	31	$\underline{m}mm$
$CoCs_3Cl_5$	0.5	$\underline{4}/\underline{m}m\underline{m}$
$CrTiNdO_5$	13	$\underline{m}mm$
$LiCuCl_3 \cdot 2H_2O$	4.5	$\underline{2}/m$
DyOOH	7.2	$2/\underline{m}$
ErOOH	4.1	$\underline{2}/m$
α-FeOOH	403	$\underline{m}mm$
γ-FeOOH	75	$2/\underline{m}$
$MnGeO_3$	16	$\underline{m}mm$
$MnNb_2O_6$	4.4	$\underline{m}mm$
UOTE	157	$4/\underline{m}\underline{m}\underline{m}$
Miscellaneous compounds		
$BiFeO_3$	375	$3m$
$FeSb_2O_4$	46	$2mm$
FeS	600	$\underline{\bar{6}}\underline{m}2$
$CuFeS_2$	815	$\bar{4}2m$
MnF_2	75	$\underline{4}/mm\underline{m}$
CoF_2	45	$\underline{4}/mm\underline{m}$
FeF_2	90	$\underline{4}/mm\underline{m}$
$MnCO_3$	32	$\underline{2}/\underline{m}$
$CoCO_3$	—	$\underline{2}/\underline{m}$
$FeCO_3$	20–35	$\bar{3}m$

APPENDIX D

Basic Quantities for Second- and Third-Order Tensors

The process of determining the linear combinations of the components of a true or an axial vector forming the carrier spaces for the irreducible representations of a given crystal class presents no difficulties. In those cases, for which finding the basic quantities is not apparent from inspection, we can apply formulas (4.23), (4.24). However, for second-, third-, and higher-order tensors and, in particular, for crystals with trigonal and hexagonal symmetry, the application of the formulas (4.23), (4.24) may become very tedious.

We devised a computer program coded in FØRTRAN IV which produces basic quantities for any order of *i*-tensors which may be true or pseudo (axial). We list below the results for second- and third-order general tensors, and for the classes of the conventional triclinic, monoclinic, rhombic, tetragonal, trigonal, and hexagonal crystal systems.

Note that if the tensor under consideration has a certain symmetry property (for example, $T_{ijk} = T_{ikj}$), the appropriate terms are selected as special cases of the results listed in the tables. Also note that if the tensor is not time-symmetric, each irreducible representation is to be multiplied by the appropriate time-reversal representation T (Chapter 4, p. 40), that is, $\Gamma_i \otimes T$.

A. Second-Order *True* (Polar) Tensors

Table A.1

$C_i(\bar{1})$ Γ_1:		T_{ij} $(i, j = 1, 2, 3)$	
C_2 Γ_1	C_s Γ_1:	$T_{11}, T_{22}, T_{23}, T_{32}, T_{33}$	
(2) Γ_2	(m) Γ_2:	$T_{12}, T_{13}, T_{21}, T_{31}$	

Table A.2

C_{2h}	Γ_1:	$T_{11}, T_{22}, T_{23}, T_{32}, T_{33}$
$(2/m)$	Γ_4:	$T_{12}, T_{13}, T_{21}, T_{31}$

D_2	Γ_1	D_{2h}	Γ_1	C_{2v}	Γ_1:	T_{11}, T_{22}, T_{33}
(222)	Γ_2	(mmm)	Γ_2	$(mm2)$	Γ_2:	T_{23}, T_{32}
	Γ_4		Γ_4		Γ_3:	T_{12}, T_{21}
	Γ_3		Γ_3		Γ_4:	T_{13}, T_{31}

Table A.3

C_4	Γ_1	C_{4h}	Γ_1	$\bar{C}_2$	Γ_1:	$T_{11} + T_{22}, T_{12} - T_{21}, T_{33}$
(4)	Γ_2	$(4/m)$	Γ_2	$(\bar{4})$	Γ_2:	$T_{11} - T_{22}, T_{12} + T_{21}$
	Γ_3		Γ_3		Γ_3:	$T_{13} + iT_{23}, T_{31} + iT_{32}$
	Γ_4		Γ_4		Γ_4:	$T_{13} - iT_{23}, T_{31} - iT_{32}$

Table A.4

D_{2v}	Γ_1	C_{4v}	Γ_1:	$T_{11} + T_{22}, T_{33}$
$(\bar{4}2m)$	Γ_2	$(4mm)$	Γ_2:	$T_{12} - T_{21}$
	Γ_3		Γ_3:	$T_{12} + T_{21}$
	Γ_4		Γ_4:	$T_{11} - T_{22}$
	$\Gamma_5 = \begin{pmatrix} 01 \\ 10 \end{pmatrix} x$		Γ_5:	$(T_{13}, T_{23}), (T_{31}, T_{32})$
D_{4h}	Γ_1	D_4	Γ_1:	$T_{11} + T_{22}, T_{33}$
$(4/mmm)$	Γ_2	(422)	Γ_2:	$T_{12} - T_{21}$
	Γ_3		Γ_3:	$T_{12} + T_{21}$
	Γ_4		Γ_4:	$T_{11} - T_{22}$
	Γ_5		Γ_5:	$(T_{23}, T_{13}), (T_{32}, -T_{31})$

Table A.5

C_3	Γ_1:	$T_{11} + T_{22}, T_{12} - T_{21}, T_{33}$
(3)	Γ_2:	$T_{11} - T_{22} + i(T_{12} + T_{21}), T_{13} - iT_{23}, T_{31} - iT_{32}$
	Γ_3:	$T_{11} - T_{22} - i(T_{12} + T_{21}), T_{13} + iT_{23}, T_{31} + iT_{32}$

Table A.6

D_3	Γ_1	C_{3v}	Γ_1:	$T_{11} + T_{22}, T_{33}$
(32)	Γ_2	$(3m)$	Γ_2:	$T_{12} - T_{21}$
	Γ_3		Γ_3:	$(T_{12} + T_{21}, T_{11} - T_{22}), (T_{13}, T_{23}), (T_{31}, T_{32})$

Table A.7

$\bar{C}_3$ Γ_1:	$T_{11} + T_{22}$, $T_{12} - T_{21}$, T_{33}
$(\bar{3})$ Γ_2:	$T_{11} - T_{22} + i(T_{12} + T_{21})$, $T_{13} - iT_{23}$, $T_{31} - iT_{32}$
Γ_3:	$T_{11} - T_{22} - i(T_{12} + T_{21})$, $T_{13} + iT_{23}$, $T_{31} + iT_{32}$

C_6 Γ_1	C_{6h} Γ_1	C_{3h} Γ_1:	$T_{11} + T_{22}$, $T_{12} - T_{21}$, T_{33}
(6) Γ_2	$(6/m)$ Γ_2	$(\bar{6})$ Γ_2:	$T_{11} - T_{22} + i(T_{12} + T_{21})$
Γ_3	Γ_3	Γ_3:	$T_{11} - T_{22} - i(T_{12} + T_{21})$
Γ_5	Γ_5	Γ_5:	$T_{13} - iT_{23}$, $T_{31} - iT_{32}$
Γ_6	Γ_6	Γ_6:	$T_{13} + iT_{23}$, $T_{31} + iT_{32}$

Table A.8

D_{3v}	Γ_1: $T_{11} + T_{22}$, T_{33}
$(\bar{3}m)$	Γ_2: $T_{12} - T_{21}$
	Γ_6: $(T_{12} + T_{21}, T_{11} - T_{22})$, (T_{13}, T_{23}), (T_{31}, T_{32})

D_{6h} Γ_1	D_6 Γ_1	C_{6v} Γ_1	D_{3h} Γ_1: $T_{11} + T_{22}$, T_{33}
$(6/mmm)$ Γ_2	(622) Γ_2	$(6mmm)$ Γ_2	$(\bar{6}m2)$ Γ_2: $T_{12} - T_{21}$
Γ_5	Γ_5	$\Gamma_5 = \begin{pmatrix} 0 & -1 \\ 1 & 0 \end{pmatrix} x$	Γ_5: $(T_{23}, -T_{13})(T_{32}, -T_{31})$
Γ_6	Γ_6	Γ_6	Γ_6: $(T_{12} + T_{21}, T_{11} - T_{22})$

B. Second-Order *Psuedo* (Axial) Tensors

Table B.1

$C_i(\bar{1})$ Γ_2:	A_{ij} $(i, j = 1, 2, 3)$

C_2 Γ_2	C_s Γ_1:	A_{12}, A_{13}, A_{21}, A_{31}
(2) Γ_1	Γ_2:	A_{11}, A_{22}, A_{33}, A_{23}, A_{32}

Table B.2

C_{2h} Γ_2:	A_{11}, A_{22}, A_{23}, A_{32}, A_{33}
$(2/m)$ Γ_3:	A_{12}, A_{13}, A_{21}, A_{31}

D_2 Γ_2	D_{2h} Γ_2'	C_{2v} Γ_1:	A_{23}, A_{32}
(222) Γ_1	(mmm) Γ_1'	$(mm2)$ Γ_2:	A_{11}, A_{22}, A_{33}
Γ_3	Γ_3'	Γ_3:	A_{13}, A_{31}
Γ_4	Γ_4'	Γ_4:	A_{12}, A_{21}

Table B.3

C_4 Γ_2	C_{4h} Γ_2'	$\bar{C}_2$ Γ_1:	$A_{11} - A_{22}$, $A_{12} + A_{21}$
(4) Γ_1	$(4/m)$ Γ_1'	$(\bar{4})$ Γ_2:	$A_{11} + A_{22}$, $A_{12} - A_{21}$, A_{33}
Γ_4	Γ_4'	Γ_3:	$A_{13} - iA_{23}$, $A_{31} - iA_{32}$
Γ_3	Γ_3'	Γ_4:	$A_{13} + iA_{23}$, $A_{31} + iA_{32}$

Table B.4

D_{2v}	Γ_4	C_{4v}	Γ_2	D_{4h}	Γ_1'	D_4	Γ_1:	$A_{11}+A_{22}, A_{33}$
$(\bar{4}2m)$	Γ_3	$(4mm)$	Γ_1	$(4/mmm)$	Γ_2'	(422)	Γ_2:	$A_{12}-A_{21}$
	Γ_2		Γ_4		Γ_3'		Γ_3:	$A_{12}+A_{21}$
	Γ_1		Γ_3		Γ_4'		Γ_4:	$A_{11}-A_{22}$
	Γ_5		Γ_5		Γ_5'		Γ_5:	$(A_{23}, -A_{13}), (A_{32}, -A_{31})$

Table B.5

C_3	Γ_1:	$A_{11}+A_{22}, A_{12}-A_{21}, A_{33}$
(3)	Γ_2:	$A_{11}-A_{22}+i(A_{12}+A_{21}), A_{13}-iA_{23}, A_{31}-iA_{32}$
	Γ_3:	$A_{11}-A_{22}-i(A_{12}+A_{21}), A_{13}+iA_{23}, A_{31}+iA_{32}$

Table B.6

C_{3v}	Γ_2	D_3	Γ_1:	$A_{11}+A_{22}, A_{33}$
$(3m)$	Γ_1	(32)	Γ_2:	$A_{12}-A_{21}$
	$\Gamma_3=\begin{pmatrix} 0 & 1 \\ -1 & 0 \end{pmatrix}\times$		Γ_3:	$(A_{12}+A_{21}, A_{11}-A_{22}), (A_{13}, A_{23}), (A_{31}, A_{32})$

Table B.7

$\bar{C}_3$	Γ_4:	$A_{11}+A_{22}, A_{12}-A_{21}, A_{33}$
$(\bar{3})$	Γ_5:	$A_{11}-A_{22}+i(A_{12}+A_{21}), A_{13}-iA_{23}, A_{31}-iA_{32}$
	Γ_6:	$A_{11}-A_{22}-i(A_{12}+A_{21}), A_{13}+iA_{23}, A_{31}+iA_{32}$

C_6	Γ_5	C_{6h}	Γ_5'	C_{3h}	Γ_2:	$A_{13}-iA_{23}, A_{31}-iA_{32}$
(6)	Γ_6	$(6/m)$	Γ_6'	$(\bar{6})$	Γ_3:	$A_{13}+iA_{23}, A_{31}+iA_{32}$
	Γ_1		Γ_1'		Γ_4:	$A_{11}+A_{22}, A_{12}-A_{21}, A_{33}$
	Γ_2		Γ_2'		Γ_5:	$A_{11}-A_{22}+i(A_{12}+A_{21})$
	Γ_3		Γ_3'		Γ_6:	$A_{11}-A_{22}-i(A_{12}+A_{21})$

Table B.8

D_{3v}	Γ_3:	$A_{11}+A_{22}, A_{33}$
$(\bar{3}m)$	Γ_4:	$A_{12}-A_{21}$
	Γ_5:	$(A_{22}-A_{11}, A_{12}+A_{21}), (A_{23}, -A_{13}), (A_{32}, -A_{31})$

D_6	Γ_2	D_{6h}	Γ_2'	C_{6v}	Γ_1	D_{3h}	Γ_3:	$A_{12}-A_{21}$
(622)	Γ_1	$(6/mmm)$	Γ_1'	$(6mm)$	Γ_2	$(\bar{6}m2)$	Γ_4:	$A_{11}+A_{22}, A_{33}$
	Γ_6		Γ_6'		$\Gamma_6=\begin{pmatrix} 0 & -1 \\ 1 & 0 \end{pmatrix}\times$		Γ_5:	$(A_{12}+A_{21}, A_{11}-A_{22})$
	Γ_5		Γ_5'		Γ_5		Γ_6:	$(A_{23}, -A_{13}), (A_{32}, -A_{31})$

C. Third-Order *True* (Polar) Tensors

Table C.1

$C_i(\bar{1})$	Γ_2:	T_{ijk} $(i, j, k = 1, 2, 3)$				
C_2 (2)	Γ_2	C_{2h} $(2/m)$	Γ_3	C_s (m)	Γ_1:	$T_{112}, T_{113}, T_{121}, T_{131}, T_{211}, T_{222}, T_{223}, T_{232}, T_{233}, T_{311}, T_{322}, T_{323}, T_{332}, T_{333}$
	Γ_1		Γ_2		Γ_2:	$T_{111}, T_{122}, T_{123}, T_{132}, T_{133}, T_{212}, T_{213}, T_{221}, T_{231}, T_{312}, T_{313}, T_{321}, T_{331}$

Table C.2

D_2 (222)	Γ_2	D_{2h} (mmm)	Γ_2'	C_{2v} $(mm2)$	Γ_1:	$T_{111}, T_{122}, T_{133}, T_{212}, T_{221}, T_{313}, T_{331}$
	Γ_1		Γ_1'		Γ_2:	$T_{123}, T_{132}, T_{213}, T_{231}, T_{312}, T_{321}$
	Γ_3		Γ_3'		Γ_3:	$T_{112}, T_{121}, T_{211}, T_{222}, T_{233}, T_{323}, T_{332}$
	Γ_4		Γ_4'		Γ_4:	$T_{113}, T_{131}, T_{223}, T_{232}, T_{311}, T_{322}, T_{333}$

Table C.3

C_4 (4)	Γ_2	C_{4h} $(4/m)$	Γ_2'	$\bar{C}_2$ $(\bar{4})$	Γ_1:	$T_{113} - T_{223}, T_{123} + T_{213}, T_{131} - T_{232}, T_{132} + T_{231}, T_{311} - T_{322}, T_{312} + T_{321}$
	Γ_1		Γ_1'		Γ_2:	$T_{113} + T_{223}, T_{123} - T_{213}, T_{131} + T_{232}, T_{132} - T_{231}, T_{311} + T_{322}, T_{312} - T_{321}, T_{333}$
	Γ_4		Γ_4'		Γ_3:	$T_{111} - iT_{222}, T_{112} + iT_{221}, T_{121} + iT_{212}, T_{122} - iT_{211}, T_{133} - iT_{233}, T_{133} - iT_{323}, T_{331} - iT_{332}$
	Γ_3		Γ_3'		Γ_4:	Seven terms complex conjugate of the elements in Γ_3 for $\bar{C}_2$.

Table C.4

D_{2v} $(\bar{4}2m)$	Γ_3 / Γ_4	D_4 (422)	Γ_2 / Γ_1	D_{4h} $(4/mmm)$	Γ_2' / Γ_1'	C_{4v} $(4mmm)$	Γ_1:	$T_{113} + T_{223}, T_{131} + T_{232}, T_{311} + T_{322}, T_{333}$
							Γ_2:	$T_{123} - T_{213}, T_{132} - T_{231}, T_{312} - T_{321}$
	Γ_1		Γ_4		Γ_4'		Γ_3:	$T_{123} + T_{213}, T_{132} + T_{231}, T_{312} + T_{321}$
	Γ_2		Γ_3		Γ_3'		Γ_4:	$T_{113} - T_{223}, T_{131} - T_{232}, T_{311} - T_{322}$
	Γ_5		Γ_5		Γ_5'		Γ_5:	$(T_{111}, T_{222}), (T_{122}, T_{211}), (T_{133}, T_{233}), (T_{221}, T_{112}), (T_{313}, T_{323}), (T_{331}, T_{332}) (T_{212}, T_{121})$

Table C.5

C_3	Γ_1:	$T_{111} - T_{122} - (T_{212} + T_{221}),\ T_{211} - T_{222} + (T_{112} + T_{121}),$
(3)		$T_{113} + T_{223},\ T_{123} - T_{213},\ T_{131} + T_{232},\ T_{132} - T_{231},$
		$T_{311} + T_{322},\ T_{312} - T_{321},\ T_{333}$
	Γ_2:	$T_{113} - T_{223} + i(T_{123} + T_{213}),\ T_{133} - iT_{233},$
		$T_{131} - T_{232} + i(T_{132} + T_{231}),\ T_{313} - iT_{323},$
		$T_{311} - T_{322} + i(T_{312} + T_{321}),\ T_{331} - iT_{332},$
		$T_{112} + T_{222} + i(T_{111} + T_{221}),\ T_{121} + T_{222} + i(T_{111} + T_{212}),$
		$T_{211} + T_{222} + i(T_{111} + T_{122})$
	Γ_3:	Nine terms complex conjugate of the elements in Γ_2.

Table C.6

D_3 Γ_2	C_{3v}	Γ_1:	$T_{211} - T_{222} + (T_{112} + T_{121}),\ T_{113} + T_{223},$ $T_{131} + T_{232},\ T_{311} + T_{322},\ T_{333}$
(32) Γ_1	(3m)	Γ_2:	$T_{111} - T_{122} - (T_{212} + T_{221}),\ T_{123} - T_{213},$ $T_{132} - T_{231},\ T_{312} - T_{321}$
$\Gamma_3 = \begin{pmatrix} 0 & 1 \\ -1 & 0 \end{pmatrix} \times$		Γ_3:	$(T_{133}, T_{233}),\ (T_{313}, T_{323}),\ (T_{331}, T_{332}),$ $(T_{123} + T_{213}, T_{113} - T_{223}),\ (T_{132} + T_{231},$ $T_{131} - T_{232}),\ (T_{312} + T_{321}, T_{311} - T_{322}),$ $(T_{111} + T_{221}, T_{112} + T_{222}),\ (T_{111} + T_{212},$ $T_{121} + T_{222})(T_{111} + T_{122}, T_{211} + T_{222}),$

Table C.7

$\bar{C}_3$	Γ_4:	$T_{111} - T_{122} - (T_{212} + T_{221}),\ T_{211} - T_{222} + (T_{112} + T_{121}),$
($\bar{3}$)		$T_{113} + T_{223},\ T_{123} - T_{213},\ T_{131} + T_{232},\ T_{132} - T_{231},\ T_{311} + T_{322},$
		$T_{312} - T_{321},\ T_{333}$
	Γ_5:	$T_{112} + T_{222} + i(T_{111} + T_{221}),\ T_{121} + T_{222} + i(T_{111} + T_{212}),$
		$T_{211} + T_{222} + i(T_{111} + T_{122}),\ T_{113} - T_{223} + i(T_{123} + T_{213}),$
		$T_{133} - iT_{233},\ T_{131} - T_{232} + i(T_{132} + T_{231}),\ T_{313} - iT_{323},$
		$T_{311} - T_{322} + i(T_{312} + T_{321}),\ T_{331} - iT_{332}$
	Γ_6:	Nine terms complex conjugate of the elements in Γ_5.

C_{6h} Γ_4'	C_6 Γ_4	C_{3h} Γ_1:	$T_{111} - T_{122} - (T_{212} + T_{221}),\ T_{211} - T_{222} +$ $(T_{112} + T_{121})$
(6/m) Γ_5'	(6) Γ_5	($\bar{6}$) Γ_2:	$T_{112} + T_{222} + i(T_{111} + T_{221}),\ T_{121} + T_{222} +$ $i(T_{111} + T_{212}),\ T_{211} + T_{222} + i(T_{111} + T_{122}),$ $T_{133} - iT_{233},\ T_{313} - iT_{323},\ T_{331} - iT_{332}$
Γ_6'	Γ_6	Γ_3:	complex conjugate of the elements in Γ_2 of C_{3h}
Γ_1'	Γ_1	Γ_4:	$T_{113} - T_{223},\ T_{123} - T_{213},\ T_{131} + T_{232},$ $T_{132} - T_{231},\ T_{311} + T_{322},\ T_{312} - T_{321},\ T_{333}$
Γ_2'	Γ_2	Γ_5:	$T_{113} - T_{223} + i(T_{123} + T_{213}),\ T_{131} - T_{232} +$ $i(T_{132} + T_{231}),\ T_{311} - T_{322} + i(T_{312} + T_{321})$
Γ_3'	Γ_3	Γ_6:	Complex conjugate of the elements in Γ_5 of C_{3h}

Table C.8

D_{6h} $(6/mmm)$	D_6 (622)	C_{6v} $(6mm)$	D_{3h} $(\bar{6}m2)$	
Γ_4'	Γ_4	Γ_4	Γ_1:	$T_{211} - T_{222} + (T_{112} + T_{121})$
Γ_3'	Γ_3	Γ_3	Γ_2:	$T_{111} - T_{122} - (T_{212} + T_{221})$
Γ_2'	Γ_2	Γ_1	Γ_3:	$T_{113} + T_{223}, T_{131} + T_{232}, T_{311} + T_{322}, T_{333}$
Γ_1'	Γ_1	Γ_2	Γ_4:	$T_{123} - T_{213}, T_{132} - T_{231}, T_{312} - T_{321}$
Γ_6'	Γ_6	$\Gamma_6 = \begin{pmatrix} 0 & -1 \\ 1 & 0 \end{pmatrix}$	Γ_5:	$(T_{223} - T_{113}, T_{123} + T_{213}), (T_{232} - T_{131}, T_{132} + T_{231}), (T_{322} - T_{311}, T_{312} + T_{321})$
Γ_5'	Γ_5	Γ_5	Γ_6:	$(T_{133}, T_{233}), (T_{313}, T_{323}), (T_{331}, T_{332}), (T_{111} + T_{221}, T_{112} + T_{222}), (T_{111} + T_{212}, T_{121} + T_{222}), (T_{111} + T_{122}, T_{211} + T_{222})$

D_{3v} $(\bar{3}m)$	
Γ_3:	$T_{111} - T_{122} - (T_{212} + T_{221}), T_{123} - T_{213}, T_{132} - T_{231}, T_{312} - T_{321}$
Γ_4:	$T_{211} - T_{222} + (T_{112} + T_{121}), T_{113} + T_{223}, T_{131} + T_{232}, T_{311} + T_{322}, T_{333}$
Γ_5:	$(T_{123} + T_{213}, T_{113} - T_{223}), (T_{132} + T_{231}, T_{131} - T_{232}), (T_{312} + T_{321}, T_{311} - T_{322}), (T_{111} + T_{212}, T_{121} + T_{222}), (T_{111} + T_{221}, T_{112} + T_{222}), (T_{111} + T_{122}, T_{211} + T_{222}), (T_{133}, T_{233}), (T_{313}, T_{323}), (T_{331}, T_{332})$

References

ASTROV, D.N. [1960]: The magneto-electric effect in antiferromagnetics, *Soviet Phys. JETP* (English transl.), **11**, 708.

ASTROV, D.N. [1961]: Magneto-electric effect in chromium oxide, *Soviet Phys. JETP* (English transl.), **13**, 729.

BELOV, N.V., NEROVONA, N.N., and SMIRNOVA, T.S. [1957]: Shubnikov groups, *Soviet Phys. Cryst.* (English transl.), **2**, 311.

BERTAUT, E.F. [1968]: Representation analysis of magnetic structure, *Acta Crystallogr.*, **A24**, 217.

BHAGAVANTAM, S. [1966]: *Crystal Symmetry and Physical Properties*, Academic Press, New York.

BIRSS, R.R. [1962]: Property tensors in magnetic crystal classes, *Proc. Phys. Soc.*, **79**, 946.

BIRSS, R.R. [1963]: Macroscopic symmetry in space–time, *Rep. Prog. Phys.*, **26**, 307.

BIRSS, R.R. [1964]: *Symmetry and Magnetism*, North-Holland, Amsterdam.

BOROVIK-ROMANOV, A.S. [1959]: Piezomagnetism in the antiferromagnetic fluorides, *Soviet Phys. JETP* (English transl.), **9**, 1390.

BOROVIK-ROMANOV, A.S. [1960]: Piezomagnetism in the antiferromagnetic fluorides of cobalt and manganese, *Soviet Phys. JETP* (English transl.), **11**, 786.

BRADLEY, C.J. and CRACKNELL A.P. [1972]: *The Mathematical Theory of Symmetry in Solids*, Clarendon Press, Oxford.

CURIE, P. [1908]: *Oeuvres de Pierre Curie*, reprinted by Gauthier-Villars, Paris, pp. 118–141.

DE GROOT, S.R. and SUTTORP L.G. [1972]: *Foundations of Electrodynamics*, North-Holland, Amsterdam.

DZYALOSHINSKII, I.E. [1960]: On the magneto-electric effect in antiferromagnetics, *Soviet Phys. JETP* (English transl.), **10**, 628.

ERINGEN, A.C. [1980]: *Mechanics of Continua*, 2nd ed., Robert E. Krieger, Florida.

ERINGEN, A.C. and MAUGIN, G.A. [1989]: *Electrodynamics of Continua*, Springer-Verlag, New York.

ERSOY, Y. and KIRAL, E. [1978]: A dynamic theory for polarizable and magnetizable magneto-electro thermo-viscoelastic, electrically and thermally conductive anisotropic solids having magnetic symmetry, *Int. J. Engng. Sci*, **16**, 483–492.

Folen, V.J., Rado, G.T., and Stalder, E.W. [1961]: Anisotropy of the magnetoelectric effect in Cr_2O_3, *Phys. Rev. Lett.*, **6**, 607.

Green, A.E. and Adkins, J.E. [1960]: *Large Elastic Deformations*, Oxford University Press (Clarendon), London and New York.

Grot, R.A., and Eringen, A.C. [1966]: Relativistic continuum mechanics, I & II, *Int. J. Engng. Sci.*, **4**, 611–638 and 639–670.

Huang, C.L. [1968]: The energy function for anisotropic materials with couple stresses—Cubic and hexagonal systems, *Int. J. Engng. Sci.*, **6**, No. 10, 609.

Huang, C.L. [1969]: The energy function for crystal materials with couple stresses, *Int. J. Engng. Sci.*, **7**, No. 12, 1221.

Indenbom, V.L. [1960]: Relation of the antisymmetry and color symmetry groups to one-dimensional representations of the ordinary symmetry groups. Isomorphism of the Shubnikov and space groups, *Soviet Phys. Crystal* (English transl.), **4**, 578.

Jackson, J.D. [1975]: *Classical Electrodynamics*, 2nd ed. Wiley, New York.

Kiral, E. [1972]: Symmetry restrictions on the constitutive relations for anisotropic materials—Polymonial integrity bases for cubic crystal systems, Habilitation Thesis, METU, Ankara, Turkey.

Kiral, E. and Mert, M. [1974]: On the number and enumeration of independent components of material tensors with intrinsic symmetry-magnetic crystal classes *Proceedings of Symposium on Symmetry, Similarity and Group Theoretic Methods in Mechanics*, University of Calgary, pp. 141–154.

Kiral, E. and Smith, G.F. [1974]: On the constitutive relations for anisotropic materials—trinclinic, monoclinic, rhombic, tetragonal, and hexagonal crystal systems, *Int. J. Engng. Sci.*, **12**, 471.

Kiral, E., Smith, M.M., and Smith, G.F. [1980]: On the constitutive relations for anisotropic materials—The crystal class D_{6h}, *Int. J. Engng. Sci*, **18**, 569–581.

Koptsik, V.A. [1966]: *Shubnikov Groups.* Handbook on the Symmetry and Physical Properties of Crystal Structures, University Press, Moscow (in Russian).

Koptsik, V.A. [1968]: A general sketch of the development of the theory of symmetry and its applications in physical crystallography over the last 50 years, *Soviet Phys. Cryst.* (English transl.), **12**, 667.

Koster, G.F., Dimmock, J.B., Wheeler, G.R. and Statz, H. [1963]: *Properties of the Thirty-Two Point Groups*, MIT Press, Cambridge, Mass.

Landau, L.D. and Lifshitz, E.M. [1960]: *Electrodynamics of Continuous Media* (English transl.) Pergamon Press, Oxford, 1960, pp. 115, 119, 313–343

Lyubimov, V.N. [1966]: The interaction of polarization and magnetization in crystals, *Soviet Phys. Cryst.* (English translation), **10**, 4, 433–436.

Maugin, G.A., and Eringen A.C. [1977]: On the equations of the electrodynamics of deformable bodies of finite extent, *J. Mécanique*, **16**, 102–147.

McMillan, J.A. [1967]: Stereographic projections of the colored crystallographic point groups, *Amer. J. Phys.*, **35**, 1049.

MERT, M. [1975]: Symmetry restrictions on linear and nonlinear constitutive equations for anisotropic materials—classical and magnetic crystal classes, Ph.D. Thesis, METU, Ankara, Turkey.

MERT, M. and KIRAL, E. [1977]: Symmetry restrictions on the constitutive equations for magnetic materials, *Int. J. Engng. Sci.*, **15**, 281–294.

NYE, J.F. [1957]: *Physical Properties of Crystals*, Clarendon Press, Oxford.

O'DELL, T.H. [1970]: The electrodynamics of magneto-electric media, in *Selected Topics in Solid State Physics*, Vol. XI, ed. E.P. Wohlforth, North-Holland/Elsevier, Amsterdam.

OPECHOWSKI, W. and GUCCIONE, R. [1965]: Magnetic symmetry, in *Magnetism*, Vol. 2A, Chap. 3, eds. G.T. Rado and H. Suhl, Academic Press, New York.

RADO, G.T. [1962]: Statistical theory of magnetoelectric effects in antiferromagnetics, *Phys. Review*, **128**, 6, 2546.

RADO, G.T. and FOLEN, V.J. [1962]: Magneto-electric effects in antiferromagnetics, *J. Appl. Phys.*, **33**, 1126.

SHUBNIKOV, A.V. [1951]: *Symmetry and Antisymmetry of Finite Figures*, USSR Academy of Sciences, Moscow. Translated into English by A.V. Shubnikov and N.V. Belov under the title *Coloured Symmetry*, Pergamon Press, Oxford, 1964.

SMITH, G.F. [1967]: Tensor and integrity bases for the gyroidal crystal class, *Quart. Appl. Math.*, **25**, 218.

SMITH, G.F. [1968]: On the generation of integrity bases, *Att. Acad. Naz. Lincei Mem. Cl. Sci. Fis. Mat. Natur. Sez.*, **Ia(8)**, 9, 51.

SMITH, G.F. and KIRAL, E. [1969]: Integrity bases for N-symmetric second-order tensors—the crystal classes, *Rendiconti, Circolo Mat. di Palermo*, **II**, XVII, 5.

SMITH, G.F. and KIRAL, E. [1978]: Anisotropic constitutive equations and Schur's lemma, *Int. J. Engng. Sci.*, **16**, 773–780.

SMITH, G.F. and Rivlin, R.S. [1958]: The strain–energy function for anisotropic elastic materials, *Trans. Amer. Math. Soc.*, **88**, 175.

SMITH, G.F. and RIVLIN, R.S. [1964]: Integrity bases for vectors—the crystal classes, *Arch. Rat. Mech. Anal.*, **15**, 169.

SMITH, G.F., SMITH, M.M., and RIVLIN, R.S. [1963]: Integrity bases for a symmetric tensor and a vector—the crystal classes, *Arch. Rat. Mech. Anal.*, **12**, 93.

TAVGER, B.A. and ZAITSEV, V.M. [1956]: Magnetic symmetry of crystals, *Soviet Phys. JETP* (English transl.), **3**, 430.

ZAMORZAEV, A.M. [1957]: Generalization of Fedorov groups, *Soviet Phys. Cryst.* (English transl.), **2**, 10.

ZAMORZAEV, A.M. [1958]: Derivation of new Shubnikov groups, *Soviet Phys.* Cryst. (English transl.), **3**, 401.

ZOCHER, H. and TÖRÖK, C. [1953]: About space–time asymmetry in the realm of classical, general and crystal physics, *Proc. Nat. Acad. Sci.*, **39**, 681.

Index